Studies in Fuzziness and Soft Computing 292

Editor-in-Chief

Prof. Janusz Kacprzyk
Systems Research Institute
Polish Academy of Sciences
ul. Newelska 6
01-447 Warsaw
Poland
E-mail: kacprzyk@ibspan.waw.pl

For further volumes:
http://www.springer.com/series/2941

Maciej Wygralak

Intelligent Counting under Information Imprecision

Applications to Intelligent Systems and Decision Support

Springer

Maciej Wygralak, Assoc. Prof.
Faculty of Mathematics and Computer Science
Adam Mickiewicz University
Umultowska 87
61-614 Poznań
Poland
E-mail: wygralak@amu.edu.pl

ISSN 1434-9922 ISSN 1860-0808 (electronic)
ISBN 978-3-642-43990-2 ISBN 978-3-642-34685-9 (eBook)
DOI 10.1007/978-3-642-34685-9
Springer Heidelberg New York Dordrecht London

Softcover reprint of the hardcover 1st edition 2013

Printed on acid-free paper

Springer is part of Springer Science+Business Media (www.springer.com)

To my four ladies:
Renata, Karolina, Agata, and Alicja

Preface

This monograph begins with a presentation of the concept of and selected issues in fuzzy sets and fuzzy logic, interval-valued fuzzy sets, and I-fuzzy sets - Atanassov's intuitionistic fuzzy sets. However, that study is only of preparatory character. The main subject of the book will be intelligent counting under imprecision of information (about the objects of counting). Why is this worth investigating and deliberating?

It seems that counting belongs to the most basic and frequent mental activities of human beings as its results are a basis for coming to a decision in a lot of situations. One should distinguish, however, between two very different cases occurring in practice. First, the objects of counting can be precisely specified and, then, the counting process collapses to the trivial task of counting in a set by means of the natural numbers. Second, those objects can be imprecisely (fuzzily) specified and just this much more advanced and sophisticated case of counting requiring intelligence will be the subject of our investigations. Speaking formally, that *intelligent counting* collapses to counting in a fuzzy set or - whenever imprecision is combined with incompleteness of information - to counting in an interval-valued fuzzy set or I-fuzzy set. Theoretical aspects as well as applications of intelligent counting will be discussed. Especially, we mean applications to intelligent systems and decision support. The emphasis will be on showing that the presented methods of intelligent counting and the resulting cardinalities are human-consistent, i.e. are reflections and formalizations of real, human counting methods under imprecision possibly combined with incompleteness of information. It is self-evident that our main interest will be in counting in finite fuzzy sets, finite interval-valued fuzzy sets and I-fuzzy sets. Nevertheless, for completeness, the infinite case will be concisely discussed, too.

The monograph is divided into two parts and eleven chapters. The first one is of introductory character, whereas Chapters 2-6, forming Part I, are devoted to those elements of fuzzy sets and fuzzy logic which are relevant from the viewpoint of the main aim of this book. We will present operations on and basic characteristics of fuzzy sets, negations, triangular norms and conorms, fuzzy numbers and linguistic variables, aggregation operators, fuzzy relations, and an introduction to approximate reasoning and fuzzy rule-based systems. Moreover, interval-valued fuzzy sets and I-fuzzy sets will be discussed as tools for modeling incompletely known fuzzy sets.

Chapters 7-11 constitute Part II, the key part of this book, devoted to methods of intelligent counting and related cardinalities of fuzzy sets, interval-valued fuzzy sets and I-fuzzy sets. Both the scalar and fuzzy approaches to these questions will be discussed in detail, including human-consistency and giving the reader a novel and up-to-date image of the subject matter.

The presentation is self-contained as much as possible and equipped with many figures, examples, and references to the source literature. Our general principle is "motivations and ideas before technical details". Nevertheless, the reader should have some basic knowledge of mathematics with special reference to set theory, mathematical logic, analysis, and general algebra.

This monograph is intended for computer and information scientists, researchers, engineers and practitioners, applied mathematicians, and postgraduate students interested in dealing with information imprecision and incompleteness.

The most pleasant moment of each book project is to write acknowledgments. I am grateful to all who supported me. Especially, I would like to thank my wife Renata and our daughters Karolina and Agata for their continuous and reliable support, understanding and patience.

As to the financial side, this book project has been partially supported by a research grant from the National Science Centre (NCN), and this support is greatly appreciated.

Last but not least, let me thank my collaborators, Dr. Krzysztof Dyczkowski and Dr. Anna Stachowiak, for kind technical assistance and valuable suggestions.

Maciej Wygralak

Contents

Chapter 1

Introduction

Fuzzy set-based methodology and techniques are firmly established in many areas of contemporary science and technology, especially in broadly conceived computer science and decision support. The reason, in brief, is that fuzzy sets are an adequate and powerful tool when trying to cope with an important and omnipresent, but troublesome, type of information, namely imprecise information. The main issue this book addresses are methods of intelligent counting under information imprecision presented and studied in Part II. We mean methods of counting in and the resulting cardinalities of fuzzy sets, a basis for coming to a decision in a lot of situations. Questions of human-consistency, theoretical aspects and multiple applications will be analyzed. Since imprecision is often combined in practice with incompleteness of information, we will also deal with counting in incompletely known fuzzy sets modeled by interval-valued fuzzy sets and I-fuzzy sets. To make this book self-contained as much as possible, Part I presents selected elements of fuzzy sets, interval-valued fuzzy sets and I-fuzzy sets. This preliminary chapter is aimed at introducing the reader to the subject matter of Parts I and II in a friendly way.

1.1 Information Imprecision and Fuzzy Sets

As we perceive it today, imprecision of information – its fuzziness, unsharpness – is actually omnipresent and unremovable. Humans have a remarkable ability to understand imprecise information and, what is more, to reason and to make right decisions on its basis. This ability seems to belong to fundamental features of human intelligence. Let us look at simple examples.

- Weather forecast: *Breezy and generally cloudy, although some cloud breaks are likely in the south. Rain, some heavy, in the north and northwest area, elsewhere mostly dry.*
- FED Press Release: ...*Readings on core inflation have improved modestly in recent months, and inflation pressures seem likely to moderate over time. However, the high level of resource utilization has the potential to sustain inflation pressures.*
- Description of the perpetrator given by eyewitness: *White male*; *about 180 cm tall*; *about 30-35 years old*; *slim build*; *high voice*; *dark, short hair*; *dark eyes.*

The imprecision in these statements is not caused by someone's lack of care or ill will, but it results from the very nature of notions and issues brought up therein. Despite that imprecision, human is able to recognize in a crowd a person similar to that

M. Wygralak: *Intelligent Counting Under Information Imprecision*, STUDFUZZ 292, pp. 1–16.
DOI: 10.1007/978-3-642-34685-9_1

from the description, to tailor plans and type of clothing to the weather forecast, and to make right financial decisions on the basis of the release.

More and more advanced applications of computers – applications to solving more and more complex problems in an intelligent way so far reserved for human beings – require an ability to overcome difficulties bound up with imprecision of information. This ability must be built on having a suitable theoretical apparatus at our disposal. A really effective response to that need became fuzzy sets introduced by Lotfi A. Zadeh in 1965 and the resulting methodology (see ZADEH (1965)). Their aim is just the formalization of information imprecision. The reader is referred to SEISING (2007b) for a detailed study devoted to the genesis of the theory of fuzzy sets and its early development (see also DUBOIS/PRADE (2000)).

Speaking generally, a *fuzzy set* is a vague, nebulous collection of elements with unsharp boundary. That an element of a fixed universe of discourse belongs to a given fuzzy set is a matter of degree, which is expressed by means of a *membership degree*, a real number from the interval [0, 1]. So, the status of each element of the universe can be *nonmembership* (0), *partial membership* (a number from (0, 1)), or *full membership* (1). In the case of sets, that status is described by the characteristic function of a set, and is reduced to just two possible options: either *membership* (1) or *nonmembership* (0). Fuzzy sets are thus clearly more general constructions than ordinary sets. This greater generality, however, is only a side effect, not the aim. In other words, fuzzy sets should not be viewed as a "new set theory".

The concept of a set is based on classical, two-valued logic. A suitable foundation of fuzzy sets is many-valued logic, logic accepting intermediate truth values from between 0 (false) and 1 (true).

Just thanks to the gradation of membership from 0 through intermediate values to 1, fuzzy sets with their conceptual apparatus and methodology make it possible to

- model and process imprecise information,
- work out effective solutions to problems involving imprecise information.

The adjective "imprecise" is then frequently replaced with "fuzzy". What we discuss can thus be called fuzzy information and pieces of fuzzy information composed of fuzzy concepts, fuzzy relationships, etc.

Worth recollecting in the context of fuzzy sets is logico-philosophical research on what one calls *vagueness of concepts*. It was initiated in the 1920s and 1930s by RUSSELL (1923), KOTARBIŃSKI (1929) and BLACK (1934) (see also BLACK (1963) and AJDUKIEWICZ (1935)). That vagueness can also be formalized by means of fuzzy sets; see NOVÁK (2005), SEISING (2007a), TAMBURINI/TERMINI (1982) and TERMINI (1984) for details. The above works from the first half of the 20th century, on the other hand, may be viewed as pre-beginnings of research on information imprecision done in the pre-computer era.

Nowadays, successful applications of fuzzy sets encompass almost all branches of computer science as well as systems and control engineering, decision support, planning, management, production research, modeling and simulation in social sciences and psychology, etc. Some of those applications are groundbreaking as they concern problems which resisted a successful solution by means of standard mathematical tools for a long time. A spectacular example are here applications in control.

Speaking about the role and applications of fuzzy sets, one cannot forget about the concept and machinery of *fuzzy logic*, FL, introduced in ZADEH (1975d); see also ZADEH (1979, 1994). Following Zadeh, we like to make a distinction between two different meanings of that term. FL in its wide sense, FL_w, is actually a synonym of the whole field of fuzzy sets. FL in the narrow sense, FL_n, forms a branch of FL_w aiming at the formalization and modeling of human approximate reasoning which is usually carried out in the presence of imprecise information. FL_n is a very essential extension of many-valued logic (see e.g. HÁJEK (1998)). It is equipped with such additional key concepts as those of a linguistic variable, a fuzzy conditional statement, a linguistic quantifier, a linguistic truth value, a fuzzy relation, the extension principle, the compositional rule of inference, etc. Part I of this book will present elements of fuzzy logic, both FL_w and FL_n.

Fields of applications of fuzzy sets are usually related to possible sources of information imprecision. Let us take a closer look at four basic instances of those sources.

A. Natural Language and Its Nature

It is rather a truism to say that natural language forms a basic, very effective and perfect tool for communication between people. Verbal information seems to be a fundamental type of information. In each particular natural language, however, one can point to two kinds of properties which are expressible in it.

- *Precise* (*sharp*) *properties*, e.g. "to be a Polish citizen", "to be a prime number", "to be born in 1987", "to be a university student".

They are properties to which a precise definition or criterion is assigned and, thus, there are no intermediate states between their fulfilment and nonfulfilment. Indeed, someone is either a Polish citizen or not, a number is either prime or not, and so on. *Tertium non datur*. Precise properties can be identified with sets. For instance, "Adam is a Polish citizen" and "17 is prime", respectively, are sentences formally equivalent to "Adam belongs to the set of Polish citizens" and "17 belongs to the set of prime numbers".

- *Imprecise* (*unsharp, fuzzy*) *properties*, e.g. "to be a tall man", "to be about 25", "to be similar to Adam".

There are no precise and objective definitions or criteria related to these properties. Their *partial* fulfilment, fulfilment *to a degree* is possible. For instance, a man being 160 cm in height is surely not tall, whereas someone being 195 cm is regarded as tall without any hesitation. Intermediate heights are labeled in everyday language as "rather tall", "pretty tall", "not very tall", etc. Imprecise properties can be identified with fuzzy sets and then, say, "to be a tall man" is understood as "to belong to the fuzzy set of tall men". Someone's membership in that fuzzy set is always a matter of degree, e.g. 0.8. Simultaneously, this is the degree of fulfilment of the property "tall" by that person.

Trying to interpret an imprecise property in the language of sets by force, we encounter essential difficulties. Indeed, looking at a population, if one likes to represent tall men or, say, men about 25 years old as sets, precise definitions of "tall" and "about 25" must be established, e.g.

$$\text{"tall"} = \text{"}\geq 180 \text{ cm in height"}$$

and

$$\text{"about 25"} = \text{"}23-27 \text{ years old".}$$

As we point out, however, this way of doing by establishing strict threshold values is highly artificial and deforming since 1 mm or 1 day make the difference, which is absurd and clashes with human, flexible understanding of those concepts.

It seems that most terms of natural languages are more or less imprecise: they collapse or refer to imprecise properties. Communicating in a natural language, humans use those imprecise terms much more frequently than precise ones. So, human beings usually refer to fuzzy sets rather than ordinary sets. This intrinsic imprecision (fuzziness) of natural languages, on the other hand, seems to be unremovable as an attempt at precisiation of the meanings of all imprecise terms would lead to absurd, whereas the removal of these terms from the language would make efficient and intelligent communication impossible. It is also worth noticing that imprecision enables us to communicate quickly and laconically ("I will be back in a moment", "Buy a few apples", etc.). Moreover, the higher the decision level, the stronger the need for laconic, fuzzy information in natural language as it is easily absorbable and quite sufficient for making right decisions. The message "We have about 200 pieces of the product in the warehouse" is more expected by the logistics director than "We have 214 pieces" as the latter is too precise at that decision level.

Concluding, fuzzy sets are an effective and adequate tool for modeling the meanings of natural language terms and statements (see e.g. ZADEH (1978b)). Worth adding is that this feature was a key to introducing by Zadeh a methodology closely related to fuzzy logic FL_n and called *computing with words*, CW. The objects of computations in CW are words and propositions of a natural language (e.g. *small, large, oil prices are medium and slowly declining*) rather than numbers and symbols. The reader is referred to ZADEH (1999a, b), ZADEH/KACPRZYK (1999a, b), KACPRZYK/ZADROŻNY (2001b, 2010b) and ZADROŻNY/KACPRZYK (2006) for details; cf. also ZADEH (1997).

B. Perceptions and Their Results

Pieces of information human beings are able to use and operate on may come from measurements, e.g.

"It is −20°C", "John is 25", "The inflation rate is 1.2%".

On the other hand, humans have a remarkable ability to operate on, reason with, and make decisions on the basis of perception-based information. We mean perception of distance, size, colour, age, temperature, speed, force, direction, likelihood, etc. For instance,

"It is very frosty", "John is young", "The inflation rate is low".

Driving a car forms a typical example of a fully perception-based process. A basic difference between measurements and perceptions is that, generally, results of measurements are crisp (sharp, nonfuzzy), whereas results of perceptions are fuzzy and, thus, have to be modeled as fuzzy sets. One should mention here the *computational theory of perceptions*, CTP, a CW-based methodology for computing and reasoning with perceptions (see ZADEH (1999a, 2000, 2001, 2002) and BATYRSHIN *et al.* (2007) for details). Going further, one uses a joint methodology of *computing with words and perceptions*, CWP (see e.g. KACPRZYK/ZADROŻNY (2005b)).

C. Commonsense and Experience-Based Knowledge

Important ingredients of human knowledge, specialized and general as well, are commonsense knowledge and knowledge coming from experience. Their familiar examples are statements such as

"The flu is often accompanied by a high fever",
"Traffic jams are very rare in the late evening",
"Dynamic increases of stock indices are usually followed by fast corrections".

The main feature of those types of knowledge is again imprecision. Fuzzy sets are thus a suitable tool for their modeling and incorporating, say, into expert systems and decision support systems (see e.g. ZADEH (1983b, c), PETROVIC *et al.* (2006)).

D. Rules, Real Processes and Phenomena

First, we like to point to systems and processes whose mathematical model is not available as its construction would be too costly, the system/process is too complex or enigmatic, etc. A way out is then to create a *linguistic model*, a collection of *linguistic rules* of the form

IF *condition/premise* THEN *action/conclusion*

with imprecisely specified "IF" and "THEN" parts involving fuzzy sets, e.g.

IF pressure is *high* THEN lower the temperature *a bit*.

They are known as *fuzzy conditional statements*. What we get are *fuzzy rule-based systems*. Their applications go far beyond engineering and control and encompass areas such as expert systems and decision support, modeling and simulation in economics, management, biology, ecology, sociology, etc. (see e.g. AHN *et al.* (2005), PAŁUBICKI (2007), PETROVIC *et al.* (2006), SILER /BUCKLEY (2005)).

Second, let us notice that many real phenomena and processes have an imprecise character in the sense that they are not two-state switches. Instead, their feature are various possible intensities and a gradual transition from one extreme state to the other, through many intermediate states, which can be modeled using fuzzy sets. Familiar examples are processes and phenomena from the areas of technology (wearing down, drying, clarification, ...), biology (maturing, growth, aging, ...), economics (economic conditions, inflation, ...), sociology (structural transformations, changes in views, ...), ecology (climate changes, changes in ecosystems, ...).

Concluding, fuzzy sets and their methodology considerably enhance our capabilities to formalize and solve problems from many fields of science and technology. They also broaden those capabilities by areas which seem to be inaccessible for conventional mathematical tools and methods. In comparison with the classical set-based look, fuzzy sets and their methodology offer a richer and more adequate optics of phenomena, processes and problems. This makes it possible to build better models, find better solutions and, finally, to create better and more advanced (more intelligent, more efficient, ...) systems, technologies and appliances.

* * *

The consecutive chapters of Part I will present those elements of fuzzy sets theory which are relevant to the main aim of this book in a more or less direct way. A brief survey of the contents of those chapters is placed below.

In Chapter 2, basic notions of the language of fuzzy sets will be introduced and studied. The first section defines the concept of a fuzzy sets and discusses various possible interpretations of membership degrees. Moreover, for completeness, it presents many-valued logical roots of fuzzy sets with special reference to Łukasiewicz logic (see also Subsections 2.2.3 and 2.4.6), although logical considerations are by no means the aim of this book. Section 2.2 deals with the standard algebra of fuzzy sets based on the minimum and maximum operations. Main characteristics of a fuzzy set will be presented in Section 2.3. In particular, convexity and fuzziness measures are discussed therein. According to the decomposition property, each fuzzy set can be viewed as a family of ordinary sets. This leads in a straightforward way to creating maps of fuzzy sets (Subsection 2.3.2). Section 2.4 shows a more flexible and sophisticated approach to operations on fuzzy sets in which triangular norms and conorms as well as arbitrary negations are involved. Also, the issue of implication operators will be outlined. The last section of Chapter 2 is devoted to fuzzy numbers and linguistic variables, two very important concepts within fuzzy logic. In particular, the powerful center-of-gravity (COG) method is described there.

Chapter 3 begins with the issue of flexible querying in databases. In a way, this is a pretext for presenting in Section 3.2 some more advanced notions and questions from the area of triangular norms, namely Archimedean triangular norms, strictness and nilpotency, and ordinal sums. The issue of (additive) generators of triangular norms and complementarity of triangular norms and conorms will also be discussed (Subsections 3.2.3 and 3.2.4).

The subject of Chapter 4 are questions of information aggregation, a process in which many different pieces of information are integrated into one piece. The cases of conjunctive, disjunctive, averaging (including OWA), and mixed aggregation operators will be presented. Section 4.6 shows applications of aggregation operators to the Bellman-Zadeh model of decision making in a fuzzy environment. Finally, the last section is devoted to mean values of aggregations and averaging with respect to triangular norms. A direct relationship between the latter issue and quasi-arithmetic means is indicated.

Chapter 5 begins with a study of fuzzy relations and their classes. In particular, similarity measures and similarity classes will be defined. Subsection 5.1.5 deals with cardinality-based similarity measures of sets (Jaccard, matching, and overlap coefficients) and the inclusion coefficient. In Subsection 5.1.6, inclusion and equality measures of fuzzy sets inspired by Łukasiewicz logic will be constructed. Finally, Section 5.2 presents the basics of approximate reasoning, including the compositional rule of inference. In Section 5.3, they will be used when showing the idea of fuzzy control and, more generally, of fuzzy rule-based systems.

1.2 Adding the Incompleteness Factor

The concept of a fuzzy sets abounds with numerous generalizations and extensions. Main instances are:

- *type-2 fuzzy sets* introduced in ZADEH (1975a); the degrees of membership in these objects are themselves fuzzy sets of numbers from the interval [0, 1] (see also MIZUMOTO/TANAKA (1976, 1981));
- *L-fuzzy sets* proposed by GOGUEN (1967); the degrees of membership are then taken from a partially ordered set forming a lattice;
- *interval-valued fuzzy sets*;
- *I-fuzzy sets* also known as *Atanassov's intuitionistic fuzzy sets*.

We like to focus in this book on the ideas of an interval-valued fuzzy set and I-fuzzy set. Resulting from different motivations, both of them are tools for representing incompletely known fuzzy sets, i.e. tools for modeling imprecision combined with incompleteness of information. What we then deal with is thus a fusion of two important forms of imperfect information.

The membership degree of an element x of the universe in an interval-valued fuzzy set is a closed interval of numbers rather than a single number from [0, 1]. The endpoints of that interval form a lower bound and an upper bound, respectively, on the membership degree of x in an incompletely known fuzzy set modeled by the interval-valued fuzzy set. Each interval-valued fuzzy set is thus a pair of fuzzy sets in which the first component is contained in the second one. Interval-valued fuzzy sets are, on the other hand, a special case of type-2 fuzzy sets and L-fuzzy sets as well.

The concept of an interval-valued fuzzy set was introduced simultaneously and independently by GRATTAN-GUINESS (1975), JAHN (1975), SAMBUC (1975) and ZADEH (1975a). Its applications encompass the areas of decision making, image processing, approximate reasoning, linguistic summarization in databases, etc. (see e.g. BUSTINCE (2000), BUSTINCE *et al.* (2008, 2010), HERRERA *et al.* (2005), NIEWIADOMSKI (2008) for further references and details).

An I-fuzzy set is also a pair of fuzzy sets. The first component, however, must be contained in the complement of the second one. These two fuzzy sets, respectively, form pieces of *positive* and *negative information* about an incompletely known fuzzy set we like to represent by means of that I-fuzzy set. To each element x a pair of degrees is now assigned: its *membership degree* and *non-membership degree.* In many practical situations, the emphasis on defining objects in this *bipolar way* through positive and negative information makes I-fuzzy sets a more useful and adequate tool than interval-valued fuzzy sets, although these two concepts are mathematically equivalent (see e.g. ARIELI *et al.* (2004) and DESCHRIJVER/KERRE (2003)). To be more specific, I-fuzzy sets make it possible and even force us to think about and look at possible options and decisions in terms of their advantages (positive features) and disadvantages (negative features), satisfaction and dissatisfaction, trust and distrust, and so on (see e.g. DE COCK/PINHEIRO DA SILVA (2006), VICTOR *et al.* (2009); see also DUBOIS (2008), DUBOIS/PRADE (2009)). This cannot be done using interval-valued fuzzy sets. The suitability of using the double, bipolar optics offered by I-fuzzy sets when solving problems is stressed by psychological investigations because decision makers have a tendency to forget about negative sides of decisions they consider. Moreover, I-fuzzy sets make it possible to take into consideration and model in a convenient way such issues as unclassifiability and abstention. All this justifies a more and more growing interest in and an intensification of theoretical and applied research on I-fuzzy sets in recent years. A natural area of applications of I-fuzzy sets is knowledge representation. Other examples of such areas are decision making, including group decision making (see e.g. LI/YANG (2003), LIU/WANG (2007), PANKOWSKA (2005, 2007, 2008), PANKOWSKA/WYGRALAK (2005, 2006), SZMIDT/KACPRZYK (1998, 1999), WU/ZHANG (2011), YE (2010)), image processing (see e.g. BUSTINCE *et al.* (2005), DYCZKOWSKI (2010), VLACHOS/ SERGIADIS (2007b)), classification (see e.g. SZMIDT/KUKIER (2006, 2008)), social networks (see STACHOWIAK (2009, 2010)).

I-fuzzy sets are younger than interval-valued fuzzy sets and were introduced by K. T. Atanassov in 1983 under the name "intuitionistic fuzzy sets" (see ATANASSOV/STOEVA (1983), ATANASSOV (1986, 1999, 2003)). A few years ago there was a debate on the suitability of that original name as, actually, there is a terminological clash between Atanassov's objects and "true" intuitionistic fuzzy sets, namely intuitionistic logic-based fuzzy sets from TAKEUTI/TITANI (1984). Details of that debate are given in DUBOIS/GOTTWALD *et al.* (2005) and GRZEGORZEWSKI/MRÓWKA (2005).

Consequently, to avoid any confusion, some alternative, modified names have been proposed, e.g. "Atanassov's intuitionistic fuzzy sets", "IF-sets" and just "I-fuzzy sets" used in this book. In retrospect and in view of the usefulness of Atanassov's objects, that debate - though necessary - seems to be of secondary importance.

The original Atanassov's formulation of I-fuzzy set theory does not allow for the use of arbitrary triangular norms. A triangular norm-based generalization of operations on I-fuzzy sets was proposed in DESCHRIJVER/KERRE (2002). In this book, we will use the approach developed in PANKOWSKA/WYGRALAK (2003, 2004a, b, 2006) and WYGRALAK (2007, 2010). The hesitation factor connected with an I-fuzzy set and resulting from incompleteness of information is then also modeled in a triangular norm-dependent way.

* * *

Chapter 6, the last one in Part I, presents the issue of incompletely known fuzzy sets. We begin with the prototypical and inspiring case of incompletely known sets and their modeling in a natural, three-valued way. It involves the *uncertainty area* of such a set which is composed of elements whose status remains uncertain or unknown (Section 6.1). Next we will move on to incompletely known fuzzy sets and their modeling by interval-valued fuzzy sets and I-fuzzy sets.

Interval-valued fuzzy sets are studied in Section 6.2. We will define and present properties of their uncertainty areas (being fuzzy sets) and uncertainty degrees. Clearly, we will also discuss basic operations, including the case of operations based on interval-valued triangular norms and conorms.

Section 6.3 is devoted to I-fuzzy sets and their uncertainty areas. They are then called *hesitation areas*, and are defined in a flexible way via a strong negation and a triangular norm. Our approach reflects a common intuition saying that hesitation consists in "not *yes* and not *no*", "not *pro* and not *contra*", etc. We will study general properties of hesitation degrees, i.e. membership degrees in the hesitation area of an I-fuzzy set. It turns out that if a nilpotent triangular norm is involved, the hesitation degree of x is (a sort of) a "pure" size of ignorance as to the membership degree of x in an incompletely known fuzzy set under modeling. If a strict triangular norm or the triangular norm minimum is used, that hesitation degree becomes the size of ignorance combined with a fuzziness index. This will be illustrated by practical, model examples in Subsection 6.3.4 using the three basic triangular norms. Moreover, connections with classification issues are discussed.

1.3 Counting under Information Imprecision

One of the most important types of information about a given collection of certain elements is its cardinality, the number of elements constituting the collection. It forms a basis for making a decision in a lot of situations. This is why counting belongs to the most basic and frequent mental activities of human beings. However, one should distinguish here between two very different cases.

- The objects of counting are *precisely specified*, e.g.

"How many schoolboys are there in the class?".

The counting process then collapses to the usual counting by means of the natural numbers in a collection forming a set. Clearly, the natural numbers, i.e. finite cardinal numbers, are also used by humans in the strongly related process of calculation. The process of counting in sets can be extended to arbitrary infinite sets by using (transfinite) cardinal numbers offered by set theory.

- In many practical situations, however, the objects of counting are *imprecisely (fuzzily) specified*, e.g.

"How many tall schoolboys are there in the class?".

We then deal with queries/tasks of the form "How many x's are A?" with A denoting an imprecise property. The counting process is now a process going on in a vague collection forming a fuzzy set. So, we try to perform *counting under imprecision of information about the objects of counting* or, in short, *counting under information imprecision*, or simply *counting under imprecision.*

As one sees, counting in a set is a comfortable and trivial situation as there is a clear distinction between those objects which have to be counted and, on the other hand, those ones which have to be skipped in the counting process. When having to count in a fuzzy set, the situation becomes more complex and sophisticated, and requires much more intelligence since to be in a fuzzy set is a matter of degree. Indeed, referring to the above example, each schoolboy in the class is tall to a degree. One must then face a qualitatively new problem: *what and how to count*? What we now deal with can thus be called and understood as *intelligent counting.* This is additionally illustrated below.

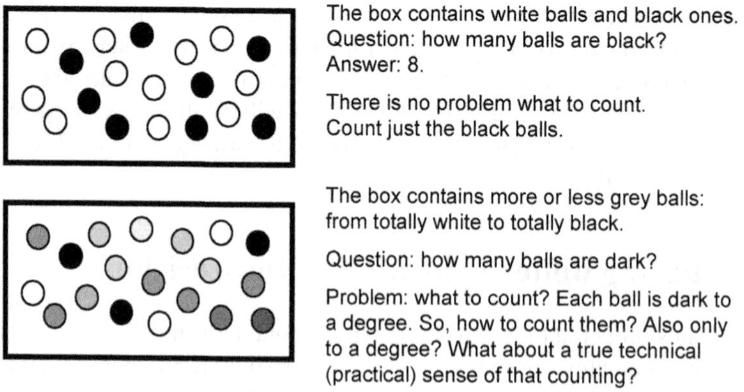

The box contains white balls and black ones.
Question: how many balls are black?
Answer: 8.
There is no problem what to count.
Count just the black balls.
The box contains more or less grey balls:
from totally white to totally black.
Question: how many balls are dark?
Problem: what to count? Each ball is dark to
a degree. So, how to count them? Also only
to a degree? What about a true technical
(practical) sense of that counting?

Fig. 1.1 Counting (top) in a set and (bottom) in a fuzzy set

Generally, counting in a fuzzy set is in a very natural way a more or less subjective process which may lead to different numerical results. Each of them can, however, be treated as a reasonable and correct one. This non-uniqueness results from using different reasonable optics in the counting process.

An even more sophisticated situation is when counting has to be performed in an interval-valued fuzzy set or I-fuzzy set, i.e. in an incompletely known fuzzy set. We then deal with *counting under information imprecision and incompleteness*.

Counting in fuzzy sets and their extensions as well as the resulting cardinalities will be the subject of Part II, the key part of this book. We like to look at that issue from the viewpoint of practice and human-consistency. Cardinalities of fuzzy sets and their extensions can be studied in a formal, purely mathematical way. This is by no means the aim or the case of this book. On the contrary, presenting various counting procedures and the resulting forms of cardinality, we will show their true technical, practical sense. What is more, we will emphasize that they reflect and formalize real counting procedures performed by human beings when counting under information imprecision or imprecision combined with incompleteness of information. We like to show applications to intelligent systems and decision support. In many cases, we will look at old constructions, well-known in the subject literature, in a fresh way: just from the viewpoint of human counting. By the way, reviews of various, more or less application-oriented approaches to cardinalities of fuzzy sets can be found in WYGRALAK (1996a, 2003a); see also DUBOIS/PRADE (1985) and WYGRALAK (1986, 2002). The dominating issue in our study is counting in finite fuzzy sets and their extensions since it plays a key role in applications. Nevertheless, for completeness, Chapter 11 is devoted to cardinalities of infinite fuzzy sets.

As to human counting procedures under information imprecision, two of them seem to be absolutely fundamental and, therefore, must be distinguished.

- Counting by thresholding (CAC, *cut-and-count method*). This basic procedure consists in

  – establishing a threshold membership degree t,
  – counting up all the elements whose membership degree is $\geq t$ or $> t$.

For instance, we try to answer the query "How many tall schoolboys are there in the class?". First, we establish a threshold height for a schoolboy to be counted as tall, say, 170 cm. Next, we count all schoolboys whose height is ≥ 170 cm or > 170 cm.

- Counting by multiple thresholding (MCAC, *multiple cut-and-count method*), exemplified by the news item

"The explosion injured 25 passersby, 10 of them seriously"

with two threshold values.

The CAC method leads to the *scalar approach* in which the cardinality of a (finite) fuzzy set is *scalar*, is a single nonnegative real number. On the other hand, the MCAC method leads to the *fuzzy approach*. The cardinality of a fuzzy set is then a *fuzzy cardinality*, a fuzzy set of nonnegative integers. This approach agrees with a simple intuition suggesting that graduated membership in a fuzzy set should imply graduated description of its cardinality by means of different numerical values. Both of the approaches will be extended to interval-valued fuzzy sets and I-fuzzy sets.

The scalar approach is conceptually and computationally simpler and, consequently, it is not surprising that one uses it more frequently than the fuzzy one. We will show, however, that fuzzy cardinalities are not artificial, sophisticated constructions. In common with scalar cardinalities, they also reflect the results of human counting procedures under information imprecision, and have a true practical sense. Constructions, issues and aspects of counting and cardinality under information imprecision we like to present in Part II are collected and summarized below.

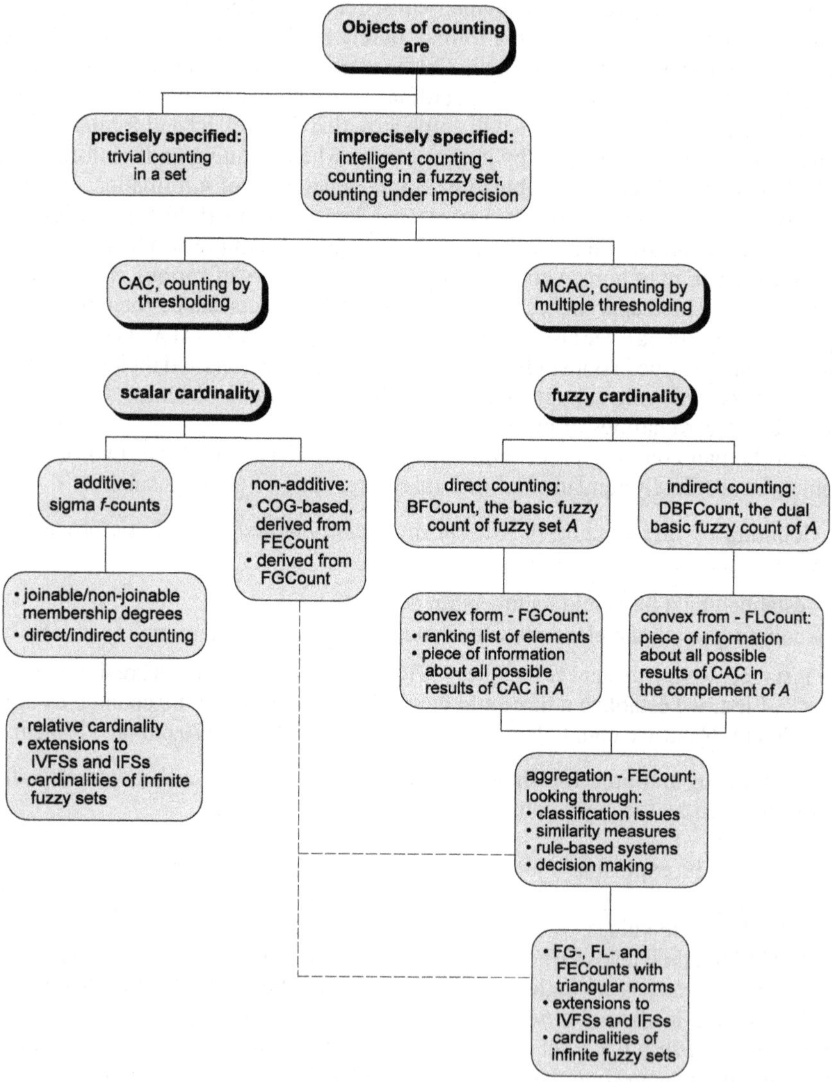

Fig. 1.2 Counting under information imprecision – tools, issues and aspects

Chapter 7 will present in detail introductory remarks and general motivations concerning the question of counting under imprecision possibly combined with incompleteness of information.

* * *

Chapter 8 is devoted to the scalar approach to cardinalities of fuzzy sets and their extensions. We begin with the concept of *sigma f-counts* of fuzzy sets proposed and developed in WYGRALAK (1998b, 1999b, c, 2000a, b, 2003a, 2008b, 2009). The sigma f-count of a fuzzy set A is the sum of weighted membership grades $f(A(x))$ in A. f denotes a weighting function. It determines the degree of participation of element x – having membership degree $A(x)$ – in the counting process. Sigma f-counts are flexible constructions bringing together all concepts of scalar cardinality used in the existing subject literature (see Section 8.1). The choice of f determines the optics of counting. Clearly, the identity function gives the usual *sigma count* of a fuzzy set. It is, speaking historically, the first concept of scalar cardinality and, generally, cardinality of a fuzzy set, introduced in DE LUCA/TERMINI (1972) and developed in ZADEH (1983a). We will show in a simple way that sigma f-counts have a true technical sense and reflect human procedures of counting under information imprecision (Subsections 8.1.3 and 8.4.3). An essential distinction between the cases of joinable and non-joinable membership degrees will be made in this context. In Section 8.2, we study sums and products of sigma f-counts. In particular, the complementarity rule is discussed. That rule makes it possible to replace direct counting in A with indirect counting performed via counting in the complement of A.

Section 8.3 will be devoted to the *relative sigma f-count* of a fuzzy set. It is an important concept from the viewpoint of applications presented in Chapter 10. For instance, we mean applications in a computational approach to linguistic quantifiers.

In Section 8.4, extensions of sigma f-counts to interval-valued fuzzy sets and I-fuzzy sets are defined and investigated, including the case of relative cardinalities.

Although the concept of sigma f-counts is very general, it does not encompass all forms of scalar cardinality which are possible to imagine and might be useful in practice. Let us mention non-additive scalar cardinalities. At least in some situations, non-additivity seems to be acceptable when counting under information imprecision. Non-additive scalar cardinalities derived from fuzzy cardinalities are presented in Subsection 9.2.2 (COG-based method) and Section 10.3.

* * *

The fuzzy approach will be studied in Chapter 9. Section 9.1 presents various counting methods leading to a fuzzy cardinality of a fuzzy set A. Our starting point is MCAC, counting by multiple thresholding. Going to extreme, MCAC can involve all possible threshold values. Combining all the results, one gets a fuzzy set of nonnegative integers which will be called the *basic fuzzy count of A*, BF(A). It seems to collapse to the historically first concept of fuzzy cardinality which was proposed in ZADEH (1979). BF(A) can be viewed as an encoded and compact piece of information about all possible values of cardinality when counting in A is performed via thresholding. We will show that the basic fuzzy count is used by humans in many situations as a convenient basis for coming to a decision.

The next concept of fuzzy cardinality we like to present is inspired by a simple trick used when having to count up the elements of an ordinary set $\boldsymbol{A}$. If more convenient, direct counting in $\boldsymbol{A}$ is then replaced with *indirect counting*, i.e. counting in the complement $\boldsymbol{A}'$. In the second step, the result is subtracted from the cardinality of the whole universe. Doing the same for fuzzy sets, counting in A can be performed *indirectly* via determining BF(A'), the basic fuzzy count of the complement of A. However, after subtraction, one gets a result differing from BF(A), and called the *dual basic fuzzy count of A*, DBF(A).

Both BF(A) and DBF(A) are generally non-convex fuzzy sets. By a simple modification, one can make them convex. The resulting fuzzy cardinalities are known in the subject literature as the *FGCount* and *FLCount of A*, respectively. We will denote them in this book by FG(A) and FL(A), respectively. The concept of FG(A) has been proposed in BLANCHARD (1981, 1982) and ZADEH (1981b), whereas the first formulation of FL(A) is given in ZADEH (1983a).

FG(A)(k), the membership degree of k in FG(A), can be interpreted as a degree to which A has *at least k* elements. However, we will also look at FGCounts from the viewpoint of applications and human counting under information imprecision (Subsection 9.1.2). Two facts are worth emphasizing in this context. First, FGCounts seem to be equivalent to ranking lists of objects. Second, FG(A) forms a dynamic and compact piece of information about all possible results of counting by thresholding (sharp or not) in A. Let us add that FG(A) is a convenient basis for determining the result of human counting by thresholding and joining discussed in Subsection 8.1.2.

Returning to FLCounts, FL(A)(k) forms a degree to which A has *at most k* elements. From the viewpoint of practice, however, a more interesting fact seems to be that FL(A) is in essence an encoded and compact piece of information about all possible results of counting by thresholding in the complement A'.

Subsection 9.1.3 will present and investigate another, more advanced type of fuzzy cardinality, namely FE(A), the *FECount of A* introduced in ZADEH (1983a) and WYGRALAK (1983a). Formally, FE(A) is the intersection of FG(A) and FL(A) and, thus, forms an aggregation of information carried by the FGCount and FLCount. FE(A)(k) is a degree to which A has *exactly k* elements. We will show that FE(A) is a compact piece of information about all possible results of counting by thresholding in the intersection $\boldsymbol{A} \cap \boldsymbol{A}'$, i.e. in the fuzzy set of "embarrassing" elements. They are elements having to a degree both an imprecise property and its opposite. This suggests a strong connection between FECounts and classification issues. That connection is investigated in Subsection 9.1.4. We introduce a simple and natural concept of t-classification of elements of the universe, where t from (0, 0.5] is a threshold value. Next, we show that FE(A) is then an encoded piece of information about the number of unclassifiable elements in each possible t-classification.

FECounts are also the subject of Section 9.2 in which we try to look at those fuzzy cardinalities in three novel ways. First, we look through similarity measures. We will show that FE(A)(k) is the degree of similarity of A to a k-element set with respect to a natural similarity measure. In Subsection 9.2.2, FECounts are results of a simple rule-based approach. Finally, in Subsection 9.2.3, we use the optics of the Bellman-Zadeh model of decision making in a fuzzy environment from Section 4.6.

Determining the cardinality of a fuzzy set A is then a decision process with FE(A) becoming the fuzzy decision. Its defuzzification leads to the COG-based scalar cardinality of A derived from FE(A), a non-additive type of scalar cardinality.

In Section 9.3, we will present and study triangular norm-based generalizations of FG-, FL-, and FECounts introduced in WYGRALAK (2001, 2003a). Those of FGCounts can be a basis for deriving further types of scalar cardinalities in Section 10.3. Moreover, we will extend FG-, FL-, and FECounts to interval-valued fuzzy sets and I-fuzzy sets. The emphasis will be placed on relationships between those extensions and human counting under imprecise and incomplete information, and on issues of unclassifiability. Our discussion will be illustrated by practical examples.

Section 9.4 is devoted to comparisons between and arithmetic operations on fuzzy cardinalities of fuzzy sets, especially if FG-, FL- or FECounts from Section 9.1 are involved. We begin with a study of the relation of *equipotency* of two fuzzy sets, i.e. of being of the same cardinality. Various characterizations of equipotency corresponding to various types of fuzzy cardinalities will be given. An analogous study of inequalities between fuzzy cardinalities is placed in Subsection 9.4.2. One should mention that inequality relations for fuzzy cardinalities are partial order relations in general and, consequently, one must reckon with the existence of fuzzy sets which are incomparable with respect to their cardinalities.

The subject of Subsection 9.4.3 are arithmetic operations on fuzzy cardinalities with special reference to addition and multiplication. We define these two operations in a classical-like way via the cardinality of the sum and cartesian product of fuzzy sets. It will be shown in which cases addition and multiplication can be equivalently performed by means of the extension principle. We will present a long list of laws and properties of sums and products of fuzzy cardinalities, e.g. distributivity, cancellation laws, side-by-side addition and multiplication of inequalities, and inequalities between sums and products. Differences between the arithmetics of fuzzy cardinalities and nonnegative integers will be emphasized, too. It seems that one of the most important differences is the failure of the compensation property. If α and γ are fuzzy cardinalities – say, two FGCounts – and $\alpha < \gamma$, then a fuzzy cardinality β satisfying the equality $\alpha + \beta = \gamma$ does not exist or is not unique in general. Our study of operations will end by giving remarks on subtraction and division of fuzzy cardinalities.

Finally, Section 9.5 refers to extensions of FG-, FL- and FECounts to I-fuzzy sets and interval-valued fuzzy sets. Questions of equipotency relations for and arithmetic operations on those extensions will be briefly presented.

* * *

Chapter 10 shows selected applications of counting in and cardinalities of fuzzy sets and their extensions. We mean applications to some areas of intelligent systems and decision support. Bibliographical references encompassing these and other applications in broadly conceived computer science are collected in the last section of the chapter.

The first section deals with cardinality-based coefficients and similarity measures for fuzzy sets. Drawing inspiration from the constructions in Subsection 5.1.5, we define and investigate the inclusion, overlap, Jaccard, and matching coefficients involving sigma *f*-counts and COG-based scalar cardinalities. Their applications to describing similarities between numerical data will be shown. Moreover, we define distances and fuzziness measures based on the Jaccard and matching coefficients. Extensions to interval-valued fuzzy sets and I-fuzzy sets are also mentioned.

Section 10.2 outlines applications of cardinalities of fuzzy sets in time series analysis. We discuss applications to selection of a suitable probabilistic model, and to evaluation of forecast accuracy.

The next section presents applications in the modeling of linguistic quantifiers like *most, a few, about 20,* etc., and in the computational approach to linguistically quantified propositions from ZADEH (1983a). These applications are already classical, but their significance is difficult to overestimate. They still open the door to new applications of fuzzy set cardinality in intelligent systems, intelligent computing, and decision support.

In Section 10.4, applications to decision making in a fuzzy environment are proposed. We present an alternative formulation of the original Bellman-Zadeh model in which, speaking generally, aggregation is replaced with counting up. First, both fuzzy goals and fuzzy constraints are treated in a uniform way as fuzzy constraints. Second, those fuzzy constraints are understood as targets, whereas the decision alternatives are viewed as shooters shooting at consecutive targets. Finally, one counts up the hits, which may be more or less distant from the center of a target, and one looks for the best shooter.

Section 10.5 is devoted to applications in group decision making, i.e. in finding a decision alternative that "best" suits to individual preferences of the members of a group of decision makers. The point is that, in practice, those preferences are generally divergent and fuzzy. We like to discuss two approaches. The *direct approach* is based on the very individual preferences. Our task is then to find a fuzzy set of decision alternatives such that a soft majority of decision makers, modeled by means of a *most*-type linguistic quantifier, is not against them. In the *indirect approach*, the individual preferences are first aggregated into a matrix of group preferences and, next, we look for a fuzzy set of decision alternatives preferred by the group over a soft majority of the remaining options. For both of the approaches, we propose improved algorithms of group decision making involving triangular norms and arbitrary scalar cardinalities. Two cases are considered: those of complete and incomplete information which is necessary to specify the intensity of individual preferences.

* * *

Chapters 8-10 deal with counting in and cardinalities of finite fuzzy sets and their extensions, which is certainly the key case in applications. For the sake of completeness, Chapter 11 will be devoted to cardinalities of infinite fuzzy sets (see WYGRALAK (1991a, 1992, 1993a, b, 1996a)). As to scalar cardinalities, it seems that only counting by thresholding (sharp or not) can be extended to infinite fuzzy sets without any additional preassumptions. As to the fuzzy approach, we define and study natural extensions of FG-, FL-, and FECounts to infinite fuzzy sets, including questions related to the Continuum Hypothesis. Moreover, axiomatic approaches to fuzzy cardinalities of those fuzzy sets are mentioned. Extensions to infinite interval-valued fuzzy sets and I-fuzzy sets are also outlined.

Part I

Elements of Fuzzy Sets and Their Extensions

Chapter 2

Basic Notions of the Language of Fuzzy Sets

The purpose of this chapter is to present the basics of fuzzy sets. We start with defining the very concept of a fuzzy set, discuss simple examples, and emphasize that many-valued logic is a suitable logical basis for fuzzy sets. Further, the standard approach to operations on fuzzy sets is described. Useful characteristics of a fuzzy set are defined and studied in Section 2.3, including questions of convexity and fuzziness measures. We also discuss the decomposition property of a fuzzy set and the resulting maps of fuzzy sets. A flexible approach to operations on fuzzy sets involving arbitrary negations and triangular norms and conorms is presented in Section 2.4. Implication operators, too, are briefly analyzed therein. The subject of the last section are two interrelated concepts, namely fuzzy numbers and linguistic variables. Fuzzy numbers form a tool for modeling imprecise numerical data. We will define basic types of fuzzy numbers. Second, the extension principle, arithmetic operations on and inequalities between fuzzy numbers will be discussed. Finally, we will move on to the question of linguistic variables, i.e. variables attaining linguistic values interpreted by means of fuzzy numbers.

2.1 What Are Fuzzy Sets?

Throughout the book, U will denote a *universe of discourse*, also called a *universal set* or briefly a *universe*. It will be understood as a given nonempty set, finite or not, of some elements.

2.1.1 The Concept of a Fuzzy Set

Let us recollect that a set in U can be presented graphically by means of its Venn diagram (see Figure 2.1).

Imagine a more complex object in U, a nebula or – in other words – a vague, nebulous collection of elements with unsharp boundary. We will call it a *fuzzy set*. Painting Venn diagrams of sets, one uses a palette composed of just two colors: white and black. Painting (counterparts of) Venn diagrams for fuzzy sets, one must use a richer palette (see again Figure 2.1). It should contain all shades of greyness: from total white to total black.

M. Wygralak: *Intelligent Counting Under Information Imprecision*, STUDFUZZ 292, pp. 19–70.
DOI: 10.1007/978-3-642-34685-9_2

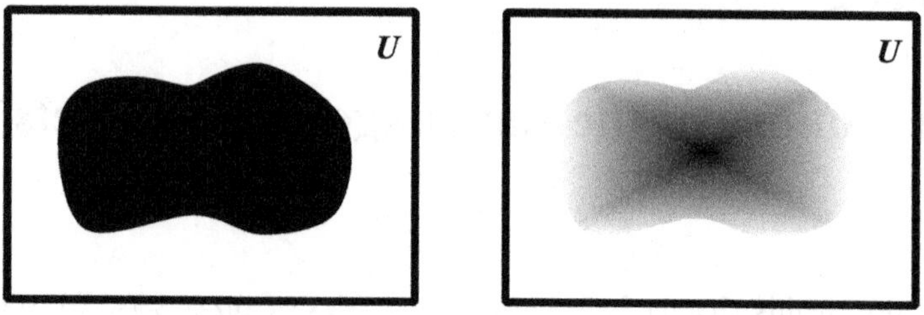

Fig. 2.1 Venn diagram (left) of a set and (right) of a fuzzy set in $\boldsymbol{U}$

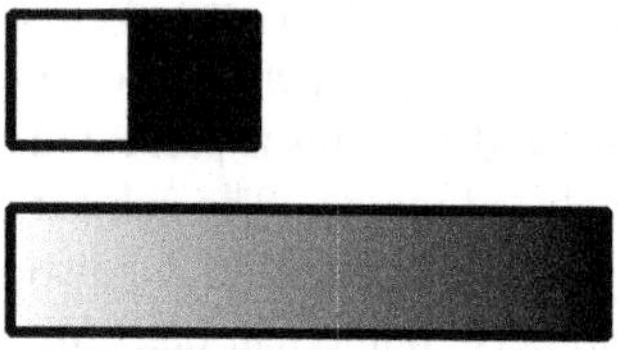

Fig. 2.2 Color palettes (top) for sets and (bottom) for fuzzy sets

In contrast to the case of sets, a fuzzy set "disappears" gradually as its boundary is unsharp (an analogy with a comet and its tail). Some points of the Venn diagram of a fuzzy set can be totally black. They form the *core* of that fuzzy set.

How to formally describe a fuzzy set, a nebula of elements? A good idea seems to be to model oneself on the case of simpler objects which are sets. As one knows, each set $\boldsymbol{D} \subset \boldsymbol{U}$ can be uniquely described by means of its *characteristic function*, a function $\mathbf{1}_{\boldsymbol{D}}: \boldsymbol{U} \to \{0, 1\}$ such that

$$1_D(x) = \begin{cases} \mathbf{1}, & \textbf{if } \boldsymbol{x} \in \boldsymbol{D}, \\ \mathbf{0}, & \textbf{otherwise.} \end{cases}$$

What is more, sets may be identified with their characteristic functions. Our intuition thus suggests to describe fuzzy sets by means of some generalizations of characteristic functions and, going further, to identify fuzzy sets with those generalizations. Good candidates for such generalizations are functions of the form

$$\boldsymbol{U} \to [0, 1]$$

as the replacement of $\{0, 1\}$ with the interval $[0, 1] \subset \mathbb{R}$ of real numbers suitably reflects the transition from the two-color palette to the full rainbow of greyness. These functions are called *generalized characteristic functions* or *membership functions* of fuzzy sets. Fuzzy sets, still understood as nebulas of elements, are then identified with their membership functions and, finally, one begins to call the very membership functions *fuzzy sets*. This reasoning leads to the following formal definition:

a *fuzzy set* is a function $U \to [0, 1]$.

One also uses the terms "a fuzzy set in $\boldsymbol{U}$", "a fuzzy subset", "a fuzzy subset in U". Throughout the book, single italicized capitals A, B, C, ... will denote fuzzy sets (viewed already as functions $\boldsymbol{U} \to [0, 1]$). Single bold italicized capitals $\boldsymbol{A}$, $\boldsymbol{B}$, $\boldsymbol{C}$, ... denote sets in $\boldsymbol{U}$, also termed *ordinary sets* or *crisp sets*. As usual, the cardinality of a set $\boldsymbol{A}$ will be denoted by $|\boldsymbol{A}|$.

The concept of a fuzzy set is clearly much more general than that of a set. Each set $\boldsymbol{D} \subset \boldsymbol{U}$ can be presented in the language of fuzzy sets as 1_D. In particular, functions $1_\emptyset$ (identically equal to 0) and 1_U (identically equal to 1) are thus the empty set and the whole universe, respectively, expressed as fuzzy sets. Fuzzy sets which are not sets, i.e. membership functions which do not collapse to characteristic functions, will be called *proper fuzzy sets*. On the other hand, each fuzzy set A: $\boldsymbol{U} \to [0, 1]$ can be expressed in the language of sets as a set of ordered pairs

$$\{(x, A(x)): x \in \boldsymbol{U}\},$$

i.e. as a set of pairs

(*element*, *its weight*).

Referring to the graphical representation in Figure 2.1, these pairs can be viewed as pairs of the form

(*pixel*, *its degree of greyness*).

The number $A(x) \in [0, 1]$ is called the *membership degree* or *membership grade of x in A*. For instance, $A(x) = 0.8$ means that x belongs to (or is in) the fuzzy set A to degree 0.8. To belong to a fuzzy set is thus a matter of degree. In particular, that degree can be equal to 0 (full nonmembership) or 1 (full membership). This graduation for sets is reduced to the scale $0 - 1$.

2.1.2 Examples and Interpretations

Fuzzy sets are a suitable tool for modeling imprecise information and, in particular, imprecise terms of natural language (Section 1.1). Let u look at some examples.

(a) The following figure presents a fuzzy set A of tall men in the universe (0, 300] of heights in centimeters. We use this redundant interval to avoid any discussion about a minimum and maximum possible height of an adult male. So, A: (0, 300]→[0, 1]. The x-axis of the graph is restricted to [160, 200] for convenience.

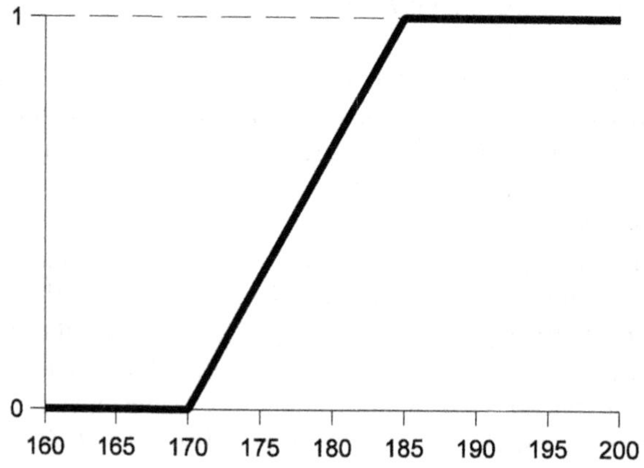

Fig. 2.3 Fuzzy set "tall men"

The values of function A are generally more or less subjective, whereas its general shape seems to be objective. A is a result of a simple way of doing:

- fix a height $h1$ up to which you like to consider a man to be absolutely not tall,
- fix a height $h2$ from which a man will be treated as absolutely tall,
- connect the points ($h1$, 0) and ($h2$, 1).

What we see in Figure 2.3 is just the case when $h1 = 170$, $h2 = 185$, and a straight line connects (170, 0) with (185, 1).

(b) Let $\boldsymbol{U}$ denote a population of men in which Bob, John, Peter and Paul, respectively, are 178, 167, 195 and 182 cm tall, respectively. Applying the interpretation of "tall" from (a), we get the following membership degrees in a fuzzy set B: $\boldsymbol{U}$→[0, 1] of tall men in $\boldsymbol{U}$:

$$B(\text{Bob}) = 0.53, \quad B(\text{John}) = 0, \quad B(\text{Peter}) = 1, \quad B(\text{Paul}) = 0.8.$$

(c) Figure 2.4 shows fuzzy sets of real numbers "about 4" and "approximately between 6 and 8" modeled as a *triangular* and *trapezoidal* fuzzy set, respectively.

(d) Finally, Figure 2.5 presents a *bell-shaped* fuzzy set of comfortable temperatures in the flat (in °C).

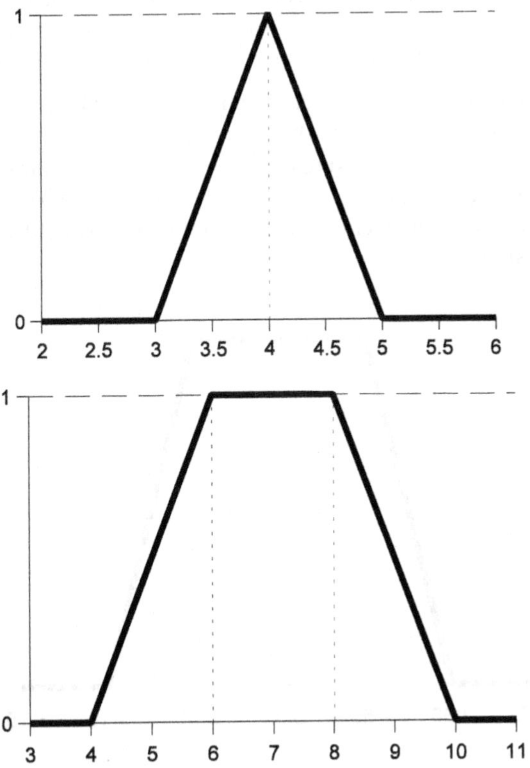

Fig. 2.4 Fuzzy sets of real numbers (top) "about 4" and (bottom) "approximately between 6 and 8"

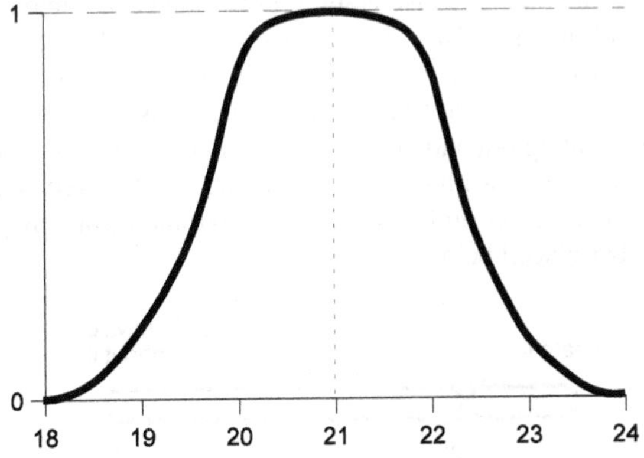

Fig. 2.5 Fuzzy set "comfortable temperature"

A question of importance for practice is how to interpret the membership degree of an element $x \in \boldsymbol{U}$ in a given fuzzy set. The above examples are a good basis for presenting a variety of related options. Indeed, that degree can be viewed as a

- degree of fulfilment, intensity, or saturation of an imprecise property modeled by the fuzzy set (see examples (a)-(d));
- measure of quality of x or measure of similarity of x to the ideal, e.g. a measure of similarity of a temperature to an absolutely comfortable one (see (a)-(d));
- degree of preference, satisfaction, or acceptability (see esp. (d));
- degree of possibility.

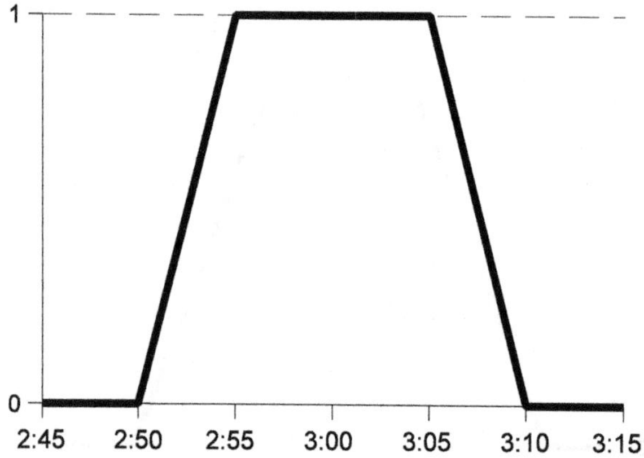

Fig. 2.6 Membership degrees as possibility degrees

As regards the last option, we then mean a degree of possibility that the value x will be attained by a variable under consideration. For instance, assume one knows that Mark is punctual and announced his arrival at about 3:00 pm. The real time of Mark's arrival can be modeled by means of a fuzzy set T presented in Figure 2.6. $T(x)$ is then the degree of possibility that Mark will arrive exactly at x.

Finally, worth pointing out and emphasizing is that membership degrees are not probabilities in general. Let us illustrate the difference using a very simple example. Assume Alice has marked her preferences, say, as to tomato soup by putting a cross on the preference bar placed below.

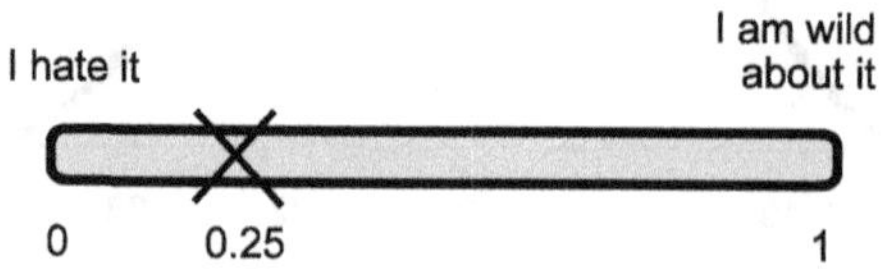

Fig. 2.7 Alice's preferences as to tomato soup

So, Alice likes tomato soup not very much, only to degree 0.25. Tomato soup belongs to a fuzzy set of Alice's favourite dishes to degree 0.25, in other words. This membership degree is a preference degree, a degree of satisfaction, a degree of similarity (according to Alice) to a perfect dish, etc. It is not a probability. Indeed, any interpretation of the 0.25 through frequency ("Alice likes tomato soup every four days, every four spoons, etc.") is difficult to accept. Eating tomato soup, Alice's level of satisfaction is always low, is just equal to 0.25.

In some cases, however, a probabilistic approach to membership degrees is possible. For instance, if 83% of the persons surveyed say that Paul is tall, then 0.83 can be taken as Paul's membership degree in a fuzzy set of tall men.

Concluding, fuzziness and randomness connected with probability are generally different notions. Randomness applies to precisely specified events whose occurrence is more or less uncertain. Fuzziness refers to imprecisely specified events which can occur with and are characterized by different degrees of intensity or saturation.

2.1.3 Remarks on Many-Valued Roots of Fuzzy Sets

The conventional notion of a set is based on classical two-valued logic. It offers only two truth values, logical values: 1 (true) and 0 (false). The truth value of a sentence $\boldsymbol{p}$ will be denoted by $[\![\boldsymbol{p}]\!]$. In classical logic, we thus have $[\![\boldsymbol{p}]\!] \in \{0, 1\}$. Consequently, the predicate $\in$ of membership in a set is also two-valued: either $x \in \boldsymbol{D}$ or $x \notin \boldsymbol{D}$ for each $x \in \boldsymbol{U}$ and $\boldsymbol{D} \subset \boldsymbol{U}$. Worth pointing out is that

$$1_D(x) = [\![x \in \boldsymbol{D}]\!]. \tag{2.1}$$

The predicate of membership in a fuzzy set is evidently many-valued, graduated since to belong to a fuzzy set is a matter of degree. Let us denote this predicate by $\in_\mathrm{m}$ with the index "m" standing for "many-valued" (see further discussion). We immediately notice that two-valued logic cannot be a logical basis for fuzzy sets. That basis is *many-valued logic*, logic accepting intermediate truth values lying between 0 and 1 (see e.g. GOTTWALD (1999, 2001)). Truth values are then called *truth degrees*. Many-valued connectives of negation, conjunction, inclusive disjunction, implication and equivalence used in many-valued logic will be denoted by $\neg_\mathrm{m}$, $\&_\mathrm{m}$, $\#_\mathrm{m}$, $\Rightarrow_\mathrm{m}$ and $\Leftrightarrow_\mathrm{m}$, respectively. The corresponding classical two-valued connectives are denoted in this book in exactly the same way, but without the subscript "m". $\forall_\mathrm{m}$ and $\exists_\mathrm{m}$, respectively, will denote many-valued general and existential quantifiers.

There is a wide variety of possible many-valued logical systems in which finitely or infinitely (even uncountably) many truth degrees are offered. As a logical foundation of fuzzy sets one can take any many-valued logic in which the set of all truth degrees collapses to the interval [0, 1] of real numbers. The following numerical interpretation of the many-valued membership predicate $\in_\mathrm{m}$ is then used:

$$A(x) = [\![x \in_{\mathrm{m}} A]\!]. \tag{2.2}$$

It forms a natural generalization of (2.1). The membership degree $A(x)$ of x in a fuzzy set $A: \boldsymbol{U} \to [0, 1]$ is thus viewed as a truth degree of the sentence saying that x belongs to A. On the left-hand side of (2.2), A is treated as a function, whereas the notation involving A on the right-hand side is set-theoretic. This duality is a consequence of identifying vague collections of elements with their membership functions. Summing up, the number $A(x) \in [0, 1]$ can be equivalently interpreted at three levels:

- logical level – $A(x)$ is the truth degree of the sentence "x belongs to A",
- set-theoretic level – $A(x)$ then forms the degree of membership of x in A,
- practical level – $A(x)$ is now a degree of fulfilment, intensity, or saturation of an imprecise property, a degree of similarity, satisfaction, acceptability, etc. (see Subsection 2.1.2).

The standard logical system used as a foundation of fuzzy sets is a particular type of many-valued logic with the real interval [0, 1] as set of truth degrees, namely *infinitely many-valued Łukasiewicz logic* $Ł_\infty$ or, in brief, *Łukasiewicz logic*. It was introduced by Jan Łukasiewicz in 1922 (see ŁUKASIEWICZ (1922); see also GILES (1976), GOTTWALD (2001) and WYGRALAK (1991b)). His three-valued logic announced in 1920, a starting point for defining $Ł_\infty$ two years later, was historically the first many-valued logical system constructed by man (see BORKOWSKI (1970)).

Before moving on to the basics of $Ł_\infty$, we have to establish some additional notation. Throughout the book, let $\wedge$, $\vee$, $\bigwedge$ and $\bigvee$, respectively, denote the operations of minimum, maximum, infimum and supremum. So, we have

$$a \wedge b = \min(a, b) \quad \text{and} \quad a \vee b = \max(a, b).$$

Moreover, let

$$a \to_Ł b = 1 \wedge (1 - a + b) = \begin{cases} 1, & \text{if } a \le b, \\ 1 - a + b, & \text{otherwise} \end{cases} \tag{2.3}$$

for $a, b \in [0, 1]$. $\to_Ł$ is called the *Łukasiewicz implication operator* (see also Subsection 2.4.5). The following numerical interpretation of many-valued connectives and quantifiers is used in Łukasiewicz logic understood as many-valued sentential calculus:

$$[\![\neg_{\mathrm{m}} p]\!] = [\![p]\!] \to_Ł 0, \tag{2.4}$$

$$[\![p \,\&_{\mathrm{m}}\, q]\!] = [\![p]\!] \wedge [\![q]\!], \tag{2.5}$$

$$[\![p \,\#_{\mathrm{m}}\, q]\!] = [\![p]\!] \vee [\![q]\!], \tag{2.6}$$

$$[\![p \Rightarrow_{\mathrm{m}} q]\!] = [\![p]\!] \to_Ł [\![q]\!], \tag{2.7}$$

$$[\![p \Leftrightarrow_{\mathrm{m}} q]\!] = [\![p \Rightarrow_{\mathrm{m}} q \,\&_{\mathrm{m}}\, q \Rightarrow_{\mathrm{m}} p]\!], \tag{2.8}$$

$$[\![\forall_{\mathrm{m}} x \in \boldsymbol{D}\colon \boldsymbol{r}(x)]\!] = \bigwedge_{a \in \boldsymbol{D}} [\![\boldsymbol{r}(x \mid a)]\!], \tag{2.9}$$

$$[\![\exists_{\mathrm{m}} x \in \boldsymbol{D}\colon \boldsymbol{r}(x)]\!] = \bigvee_{a \in \boldsymbol{D}} [\![\boldsymbol{r}(x \mid a)]\!], \tag{2.10}$$

where $\boldsymbol{p}$ and $\boldsymbol{q}$ denote two sentences, $\boldsymbol{r}(x)$ is a sentential formula with variable x, and $\boldsymbol{r}(x \mid a)$ is the ordinary substitution notation symbolizing a substitution of variable x by constant a in $\boldsymbol{r}(x)$. We thus get

$$[\![\neg_{\mathrm{m}} \boldsymbol{p}]\!] = 1 - [\![\boldsymbol{p}]\!], \tag{2.11}$$

$$[\![\boldsymbol{p} \Rightarrow_{\mathrm{m}} \boldsymbol{q}]\!] = 1 \wedge (1 - [\![\boldsymbol{p}]\!] + [\![\boldsymbol{q}]\!]), \tag{2.12}$$

$$[\![\boldsymbol{p} \Leftrightarrow_{\mathrm{m}} \boldsymbol{q}]\!] = 1 - |[\![\boldsymbol{p}]\!] - [\![\boldsymbol{q}]\!]|. \tag{2.13}$$

Hence

$$[\![\boldsymbol{p} \Rightarrow_{\mathrm{m}} \boldsymbol{q}]\!] = 1 \Leftrightarrow [\![\boldsymbol{p}]\!] \le [\![\boldsymbol{q}]\!], \tag{2.14}$$

$$[\![\boldsymbol{p} \Leftrightarrow_{\mathrm{m}} \boldsymbol{q}]\!] = 1 \Leftrightarrow [\![\boldsymbol{p}]\!] = [\![\boldsymbol{q}]\!]. \tag{2.15}$$

A unique *positively distinguished* truth degree in $Ł_\infty$ is 1, i.e. only sentences whose truth degree is 1 are regarded as true ones. It is easy to check that the system (2.4)-(2.10) collapses to the well-known rules of classical sentential calculus whenever $[\![\boldsymbol{p}]\!], [\![\boldsymbol{q}]\!], [\![\boldsymbol{r}(a)]\!] \in \{0, 1\}$. Immediate consequences of (2.4)-(2.8) are the following properties:

$$[\![\neg_{\mathrm{m}} \neg_{\mathrm{m}} \boldsymbol{p}]\!] = [\![\boldsymbol{p}]\!], \tag{2.16}$$

$$[\![\boldsymbol{p} \,\&_{\mathrm{m}} \neg_{\mathrm{m}} \boldsymbol{p}]\!] = [\![\boldsymbol{p}]\!] \wedge (1 - [\![\boldsymbol{p}]\!]) \le 0.5, \tag{2.17}$$

$$[\![\boldsymbol{p} \,\#_{\mathrm{m}} \neg_{\mathrm{m}} \boldsymbol{p}]\!] = [\![\boldsymbol{p}]\!] \vee (1 - [\![\boldsymbol{p}]\!]) \ge 0.5, \tag{2.18}$$

$$[\![\boldsymbol{p} \,\&_{\mathrm{m}} \boldsymbol{p}]\!] = [\![\boldsymbol{p}]\!], \quad [\![\boldsymbol{p} \,\#_{\mathrm{m}} \boldsymbol{p}]\!] = [\![\boldsymbol{p}]\!]. \tag{2.19}$$

Let us consider the following simple example of sentences:

$\boldsymbol{p}$: "Hotel *Adler* is affordable",

$\boldsymbol{q}$: "Hotel *Adler* lies close to the city center".

Assume $[\![\boldsymbol{p}]\!] = 0.9$ and $[\![\boldsymbol{q}]\!] = 0.6$. Using (2.11), (2.5), (2.6), (2.12) and (2.13), we get

$$[\![\neg_{\mathrm{m}} \boldsymbol{p}]\!] = 0.1,\ [\![\boldsymbol{p} \,\&_{\mathrm{m}} \boldsymbol{q}]\!] = 0{,}6,\ [\![\boldsymbol{p} \,\#_{\mathrm{m}} \boldsymbol{q}]\!] = 0.9,\ [\![\boldsymbol{p} \Rightarrow_{\mathrm{m}} \boldsymbol{q}]\!] = 0.7,\ [\![\boldsymbol{p} \Leftrightarrow_{\mathrm{m}} \boldsymbol{q}]\!] = 0.7.$$

Finally, let $\boldsymbol{D} = \{a, b, c, d, e\}$ be a set of hotels which are affordable to degree 0.3, 0.9, 0.5, 0.8 and 0.4, respectively. Let us look at the sentences

"Each hotel from $\boldsymbol{D}$ is affordable",

"At least one hotel from $\boldsymbol{D}$ is affordable".

By (2.9) and (2.10),

$$[\![\forall_{\mathrm{m}} x \in \boldsymbol{D}\colon x \text{ is affordable}]\!] = 0.3 \wedge 0.9 \wedge 0.5 \wedge 0.8 \wedge 0.4 = 0.3$$

and

$$[\![\exists_{\mathrm{m}} x \in \boldsymbol{D}\colon x \text{ is affordable}]\!] = 0.3 \vee 0.9 \vee 0.5 \vee 0.8 \vee 0.4 = 0.9.$$

2.2 Operations on Fuzzy Sets – The Standard Approach

This section presents the most frequently used approach to operations on fuzzy sets. The operations of minimum and maximum as well as the Łukasiewicz implication operator are then involved.

2.2.1 Basic Definitions

What we have to do first is to define two fundamental relationships between fuzzy sets $A, B\colon \boldsymbol{U} \to [0, 1]$. One says that *A is included in B*, and one writes $A \subset B$, if

$$A(x) \le B(x) \quad \text{for each element } x \in \boldsymbol{U}.$$

We say that *A is equal to B*, and we write $A = B$, whenever

$$A(x) = B(x) \quad \text{for each } x \in \boldsymbol{U}.$$

Hence

$$A = B \iff A \subset B \;\&\; B \subset A.$$

If $A \subset B$ and $A \neq B$, we will say that *A is properly included in B* or that *A is a proper fuzzy subset of B*. Clearly, $\subset$ is only a partial order relation and, thus, two fuzzy sets may be incomparable with respect to $\subset$.

Let us move on to operations on fuzzy sets. We like to present the standard and most frequently used approach proposed in ZADEH (1965).

The *sum of fuzzy sets A and B* is a fuzzy set $A \cup B$ such that

$$(A \cup B)(x) = A(x) \vee B(x) \quad \text{for each } x \in \boldsymbol{U}.$$

The *intersection of A and B* is a fuzzy set $A \cap B$ defined as

$$(A \cap B)(x) = A(x) \wedge B(x) \quad \text{for each } x \in \boldsymbol{U}.$$

The *complement of A* is a fuzzy set A' with

$$A'(x) = 1 - A(x) \quad \text{for each } x \in \boldsymbol{U}.$$

The *difference of A and B* can be defined as $A \setminus B = A \cap B'$, i.e.

$$(A \setminus B)(x) = A(x) \wedge (1 - B(x)) \quad \text{for each } x \in \boldsymbol{U}.$$

Alternatively, one defines

$$(A \ominus B)(x) = 0 \vee (A(x) - B(x)) \quad \text{for each } x \in \boldsymbol{U},$$

which forms a degree to which x is more in A than in B. $A \ominus B$ is called the *bounded difference of A and B*, and one has $A \ominus B \subset A \setminus B$.

Finally, the *cartesian product of fuzzy sets A and B* in possibly different universes, A: $\boldsymbol{U} \to [0, 1]$ and B: $\boldsymbol{V} \to [0, 1]$, is a fuzzy set $A \times B$: $\boldsymbol{U} \times \boldsymbol{V} \to [0, 1]$ of ordered pairs from $\boldsymbol{U} \times \boldsymbol{V}$, where

$$(A \times B)(x, y) = A(x) \wedge B(y) \quad \text{for each } (x, y) \in \boldsymbol{U} \times \boldsymbol{V}.$$

Graphical examples of $A \cup B$, $A \cap B$ and A' are presented in Figure 2.8. Simple numerical examples in the singleton notation, including the case of $A \times B$, will be given in Example 2.1(b) (see Subsection 2.3.2).

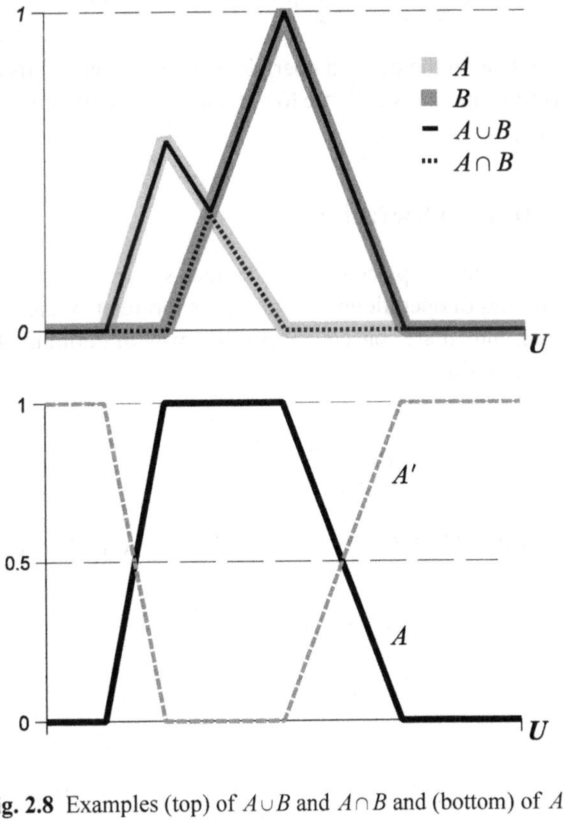

Fig. 2.8 Examples (top) of $A \cup B$ and $A \cap B$ and (bottom) of A'

We immediately notice that

$$A \cap B \subset A, B \subset A \cup B \tag{2.20}$$

and

$$A \subset B \Leftrightarrow A \cap B = A \Leftrightarrow A \cup B = B. \tag{2.21}$$

Moreover, the *monotonicity law* is valid, i.e.

$$A \cap B \subset C \cap D \text{ and } A \cup B \subset C \cup D \quad \text{whenever } A \subset C \text{ and } B \subset D. \tag{2.22}$$

It is also easy to point out that $A \cup B$ is a least fuzzy set (with respect to $\subset$) containing both A and B, whereas $A \cap B$ is a greatest fuzzy set contained in both A and B.

Let $\boldsymbol{J}$ denote a nonempty set of indices, finite or not, and let $A_i: \boldsymbol{U} \to [0, 1]$ for each index $i \in \boldsymbol{J}$. The *sum* of fuzzy sets A_i with $i \in \boldsymbol{J}$ is a fuzzy set $\bigcup_{i \in \boldsymbol{J}} A_i$ such that

$$(\bigcup_{i \in \boldsymbol{J}} A_i)(x) = \bigvee_{i \in \boldsymbol{J}} A_i(x) \quad \text{for each } x \in \boldsymbol{U}.$$

The *intersection* of the A_i's is a fuzzy set $\bigcap_{i \in \boldsymbol{J}} A_i$ with

$$(\bigcap_{i \in \boldsymbol{J}} A_i)(x) = \bigwedge_{i \in \boldsymbol{J}} A_i(x) \quad \text{for each } x \in \boldsymbol{U}.$$

It is easy to check that all the defined operations give correct results of classical set algebra whenever the operands collapse to ordinary sets, i.e. if A, B, A_i are functions ranging in $\{0, 1\}$.

2.2.2 Properties of Operations

The properties of operations placed in the previous subsection are all analogous to well-known properties of operations on sets. Let us formulate more similarities. The following laws are immediate consequences of the way of defining $A \cup B$ and $A \cap B$ via max and min operations:

commutativity,

$$A \cup B = B \cup A, \quad A \cap B = B \cap A, \tag{2.23}$$

associativity,

$$A \cup (B \cup C) = (A \cup B) \cup C, \quad A \cap (B \cap C) = (A \cap B) \cap C, \tag{2.24}$$

idempotency,

$$A \cup A = A, \quad A \cap A = A, \tag{2.25}$$

absorption,

$$A \cup (A \cap B) = A, \quad A \cap (A \cup B) = A, \tag{2.26}$$

distributivity,

$$A \cup (B \cap C) = (A \cup B) \cap (A \cup C), \quad A \cap (B \cup C) = (A \cap B) \cup (A \cap C). \tag{2.27}$$

Further, one sees that the empty set and the whole universe, respectively, are *neutral elements* of $\cup$ and $\cap$:

$$A \cup 1_{\varnothing} = A, \quad A \cap 1_U = A. \tag{2.28}$$

As regards the operation of complementation, it is

involutive,

$$(A')' = A, \tag{2.29}$$

order-reversing,

$$B' \subset A' \quad \text{whenever } A \subset B, \tag{2.30}$$

and *De Morgan laws* are still valid,

$$(A \cup B)' = A' \cap B' \quad \text{and} \quad (A \cap B)' = A' \cup B'. \tag{2.31}$$

Intuitively, since fuzzy sets are more general objects than sets, there must be laws of set algebra which are no longer valid for fuzzy sets. And this is the case of

$$\boldsymbol{A} \cap \boldsymbol{A}' = \varnothing \quad \text{and} \quad \boldsymbol{A} \cup \boldsymbol{A}' = \boldsymbol{U}$$

which cannot be transferred to proper fuzzy sets as we then have (see Figure 2.9)

$$A \cap A' \neq 1_{\varnothing} \quad \text{and} \quad A \cup A' \neq 1_U. \tag{2.32}$$

Indeed,

$$A \cap A' = 1_{\varnothing} \Leftrightarrow \forall x \in \boldsymbol{U}\text{: } A(x) \wedge (1 - A(x)) = 0 \Leftrightarrow A\text{: } \boldsymbol{U} \to \{0, 1\}$$

and

$$A \cup A' = 1_U \Leftrightarrow \forall x \in \boldsymbol{U}\text{: } A(x) \vee (1 - A(x)) = 1 \Leftrightarrow A\text{: } \boldsymbol{U} \to \{0, 1\}.$$

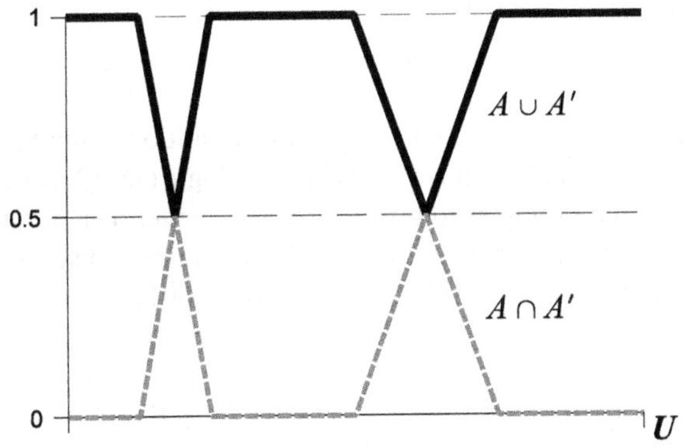

Fig. 2.9 $A \cup A'$ and $A \cap A'$ for fuzzy set A from Figure 2.8(bottom)

Undoubtedly, (2.32) forms a fundamental difference between operational properties of sets and fuzzy sets. Its consequences are far-reaching. Looking at (2.32) once more, we see that

$$A(x)\in(0,\,1) \Leftrightarrow A'(x)\in(0,\,1) \Leftrightarrow (A\cap A')(x)>0. \tag{2.33}$$

So, x can satisfy to a positive degree both an imprecise property (modeled by A) and its opposite. This reflects in a suitable way what happens in reality. For instance, sometimes a software or technological system works properly to a degree and, simultaneously, it does not work properly to a degree. We mean that it works at all, but some of its functions or modules are inaccessible, slowed down, etc. Giving another familiar example, the terms and conditions of a contract can be viewed as favourable and, on the other hand, unfavourable to a degree. Moreover, it is worth adding that, dealing with proper fuzzy sets, we not only have $A\cap A' \neq 1_\varnothing$, but even $A = A'$ is possible and holds true whenever $A(x) = 0.5$ for each $x\in \boldsymbol{U}$. A and its complement are then totally indistinguishable.

Concluding, on account of (2.32), fuzzy sets do not form a Boolean algebra. Their algebra is more general: the system $([0,\,1]^U, \cup, \cap, ', 1_\varnothing, 1_U)$ forms a De Morgan algebra or soft algebra, in other words, with $[0,\,1]^U$ denoting the family of all functions $\boldsymbol{U}\to[0,\,1]$, i.e. of all fuzzy sets in $\boldsymbol{U}$. Similarly, $\{0,\,1\}^U$ will denote the family of all sets in $\boldsymbol{U}$, i.e. all functions $\boldsymbol{U}\to\{0,\,1\}$.

2.2.3 Looking through Many-Valued Logic

There is an elementary, well-known and very close connection between basic definitions of set algebra and classical sentential calculus. For instance, we have

$$\boldsymbol{A}\subset\boldsymbol{B} \Leftrightarrow (\forall x\in \boldsymbol{U}\colon\, x\in\boldsymbol{A} \Rightarrow x\in\boldsymbol{B}),$$

$$x\in\boldsymbol{A}\cap\boldsymbol{B} \Leftrightarrow x\in\boldsymbol{A}\ \&\ x\in\boldsymbol{B},$$

$$x\in\boldsymbol{A}' \Leftrightarrow \neg x\in\boldsymbol{A}.$$

There exists an analogous connection between the definitions from Subsection 2.2.1 and many-valued sentential calculus in Łukasiewicz logic (see (2.4)-(2.10)). We like to present three related examples to give the reader a better insight into the roots of the definitions given in Subsection 2.2.1 for fuzzy sets. Further discussion and details can be found in GILES (1976) and GOTTWALD (1999, 2001).

First, let us notice that

$$A\subset B \Leftrightarrow [\![\forall_{\mathrm{m}} x\in \boldsymbol{U}\colon\, x\in_{\mathrm{m}} A \Rightarrow_{\mathrm{m}} x\in_{\mathrm{m}} B]\!] = 1. \tag{2.34}$$

Similarly to sets, $A\subset B$ thus means that the sentence "$\forall_{\mathrm{m}} x\in \boldsymbol{U}\colon\, x\in_{\mathrm{m}} A \Rightarrow_{\mathrm{m}} x\in_{\mathrm{m}} B$" is true. Indeed, by (2.9), (2.14) and (2.2),

$$[\![\forall_{\mathrm{m}} x \in \boldsymbol{U}\colon x \in_{\mathrm{m}} A \Rightarrow_{\mathrm{m}} x \in_{\mathrm{m}} B]\!] = 1 \Leftrightarrow \bigwedge_{a \in \boldsymbol{U}} [\![a \in_{\mathrm{m}} A \Rightarrow_{\mathrm{m}} a \in_{\mathrm{m}} B]\!] = 1$$

$$\Leftrightarrow \forall a \in \boldsymbol{U}\colon [\![a \in_{\mathrm{m}} A \Rightarrow_{\mathrm{m}} a \in_{\mathrm{m}} B]\!] = 1$$

$$\Leftrightarrow \forall a \in \boldsymbol{U}\colon [\![a \in_{\mathrm{m}} A]\!] \le [\![a \in_{\mathrm{m}} B]\!]$$

$$\Leftrightarrow \forall x \in \boldsymbol{U}\colon A(x) \le B(x).$$

Second, we have

$$[\![x \in_{\mathrm{m}} A \cap B \Leftrightarrow_{\mathrm{m}} x \in_{\mathrm{m}} A \,\&_{\mathrm{m}}\, x \in_{\mathrm{m}} B]\!] = 1. \tag{2.35}$$

Indeed, by (2.15) and (2.5), this equality is equivalent to

$$[\![x \in_{\mathrm{m}} A \cap B]\!] = [\![x \in_{\mathrm{m}} A \,\&_{\mathrm{m}}\, x \in_{\mathrm{m}} B]\!] = [\![x \in_{\mathrm{m}} A]\!] \wedge [\![x \in_{\mathrm{m}} B]\!].$$

Hence $(A \cap B)(x) = A(x) \wedge B(x)$. Further, let us notice that

$$[\![x \in_{\mathrm{m}} A' \Leftrightarrow_{\mathrm{m}} \neg_{\mathrm{m}}\, x \in_{\mathrm{m}} A]\!] = 1, \tag{2.36}$$

which follows from (2.15) and (2.11), and justifies the definition of the complement of a fuzzy set. Finally, it is clear that the differences in (2.32) are immediate consequences of (2.17) and (2.18).

2.3 Main Characteristics of Fuzzy Sets

We like to introduce in this section a few auxiliary notions which are useful when analyzing and speaking about fuzzy sets, and which form convenient characteristics of those objects. One very important characteristic, the cardinality of a fuzzy set, will be skipped here. Its study is the subject of Part II.

2.3.1 Core, Support, *t*-Cuts

Let $A\colon \boldsymbol{U} \to [0,1]$. The sets

$$\mathrm{supp}(A) = \{x \in \boldsymbol{U}\colon A(x) > 0\},$$

$$\mathrm{core}(A) = \{x \in \boldsymbol{U}\colon A(x) = 1\},$$

$$A_t = \{x \in \boldsymbol{U}\colon A(x) \ge t\} \text{ with } t \in (0,1]$$

and

$$A^t = \{x \in \boldsymbol{U}\colon A(x) > t\} \text{ with } t \in [0,1),$$

respectively, are called the *support*, *core*, *t-cut set* and *sharp t-cut set of A*, respectively (see Figure 2.10). The t-cut set (sharp or not) is thus a set of elements of $\boldsymbol{U}$ whose membership degree in A is sufficiently "large": is at least equal to t (t-cut) or is greater than t (sharp t-cut), where t is a given threshold value. Clearly, we have $\mathrm{supp}(A) = A^0$ and $\mathrm{core}(A) = A_1$. It is an elementary task to check that

$$A_t \subset B_t \quad \text{whenever } A \subset B, \tag{2.37}$$

$$A_u \subset A_t \quad \text{whenever } t \le u, \tag{2.38}$$

$$(A * B)_t = A_t * B_t \quad \text{for } * \in \{\cup, \cap, \times\}. \tag{2.39}$$

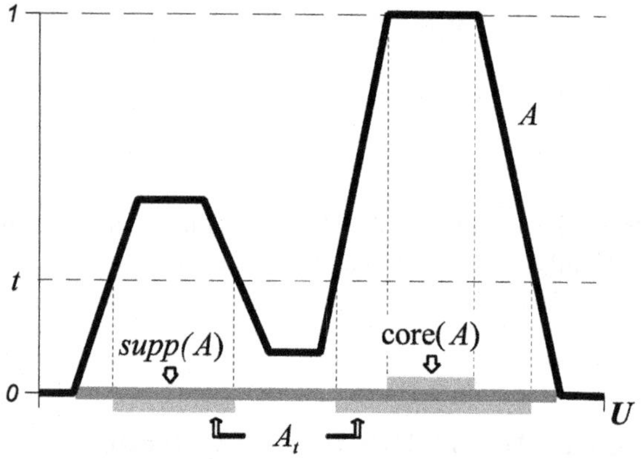

Fig. 2.10 Examples of supp(A), core(A) and A_t

The same holds true for sharp t-cuts. So, in particular, one has

$$\mathrm{supp}(A) \subset \mathrm{supp}(B) \quad \text{for } A \subset B, \tag{2.40}$$

$$\mathrm{supp}(A * B) = \mathrm{supp}(A) * \mathrm{supp}(B) \quad \text{for } * \in \{\cup, \cap, \times\}. \tag{2.41}$$

If $\mathrm{core}(A) \neq \varnothing$, one says that A is *normal*. Otherwise, A is called a *subnormal* fuzzy set. A is said to be *finite* if supp(A) is a finite set, else A is called an *infinite* fuzzy set. The number

$$\mathrm{hgt}(A) = \bigvee\{A(x)\colon x \in \boldsymbol{U}\}$$

is usually termed the *height of A*. So, hgt(A) becomes the largest membership degree in A whenever A is finite, and $\mathrm{hgt}(A) = 1$ for normal fuzzy sets.

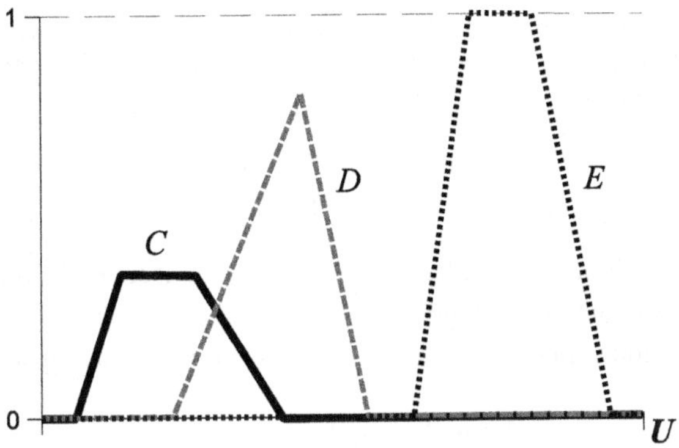

Fig. 2.11 Disjoint (C and E, D and E) and nondisjoint (C and D) fuzzy sets

Finally, similarly to the case of sets, two fuzzy sets, A and B, are said to be *disjoint* whenever $A \cap B = 1_{\varnothing}$. That A and B are disjoint thus means that the intersection of supp(A) and supp(B) is empty or, equivalently, that $A_t \cap B_t = \varnothing$ for each $t \in (0,1]$ (see Figure 2.11).

2.3.2 Decompositions and Maps of Fuzzy Sets

One says that a fuzzy set is a *singleton* whenever its support is a 1-element set. The notation a/x with $x \in \boldsymbol{U}$ and $a \in (0,1]$ will be used to describe a singleton supported by x (see Figure 2.12).

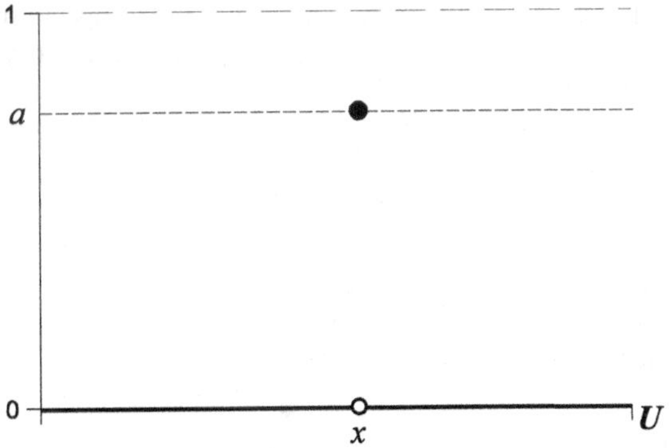

Fig. 2.12 Singleton a/x

Formally, a/x is thus a fuzzy set such that

$$(a/x)(z) = \begin{cases} a, & \text{if } z = x, \\ 0, & \text{otherwise.} \end{cases}$$

Each set is trivially the sum of its 1-element subsets. This property has a natural generalization to fuzzy sets. Each fuzzy set A can be decomposed into and viewed as a sum of singletons:

$$A = \bigcup_{x \in \mathrm{supp}(A)} A(x)/x. \tag{2.42}$$

By a slight abuse of notation, we have $0/x = 1_\emptyset$ and, consequently, the summation in (2.42) may be extended to the whole $\boldsymbol{U}$.

If A is finite and $\mathrm{supp}(A) = \{x_1, x_2, ..., x_n\}$, $n \geq 1$, (2.42) collapses to

$$A = a_1/x_1 \cup a_2/x_2 \cup ... \cup a_n/x_n,$$

where $a_i = A(x_i)$ for $1 \leq i \leq n$. Following common practice, we will use the arithmetical version of this notation in which $\cup$ is replaced with +:

$$A = a_1/x_1 + a_2/x_2 + ... + a_n/x_n = \sum_{i=1}^{n} a_i/x_i.$$

This *singleton notation* is a convenient and legible way of encoding finite fuzzy sets. It can be extended to fuzzy sets with countable and uncountable supports, and one then writes, respectively,

$$A = \sum_{x \in \mathrm{supp}(A)} A(x)/x \quad \text{and} \quad A = \int A(x)/x.$$

Example 2.1. (a) Let $\boldsymbol{U} = \{0, 1, 2, ...\}$ and $B = 0.1/0 + 1/3 + 0.8/4 + 0.5/5 + 0.6/7$. So, B is a fuzzy set such that $B(0) = 0.1$, $B(3) = 1$, $B(4) = 0.8$, $B(5) = 0.5$, $B(7) = 0{,}6$, and $B(k) = 0$ for $k \notin \{0, 3, 4, 5, 7\}$.

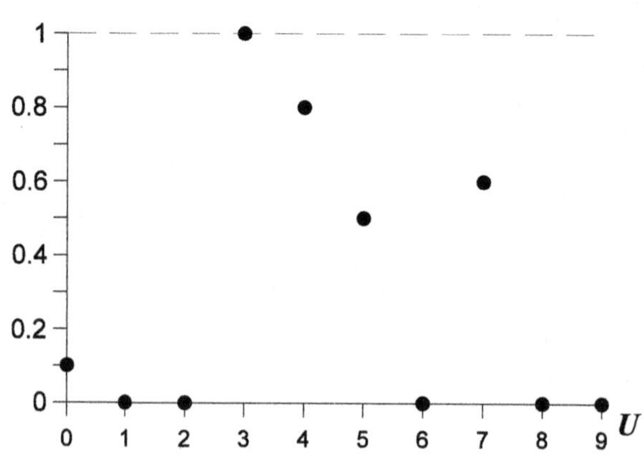

Fig. 2.13 Fuzzy set $B = 0.1/0 + 1/3 + 0.8/4 + 0.5/5 + 0.6/7$

(b) Using the singleton notation, we present results of operations on fuzzy sets. Let $A = 0.3/2 + 0.8/3 + 1/4 + 0.7/6$ and $B = 0.5/2 + 1/3 + 0.9/4 + 1/5 + 0.2/6 + 0.3/7$ in $\boldsymbol{U} = \{0, 1, ..., 8\}$. Recollect that $(A \cup B)(k) = A(k) \vee B(k)$, $(A \cap B)(k) = A(k) \wedge B(k)$ and $A'(k) = 1 - A(k)$. So,

$$A \cup B = 0.5/2 + 1/3 + 1/4 + 1/5 + 0.7/6 + 0.3/7,$$
$$A \cap B = 0.3/2 + 0.8/3 + 0.9/4 + 0.2/6,$$
$$A' = 1/0 + 1/1 + 0.7/2 + 0.2/3 + 1/5 + 0.3/6 + 1/7 + 1/8.$$

For $C = 0.8/5 + 0.9/7$ and $D = 0.4/3 + 1/5 + 0.7/6$, we get

$$C \times D = 0.4/(5, 3) + 0.8/(5, 5) + 0.7/(5, 6) + 0.4/(7, 3) + 0.9/(7, 5) + 0.7/(7, 6)$$

as $(C \times D)(i, k) = A(i) \wedge B(k)$. □

Besides the decomposition of a fuzzy set $A: \boldsymbol{U} \to [0, 1]$ into singletons in (2.42), there is another, more advanced way of decomposing A which involves t-cuts. Define $t \cdot A: \boldsymbol{U} \to [0, 1]$ with $t \in [0, 1]$ as a fuzzy set such that $(t \cdot A)(x) = tA(x)$. As usual, one can write tA instead of $t \cdot A$. We then have

$$A = \bigcup_{t \in (0, 1]} t \mathbf{1}_{A_t}. \tag{2.43}$$

An analogous formula involving sharp t-cut sets is also valid. (2.43) is known as the *decomposition property* of a fuzzy set A. Let us outline its mechanism.

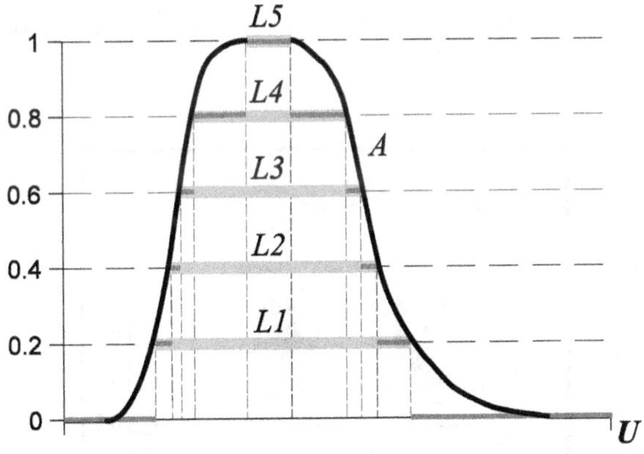

Fig. 2.14 Decomposition of a fuzzy set

The above A is partially decomposed using t-cut sets with $t = 0.2, 0.4, 0.6, 0.8, 1$. Five fuzzy sets are then generated:

$$L1 = 0.2 \cdot \mathbf{1}_{A_{0.2}},\ L2 = 0.4 \cdot \mathbf{1}_{A_{0.4}},\ L3 = 0.6 \cdot \mathbf{1}_{A_{0.6}},\ L4 = 0.8 \cdot \mathbf{1}_{A_{0.8}},\ L5 = 1 \cdot \mathbf{1}_{A_1}.$$

Their sum $L1 \cup L2 \cup L3 \cup L4 \cup L5$ is a step membership function (see dark grey line) approximating A. Clearly, the more the t-cuts, the better the approximation. Using all of them, one gets exactly A.

If A is a finite fuzzy set and $\{a_1, a_2, ..., a_k\}$ is the set of all positive membership degrees in A, then the summation in (2.43) can be restricted to $\{a_1, a_2, ..., a_k\}$.

Example 2.2. Let $A = 1/1 + 0.9/3 + 0.6/4 + 0.8/6 + 0.6/7 + 0.2/8$, $\boldsymbol{U} = \{0, 1, 2, ...\}$. Then $A_{0.2} = \{1, 3, 4, 6, 7, 8\}$, $A_{0.6} = \{1, 3, 4, 6, 7\}$, $A_{0.8} = \{1, 3, 6\}$, $A_{0.9} = \{1, 3\}$, and $A_1 = \{1\}$. By (2.43),

$$A = 0.2 \cdot 1_{\{1, 3, 4, 6, 7, 8\}} \cup 0.6 \cdot 1_{\{1, 3, 4, 6, 7\}} \cup 0.8 \cdot 1_{\{1, 3, 6\}} \cup 0.9 \cdot 1_{\{1, 3\}} \cup 1_{\{1\}}. \qquad \square$$

The decomposition property leads to a worth noticing corollary:

$$A = B \Leftrightarrow \forall t \in (0, 1]\colon A_t = B_t \Leftrightarrow \forall t \in [0, 1)\colon A^t = B^t. \tag{2.44}$$

Two arbitrary fuzzy sets are thus identical iff their corresponding t-cut sets, sharp or not, are always identical. Moreover, combining this with (2.38), each fuzzy set A can be treated as a descending family $(A_t)_{t\in(0,1]}$ or $(A^t)_{t\in[0,1)}$ of usual sets. If $\boldsymbol{U}$ is a subset of $\mathbb{R}^2$, this way of viewing A as a family of its t-cuts resembles the contour line method in cartography. Indeed, the t-cut sets of A – actually, their boundaries – then play the role of contour lines, whereas a collection of those lines forms a *map* of A (see below).

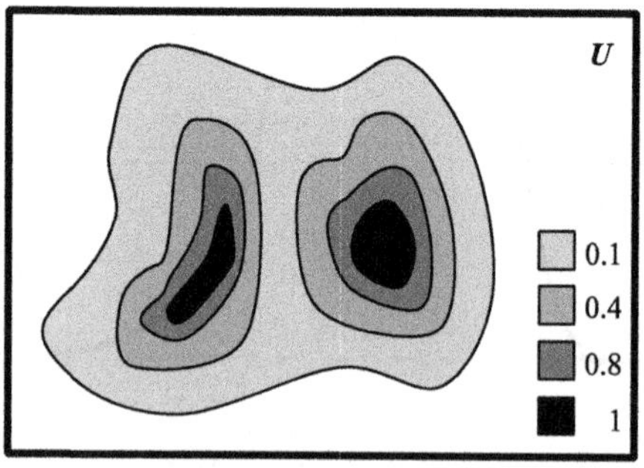

Fig. 2.15 Map of a fuzzy set based on four t-cut sets with $t = 0.1, 0.4, 0.8, 1$

Looking at the above map involving four t-cut sets of a fuzzy set, the reader can easily reconstruct an approximate 3D form of that fuzzy set.

2.3.3 Convexity

If the universe $\boldsymbol{U}$ is a convex subset of $\mathbb{R}^n$, $n \geq 1$, a fuzzy set $A\colon \boldsymbol{U} \to [0, 1]$ is called *convex* whenever

$$A(\alpha x_1 + (1-\alpha)x_2) \geq A(x_1) \wedge A(x_2) \quad \text{for each } x_1, x_2 \in \boldsymbol{U} \text{ and } \alpha \in [0, 1]. \qquad (2.45)$$

Using $\boldsymbol{U}$ that is linearly ordered by a relation $\leq$, A is said to be *convex* if

$$A(x_3) \geq A(x_1) \wedge A(x_2) \quad \text{for each } x_1, x_2, x_3 \in \boldsymbol{U} \ (x_1 \leq x_3 \leq x_2). \qquad (2.46)$$

The essence of convexity is thus that $A(y) \geq A(x_1) \wedge A(x_2)$ for each $x_1, x_2 \in \boldsymbol{U}$ and each intermediate point y lying between x_1 and x_2 (see Figure 2.16). Conditions (2.45) and (2.46) are nothing else than two formulations of this requirement adapted to $\boldsymbol{U}$ and its features.

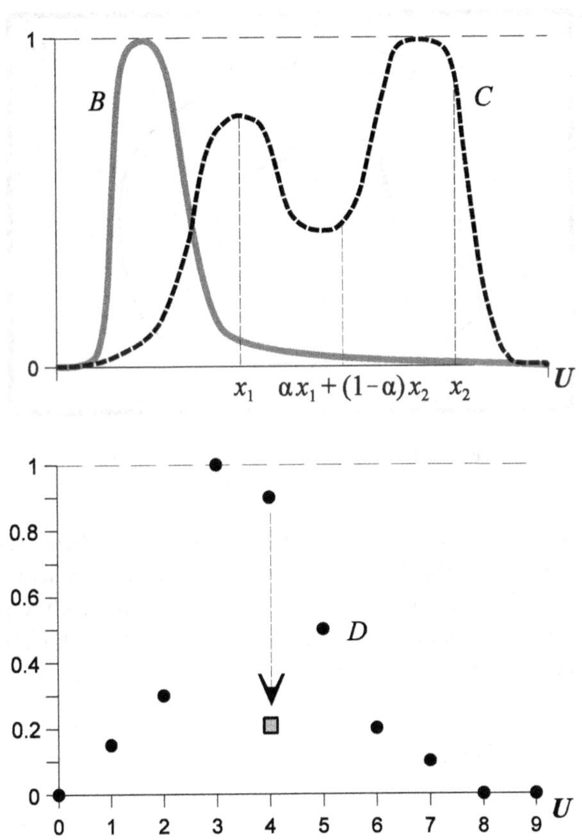

Fig. 2.16 Examples of convex and nonconvex fuzzy sets

For a fuzzy set in $\mathbb{R}^n$, y is expressed through a convex combination of x_1 and x_2. It is clear that (2.46) is suitable if $\boldsymbol{U}$ forms, say, a set of natural, integer, or cardinal numbers. (2.45) and (2.46) become identical in $\boldsymbol{U} = \mathbb{R}$. Both (2.45) and (2.46) collapse to the definition of a convex set whenever $A: \boldsymbol{U} \to \{0, 1\}$.

Figure 2.16(top) shows examples of a convex (B) and nonconvex (C) fuzzy set with respect to (2.45) and (2.46) in $\boldsymbol{U} = [0, \infty)$. Referring to Figure 2.16(bottom), it presents a fuzzy set D which is convex with respect to (2.46) in $\boldsymbol{U} = \mathbb{N} = \{0, 1, 2, ...\}$. Lowering $D(4)$ to the value marked by a square, we spoil that convexity.

Let us look a bit closer at nonconvex fuzzy sets. Recollect that a set is nonconvex whenever there exist points x_1 and x_2 such that walking from x_1 to x_2 the set "vanishes": the value of its characteristic function falls to zero (see Figure 2.17). Nonconvexity of a fuzzy set is a more general phenomenon. Referring once again to C in Figure 2.16(top) and walking from x_1 to x_2, the membership degrees in C do not have to go down to zero, but fall below $C(x_1)$ and $C(x_2)$.

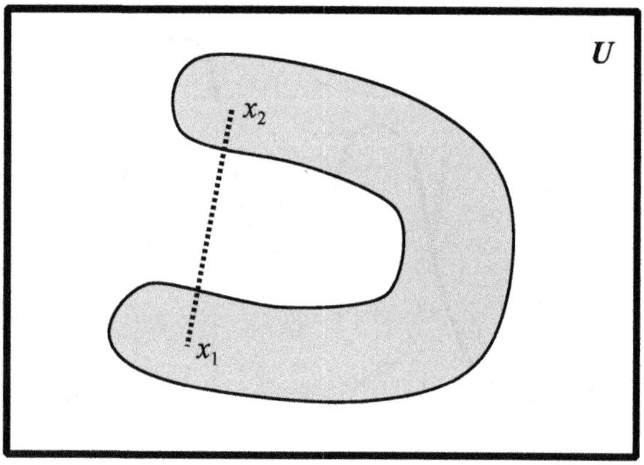

Fig. 2.17 Example of a nonconvex set

Finally, it is easy to notice that the intersection of two convex fuzzy sets is convex, too. Worth mentioning are also two criteria of convexity. A fuzzy set is convex with respect to (2.45) iff all of its t-cuts are convex sets. It is convex with respect to (2.46) iff its t-cuts are always intervals in $\boldsymbol{U}$. Depending on the form of $\boldsymbol{U}$, we mean here intervals of reals, integers, etc.

2.3.4 Fuzziness Measures

Our intuition suggests that a fuzzy set can be more or less fuzzy. Let us look at three examples in Figure 2.18. C is "not very fuzzy", D seems to be "much more fuzzy", whereas E identically equal to 0.5 forms an "extremely fuzzy" fuzzy set. All this leads

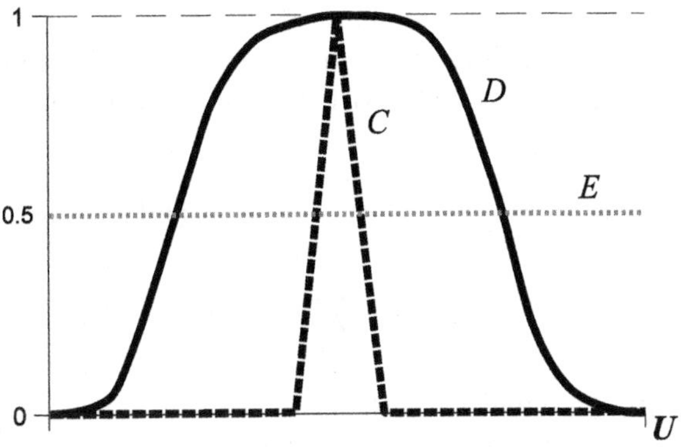

Fig. 2.18 Fuzzy sets with intuitively different degrees of fuzziness

to the idea of fuzziness measures for fuzzy sets. We like to present the classical formalization of that issue according to DE LUCA/TERMINI (1972, 1979, 1982).

First, let us introduce the following partial order relation $\leq'$ for fuzzy sets:

$$A \leq' B \Leftrightarrow \forall x \in \boldsymbol{U}\colon A(x) \leq B(x) \leq 0.5 \;\#\; 0.5 \leq B(x) \leq A(x). \tag{2.47}$$

$A \leq' B$ thus means that the membership values in A lie nearer 0 or 1 than those in B. In other words, A is more similar to a set than B. If $A \leq' B$, we will say that *A is sharper than B* or, dually, that *B is fuzzier than A*. By (2.47), the fuzziest fuzzy set is that identically equal to 0.5, which coincides with our intuition.

Definition 2.3. A mapping Fuzz: $[0, 1]^{\boldsymbol{U}} \to [0, \infty)$ is called a *fuzziness measure* if the following postulates are satisfied for each $A, B \in [0, 1]^{\boldsymbol{U}}$:

(F1) Fuzz$(A) = 0$ whenever $A \in \{0, 1\}^{\boldsymbol{U}}$,

(F2) Fuzz reaches its maximum at $A \equiv 0.5$,

(F3) $A \leq' B$ implies Fuzz$(A) \leq$ Fuzz(B),

(F4) Fuzz$(A) =$ Fuzz(A').

We will then say that Fuzz(A) is the *fuzziness measure of A*. If a fuzziness measure is defined for finite fuzzy sets only, one usually replaces (F1)-(F2) with stronger conditions, namely:

Fuzz$(A) = 0$ iff $A \in \{0, 1\}^{\boldsymbol{U}}$,
Fuzz reaches its unique maximum at $A \equiv 0.5$.

The number Fuzz(A) says how fuzzy A is. It describes a deviation of A from a crisp set. In other words, it can be viewed as a measure of imprecision of information carried by A. Fuzziness measures from Definition 2.3 are also called *entropy measures of fuzziness*. This leads to another interpretation in which Fuzz(A) is understood as a total measure of the amount of information lacking to - or distinguishing A from - a state in which there is no uncertainty (ambiguity) when classifying the elements of $\boldsymbol{U}$ (see also Subsection 6.3.4). Let us present main instances of fuzziness measures from Definition 2.3.

Example 2.4. Assume A is a finite fuzzy set and $\text{supp}(A) = \{x_1, x_2, ..., x_n\}$, $n \geq 1$. Then

$$\text{Fuzz}(A) = \sum_{i=1}^{n} \varphi(A(x_i)), \tag{2.48}$$

where $\varphi: [0, 1] \to [0, 1]$ is a function such that

- $\varphi(0) = \varphi(1) = 0$ and $\varphi(a) > 0$ for each $a \in (0, 1)$,
- $\varphi(a) < \varphi(0.5)$ for each $a \neq 0.5$,
- φ is nondecreasing on $[0, 0.5]$ and nonincreasing on $[0.5, 1]$,
- $\varphi(a) = \varphi(1 - a)$ for each $a \in [0, 1]$.

$\varphi(a)$ will be viewed as the *fuzziness index* of a and, consequently, $\text{Fuzz}(A)$ becomes the sum of all fuzziness indices of membership degrees in A. If convenient, one refers to $\varphi(A(x_i))$ as the fuzziness index of the very x_i. (2.48) has a natural extension to arbitrary fuzzy sets in $[a, b] \subset \mathbb{R}$, namely

$$\text{Fuzz}(A) = \int_a^b \varphi(A(x))\, dx. \tag{2.49}$$

An important particular case of φ seems to be

$$\varphi(a) = 2(a \wedge (1-a)) \quad \text{with } a \in [0, 1] \tag{2.50}$$

illustrated below (see also its modifications in (6.25), (6.28) and (6.34)).

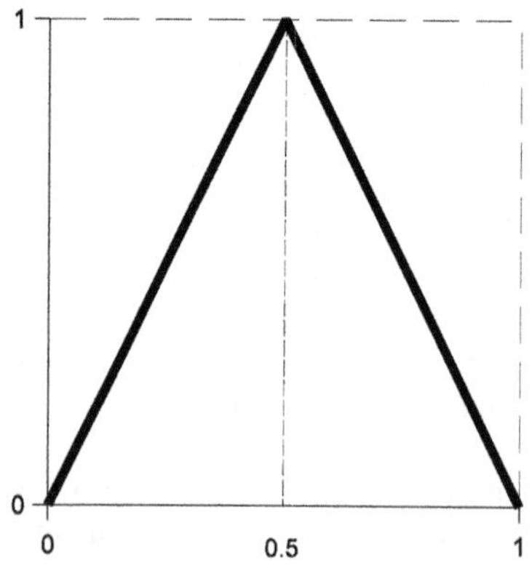

Fig. 2.19 Fuzziness index (2.50)

Then

$$\text{Fuzz}(A) = 2\sum_{i=1}^{n} A(x_i)\wedge(1-A(x_i)) = 2\sum_{i=1}^{n}(A\cap A')(x_i)$$

and (2.51)

$$\text{Fuzz}(A) = 2\int_a^b (A\cap A')(x)\,dx.$$

This Fuzz(A) thus collapses to a measure of the intersection $A\cap A'$. It can be transformed into a convenient normed form $\text{Fuzz}_n(A)\in[0,1]$ which makes it easier in practice to compare fuzzy sets with respect to their fuzziness:

$$\text{Fuzz}_n(A) = \frac{2}{|U|}\sum_{x\in U}(A\cap A')(x), \quad \text{if } U \text{ is finite,}$$

and (2.52)

$$\text{Fuzz}_n(A) = \frac{2}{b-a}\int_a^b (A\cap A')(x)\,dx, \quad \text{if } U=[a,b].$$

□

The next example offers distance-based fuzziness measures.

Example 2.5. Now, let d denote a normed metric in $[0, 1]^U$, i.e. $d(A, B)\in[0, 1]$ for each pair of fuzzy sets. Assume d is such that

$$d(A, A') = 1 \quad \text{whenever } A\in\{0, 1\}^U$$

and

$$d(B, B') \le d(A, A') \quad \text{for } A \le' B.$$

It is a routine task to check that

$$\text{Fuzz}(A) = 1 - d(A, A') \tag{2.53}$$

forms a fuzziness measure in the sense of Definition 2.3. Fuzz(A) now becomes a measure of difficulty in distinguishing between A and its complement A' expressed through distance between A and A'. An interesting particular case seems to be that with $d = d_p$ and $p\in[1,\infty)$, where $d_p(B, C)$ denotes the *normed Minkowski distance* between fuzzy sets B and C, namely

$$d_p(B, C) = \left(\frac{1}{|U|}\sum_{x\in U}|B(x)-C(x)|^p\right)^{1/p} \quad \text{whenever } U \text{ is finite,}$$

and (2.54)

$$d_p(B, C) = \left(\frac{1}{b-a}\int_a^b |B(x)-C(x)|^p dx\right)^{1/p} \quad \text{for } U=[a,b]\subset\mathbb{R}.$$

So, $d_1(B, C)$ is just the *normed Hamming distance*, whereas $d_2(B, C)$ collapses to the *normed Euclidean distance* between B and C. The removal of the factors $1/|U|$ and

$1/(b-a)$ in (2.54) leads to the usual *Minkowski*, *Hamming*, and *Euclidean distance*, respectively. Put

$$\mathrm{Fuzz}_p(A) = 1 - d_p(A, A'). \tag{2.55}$$

Since $A(x) - A'(x) = 2A(x) - 1$, (2.54) then gives

$$\mathrm{Fuzz}_1(A) = 1 - \frac{1}{|U|}\sum_{x\in U} |2A(x) - 1| \quad (U \text{ finite})$$

and (2.56)

$$\mathrm{Fuzz}_1(A) = 1 - \frac{1}{b-a}\int_a^b |2A(x) - 1|\, dx \quad (U = [a, b]),$$

whereas

$$\mathrm{Fuzz}_2(A) = 1 - \left(\frac{1}{|U|}\sum_{x\in U} (2A(x) - 1)^2\right)^{1/2} \quad (U \text{ finite})$$

and (2.57)

$$\mathrm{Fuzz}_2(A) = 1 - \left(\frac{1}{b-a}\int_a^b (2A(x) - 1)^2\, dx\right)^{1/2} \quad (U = [a, b]).$$

It is easy to notice that in both the cases of U under discussion we get

$$\mathrm{Fuzz}_1(A) = \mathrm{Fuzz}_\cap(A). \tag{2.58}$$

This equality has a simple and elegant graphical justification for $U = [a, b]$. Indeed, put

$$I_1 = \int_a^b (A \cap A')(x)\, dx \quad \text{and} \quad I_2 = \int_a^b |A(x) - A'(x)|\, dx.$$

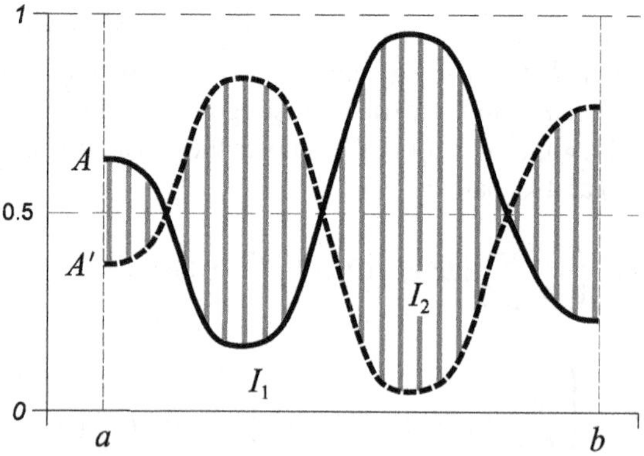

Fig. 2.20 Graphical interpretation of I_1 and I_2

Looking at the above figure, we see that $2I_1 + I_2 = b - a$, which implies (2.58). □

Finally, the reader is referred to EBANKS (1983), HIGASHI/KLIR (1982), KLIR (2000), PAL/BEZDEK (1994) and PEDRYCZ/GOMIDE (2007) for further details, examples of, and references to fuzziness measures (see also Section 10.1 of this book).

2.4 Flexible Framework for Operations on Fuzzy Sets

The standard approach to operations on fuzzy sets placed in Section 2.2 is not an only possible one. We like to present a much more flexible approach through negations and triangular norms and conorms. The user can then make a free choice of the form of basic operations on fuzzy sets from an infinite set of options. In each case, however, some desired fundamental properties of those operations are preserved.

2.4.1 Negations

A function ν: $[0,1] \to [0,1]$ is called a *negation function* or, simply, a *negation* if ν is nonincreasing, $\nu(0) = 1$ and $\nu(1) = 0$. One says that a negation is *strict* whenever it is strictly decreasing and continuous. A strict negation ν is said to be *strong* if ν is *involutive*, i.e. $\nu(\nu(a)) = a$ for each $a \in [0,1]$ and, thus, $\nu \circ \nu = id$ with *id* denoting the identity function.

Basic examples of nonstrict negations are the *threshold negations* ν_p with parameter $p \in [0,1)$, and ν^p with $p \in (0,1]$ defined as (see Figure 2.21)

$$\nu_p(a) = \begin{cases} 1, & \text{if } a \le p, \\ 0, & \text{otherwise,} \end{cases} \qquad \nu^p(a) = \begin{cases} 1, & \text{if } a < p, \\ 0, & \text{otherwise.} \end{cases} \tag{2.59}$$

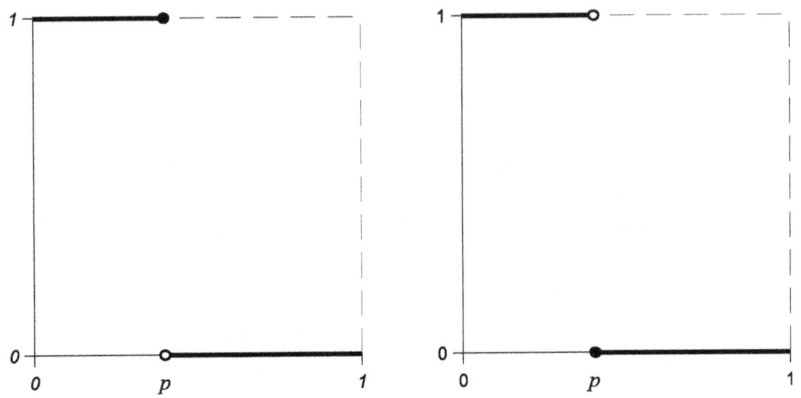

Fig. 2.21 Threshold negations (left) ν_p and (right) ν^p

Two particular cases of these constructions have to be distinguished, namely the *extreme negations* $\nu_* = \nu_0$ (known as the *Gődel negation*) and $\nu^* = \nu^1$ (see the next figure). It is clear that

$$\nu_* \leq \nu \leq \nu^*$$

for each negation ν with ≤ understood in the natural pointwise way as

$$\nu_1 \leq \nu_2 \Leftrightarrow \forall a \in [0,1]: \nu_1(a) \leq \nu_2(a).$$

ν_* is thus the least possible negation, whereas ν^* forms the greatest possible one.

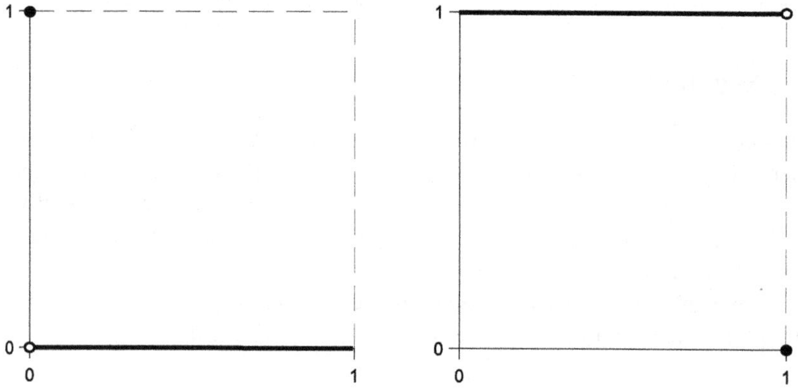

Fig. 2.22 Extreme negations (left) ν_* and (right) ν^*

The negation $\nu(a) = 1 - a^2$ exemplifies a negation that is strict, but not strong. A typical strong negation is the *Łukasiewicz negation* $\nu_Ł$ with $\nu_Ł(a) = 1 - a$, which is also termed the *standard negation* (see below).

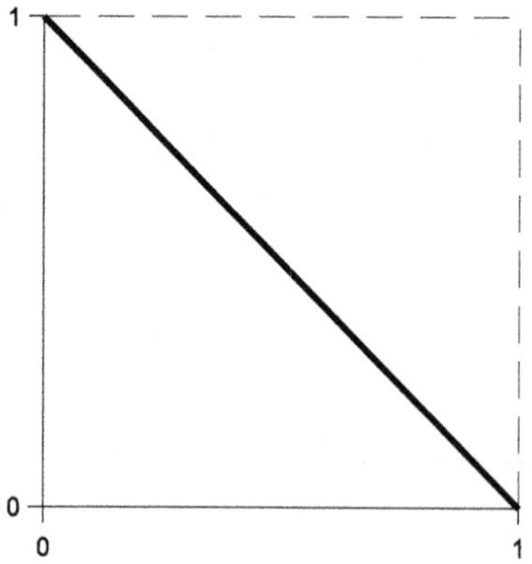

Fig. 2.23 Łukasiewicz negation $\nu_Ł$

Further important examples of strong negations are

- *Sugeno negations*: $\nu_{S,\lambda}(a) = \dfrac{1-a}{1+\lambda a}$ with $\lambda > -1$, (2.60)
- *Yager negations*: $\nu_{Y,p}(a) = (1 - a^p)^{1/p}$ with $p > 0$. (2.61)

Each strict negation as a continuous function has a unique fixed point $e(\nu) \in (0, 1)$, i.e. the equation $\nu(a) = a$ has exactly one solution $a = e(\nu)$. The number $e(a)$ can be viewed as an *equilibrium point* or a *logical center of gravity* of the scale [0, 1] with respect to ν.

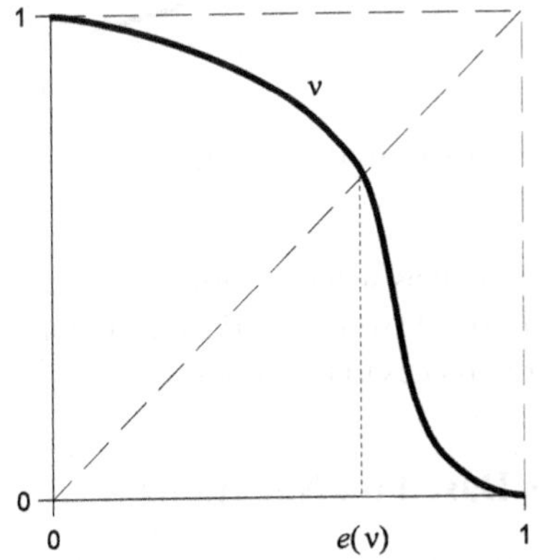

Fig. 2.24 Strict negation ν and its fixed point

Obviously, $e(\nu) = 0.5$ for $\nu = \nu_Ł$. If $\nu = \nu_{Y,p}$, one gets

$$e(\nu) = 0.5^{1/p}, \tag{2.62}$$

i.e. $e(\nu) = 0.71$ for $p = 2$.

Finally, we like to present a family of strong negations $\nu_{=c}$ with predefined fixed point $c \in (0, 1)$:

$$\nu_{=c}(a) = \begin{cases} 1 - \dfrac{1-c}{c}\, a, & \text{if } a \in [0, c], \\[2ex] \dfrac{c}{1-c}(1-a), & \text{if } a \in (c, 1]. \end{cases} \tag{2.63}$$

We will call them *broken negations*. Their examples are illustrated in Figure 2.25. One sees that

$$\nu_{=0.5} = \nu_Ł,$$
$$\nu_{=c} \to \nu_* \quad \text{for } c \to 0,$$
$$\nu_{=c} \to \nu^* \quad \text{for } c \to 1.$$

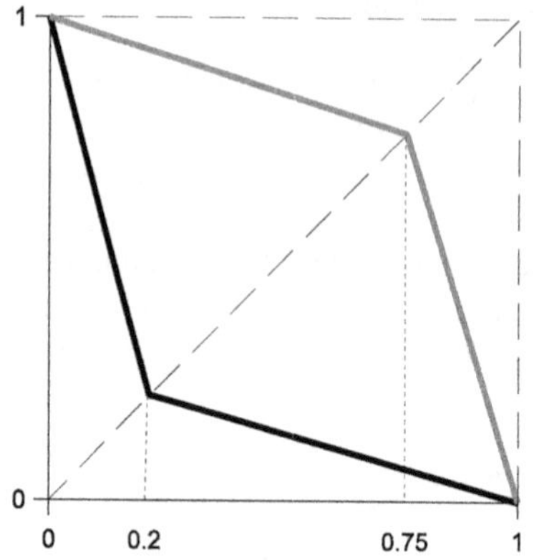

Fig. 2.25 Broken negations (black line) $\nu_{=0.2}$ and (grey line) $\nu_{=0.75}$

A discussion on negation functions will be continued in Subsections 2.4.5 and 3.2.4. The reader is also referred to GOTTWALD (2001) and FODOR/YAGER (2000) for more details about and further references to negations.

2.4.2 Complements Based on Negations

For a fuzzy set A: $\boldsymbol{U}\to[0, 1]$ and a negation ν, the *complement* A^ν *of A induced by* ν is a fuzzy set defined as

$$A^\nu(x) = \nu(A(x)) \quad \text{for each } x \in \boldsymbol{U}.$$

The standard complement A' is thus that induced by the standard negation $\nu_Ł$. An example of the complement induced by $\nu_{=0.8}$ is given below (cf. Figure 2.8(bottom)).

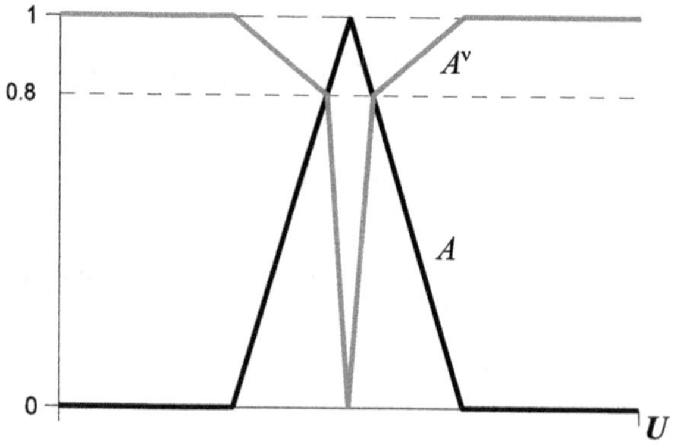

Fig. 2.26 Complement induced by $\nu = \nu_{=0.8}$

By definition, each negation ν is order-reversing, $A(x) \le B(x) \Rightarrow \nu(A(x)) \ge \nu(B(x))$, which leads to

$$A \subset B \Rightarrow B^{\nu} \subset A^{\nu}. \tag{2.64}$$

If ν is strong, then it is involutive and, consequently,

$$(A^{\nu})^{\nu} = A, \tag{2.65}$$

i.e. the double complementation law is still true. Obviously, it does not hold for non-strong negations. An interesting example is the least negation $\nu = \nu_*$. We then get $a \le \nu(\nu(a))$ and $\nu(\nu(\nu(a))) = \nu(a)$, which implies

$$A \subset (A^{\nu})^{\nu} \text{ and } ((A^{\nu})^{\nu})^{\nu} = A^{\nu}. \tag{2.66}$$

For strict negations, notice that A and its complement A^{ν} are totally indistinguishable, $A = A^{\nu}$, whenever $A(x) = e(\nu)$ for each $x \in \boldsymbol{U}$ (see Subsection 2.4.1). Using fuzziness measures of fuzzy sets with complements induced by a strong negation, the definitions and constructions from Subsection 2.3.4 should be modified by replacing A' with A^{ν}, 0.5 with $e(\nu)$, and $1 - a$ with $\nu(a)$.

2.4.3 Triangular Norms

A binary operation $\boldsymbol{t}$: $[0, 1] \times [0, 1] \to [0, 1]$ is called a *triangular norm* (*t-norm*, in short) if $\boldsymbol{t}$ is commutative, associative, nondecreasing in the first and, hence, in each argument, and has 1 as neutral element, i.e.

(T1) $a\boldsymbol{t}b = b\boldsymbol{t}a$,

(T2) $a\boldsymbol{t}(b\boldsymbol{t}c) = (a\boldsymbol{t}b)\boldsymbol{t}c$,

(T3) $a\boldsymbol{t}b \le c\boldsymbol{t}d$ whenever $a \le c$ and $b \le d$,

(T4) $a\boldsymbol{t}1 = a$

for each $a, b, c, d \in [0, 1]$. If an operation $\boldsymbol{s}$: $[0, 1] \times [0, 1] \to [0, 1]$ satisfies (T1)-(T3) and has 0 as neutral element, i.e.

(T5) $a\boldsymbol{s}0 = a$

for each $a \in [0, 1]$, then $\boldsymbol{s}$ is said to be a *triangular conorm* (*t-conorm*, in short). Triangular norms together with triangular conorms will be called *triangular operations* (*t-operations*, in short).

For completeness, we like to mention that triangular norms were originally introduced in MENGER (1942) and, later on, they were redefined in SCHWEIZER/SKLAR (1961, 1983) by strengthening their axioms into the above form used today.

Simple and very important examples of t-norms and t-conorms are the following operations (see also Figure 2.27):

- $a \wedge b$, (*t-norm minimum*)
  $a \vee b$, (*t-conorm maximum*)
- $a\boldsymbol{t}_{\mathrm{a}}b = ab$, (*algebraic* or *product t-norm*)
  $a\boldsymbol{s}_{\mathrm{a}}b = a + b - ab$, (*algebraic* or *product t-conorm*)
- $a\boldsymbol{t}_{Ł}b = 0 \vee (a + b - 1)$, (*Łukasiewicz t-norm*)
  $a\boldsymbol{s}_{Ł}b = 1 \wedge (a + b)$, (*Łukasiewicz t-conorm*)
- $a\boldsymbol{t}_{\mathrm{d}}b = \begin{cases} a \wedge b, & \text{if } a = 1 \text{ or } b = 1, \\ 0, & \text{otherwise,} \end{cases}$ (*drastic t-norm*)

  $a\boldsymbol{s}_{\mathrm{d}}b = \begin{cases} a \vee b, & \text{if } a = 0 \text{ or } b = 0, \\ 1, & \text{otherwise.} \end{cases}$ (*drastic t-conorm*)

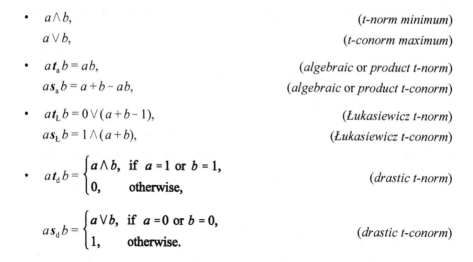

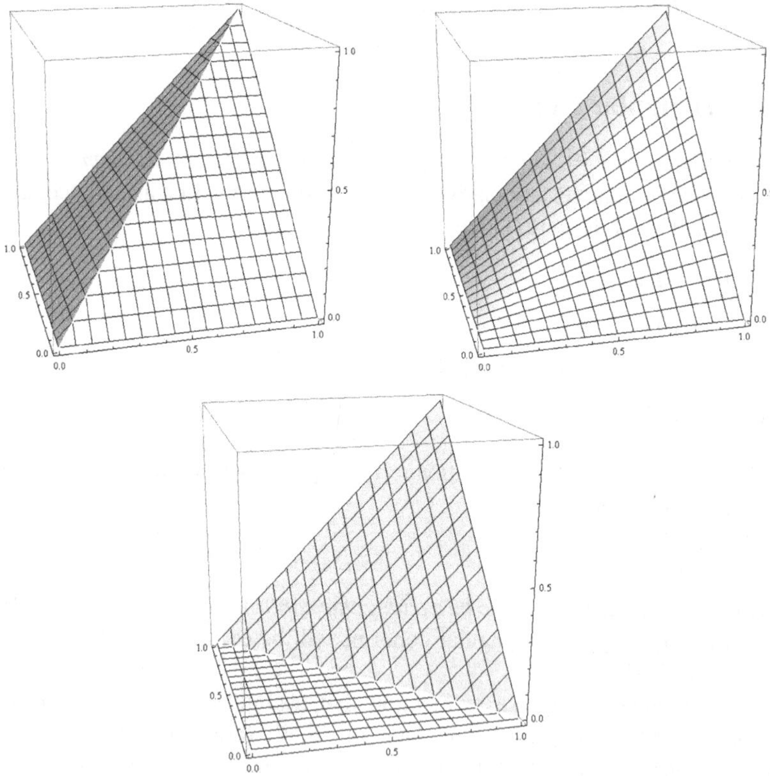

Fig. 2.27 3D plots of $\wedge$ (top left), $\boldsymbol{t}_{\mathrm{a}}$ (top right), and $\boldsymbol{t}_{Ł}$ (bottom)

As regards n-ary extensions with $n \geq 2$, they are obvious for $\wedge$ and $\boldsymbol{t}_{\mathrm{a}}$:

$$a_1 \wedge a_2 \wedge \ldots \wedge a_n = \min(a_1, a_2, \ldots, a_n),$$

$$a_1 \boldsymbol{t}_{\mathrm{a}} a_2 \boldsymbol{t}_{\mathrm{a}} \ldots \boldsymbol{t}_{\mathrm{a}} a_n = \prod_{i=1}^{n} a_i,$$

whereas (see also Subsections 3.2.3 and 3.2.4)

$$a_1 \boldsymbol{t}_{\mathrm{Ł}} a_2 \boldsymbol{t}_{\mathrm{Ł}} \ldots \boldsymbol{t}_{\mathrm{Ł}} a_n = 0 \vee \left(\sum_{i=1}^{n} a_i - (n-1) \right),$$

$$a_1 \boldsymbol{t}_{\mathrm{d}} a_2 \boldsymbol{t}_{\mathrm{d}} \ldots \boldsymbol{t}_{\mathrm{d}} a_n = \begin{cases} a_i, & \text{if } a_j = 1 \text{ for each } j \neq i, \\ 0, & \text{otherwise.} \end{cases}$$

For the corresponding t-conorms, one has

$$a_1 \vee a_2 \vee \ldots \vee a_n = \max(a_1, a_2, \ldots, a_n),$$

$$a_1 \boldsymbol{s}_{\mathrm{a}} a_2 \boldsymbol{s}_{\mathrm{a}} \ldots \boldsymbol{s}_{\mathrm{a}} a_n = 1 - \prod_{i=1}^{n} (1 - a_i),$$

$$a_1 \boldsymbol{s}_{\mathrm{Ł}} a_2 \boldsymbol{s}_{\mathrm{Ł}} \ldots \boldsymbol{s}_{\mathrm{Ł}} a_n = 1 \wedge \sum_{i=1}^{n} a_i,$$

$$a_1 \boldsymbol{t}_{\mathrm{d}} a_2 \boldsymbol{t}_{\mathrm{d}} \ldots \boldsymbol{t}_{\mathrm{d}} a_n = \begin{cases} a_i, & \text{if } a_j = 0 \text{ for each } j \neq i, \\ 1, & \text{otherwise.} \end{cases}$$

Straightforward consequences of (T1)-(T5) are the following properties of a t-norm $\boldsymbol{t}$ and t-conorm $\boldsymbol{s}$:

$$a \boldsymbol{t} 0 = 0, \quad a \boldsymbol{s} 1 = 1, \tag{2.67}$$

$$a \boldsymbol{t} b = 1 \Leftrightarrow a = b = 1, \quad a \boldsymbol{s} b = 0 \Leftrightarrow a = b = 0, \tag{2.68}$$

$$a \boldsymbol{t} a \leq a \leq a \boldsymbol{s} a, \tag{2.69}$$

$$a \boldsymbol{t}_{\mathrm{d}} b \leq a \boldsymbol{t} b \leq a \wedge b \leq a, b \leq a \vee b \leq a \boldsymbol{s} b \leq a \boldsymbol{s}_{\mathrm{d}} b, \tag{2.70}$$

where $a, b \in [0, 1]$. Moreover, we have

$$(\forall a \in [0,1]\colon a \boldsymbol{t} a = a) \Leftrightarrow \boldsymbol{t} = \wedge$$

and

$$(\forall a \in [0,1]\colon a \boldsymbol{s} a = a) \Leftrightarrow \boldsymbol{s} = \vee, \tag{2.71}$$

i.e. $\wedge$ is a unique idempotent t-norm, while $\vee$ forms a unique idempotent t-conorm.

Finally, one gets

$$[\forall a, b, c \in [0,1]: a\boldsymbol{t}(b\boldsymbol{s}c) = (a\boldsymbol{t}b)\boldsymbol{s}(a\boldsymbol{t}c)] \Leftrightarrow \boldsymbol{s} = \vee$$

and (2.72)

$$[\forall a, b, c \in [0,1]: a\boldsymbol{s}(b\boldsymbol{t}c) = (a\boldsymbol{s}b)\boldsymbol{t}(a\boldsymbol{s}c)] \Leftrightarrow \boldsymbol{t} = \wedge.$$

A t-norm is thus distributive with respect to a t-conorm only if that t-conorm is $\vee$ and, on the other hand, a t-conorm is distributive with respect to a t-norm $\boldsymbol{t}$ only if $\boldsymbol{t} = \wedge$.

Binary operations in [0, 1] are partially ordered by defining

$$\boldsymbol{u} \le \boldsymbol{w} \Leftrightarrow \forall a, b \in [0,1]: a\boldsymbol{u}b \le a\boldsymbol{w}b$$

with $\boldsymbol{u}, \boldsymbol{w}: [0,1]\times[0,1] \to [0,1]$. Property (2.70) can then be rewritten in a shorter form as

$$\boldsymbol{t}_{\mathrm{d}} \le \boldsymbol{t} \le \wedge \le \vee \le \boldsymbol{s} \le \boldsymbol{s}_{\mathrm{d}}. \tag{2.73}$$

The t-norms $\boldsymbol{t}_{\mathrm{d}}$ and $\wedge$ are thus, respectively, the least possible t-norm and the greatest one, whereas $\vee$ and $\boldsymbol{s}_{\mathrm{d}}$, respectively, form the least t-conorm and the greatest one.

For a t-norm $\boldsymbol{t}$, t-conorm $\boldsymbol{s}$ and strong negation ν, let $\boldsymbol{t}^{\nu}$ and $\boldsymbol{s}^{\nu}$ denote binary operations in [0, 1] defined by

$$a\boldsymbol{t}^{\nu}b = \nu(\nu(a)\boldsymbol{t}\nu(b)), \quad a\boldsymbol{s}^{\nu}b = \nu(\nu(a)\boldsymbol{s}\nu(b)). \tag{2.74}$$

It is a routine task to check that $\boldsymbol{t}^{\nu}$ is always a t-conorm, $\boldsymbol{s}^{\nu}$ forms a t-norm, and

$$(\boldsymbol{t}^{\nu})^{\nu} = \boldsymbol{t}, \quad (\boldsymbol{s}^{\nu})^{\nu} = \boldsymbol{s}.$$

Each strong negation ν thus establishes via (2.74) a mutual correspondence between t-norms and t-conorms.

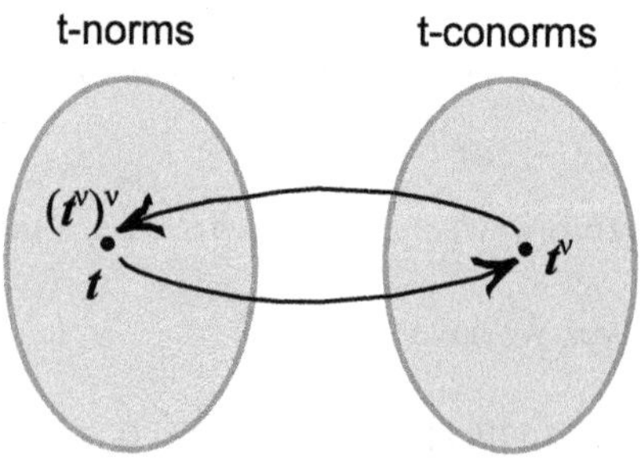

Fig. 2.28 1-1 correspondence between t-norms and t-conorms determined by strong negation ν

A t-norm $\boldsymbol{t}$ and t-conorm $\boldsymbol{s}$ are said to be ν-*dual* if $\boldsymbol{s} = \boldsymbol{t}^{\nu}$ or, equivalently, $\boldsymbol{t} = \boldsymbol{s}^{\nu}$. One can easily check that

$$\boldsymbol{t}_1 \leq \boldsymbol{t}_2 \Leftrightarrow \boldsymbol{t}_2^{\nu} \leq \boldsymbol{t}_1^{\nu} \tag{2.75}$$

for t-norms $\boldsymbol{t}_1$ and $\boldsymbol{t}_2$. The most important particular case of ν-duality is $\nu_Ł$-duality (see also Subsection 3.2.4). If $\boldsymbol{t}$ and $\boldsymbol{s}$ are $\nu_Ł$-dual, one says that they are *associated*, and one writes $\boldsymbol{s} = \boldsymbol{t}^*$ or, equivalently, $\boldsymbol{t} = \boldsymbol{s}^*$. Thus

$$a\boldsymbol{t}^* b = 1 - (1-a)\boldsymbol{t}(1-b) \quad \text{and} \quad a\boldsymbol{s}^* b = 1 - (1-a)\boldsymbol{s}(1-b). \tag{2.76}$$

$\wedge$ and $\vee$, $\boldsymbol{t}_a$ and $\boldsymbol{s}_a$, $\boldsymbol{t}_Ł$ and $\boldsymbol{s}_Ł$ as well as $\boldsymbol{t}_d$ and $\boldsymbol{s}_d$ are all examples of a t-norm and the associated t-conorm. Further instances are given below.

Example 2.6. Selected parameterized families of t-norms and associated t-conorms:

- $a\boldsymbol{t}_{S,p}b = [0 \vee (a^p + b^p - 1)]^{1/p}$, (*Schweizer t-norms*)
  $a\boldsymbol{s}_{S,p}b = 1 - [0 \vee ((1-a)^p + (1-b)^p - 1)]^{1/p}$, $p > 0$, (*Schweizer t-conorms*)
- $a\boldsymbol{t}_{Y,p}b = 0 \vee [1 - ((1-a)^p + (1-b)^p)^{1/p}]$, (*Yager t-norms*)
  $a\boldsymbol{s}_{Y,p}b = 1 \wedge (a^p + b^p)^{1/p}$, $p > 0$, (*Yager t-conorms*)
- $a\boldsymbol{t}_{H,\gamma}b = \dfrac{ab}{\gamma + (1-\gamma)(a+b-ab)}$, (*Hamacher t-norms*)

  $a\boldsymbol{s}_{H,\gamma}b = \dfrac{a+b-ab-(1-\gamma)ab}{1-(1-\gamma)ab}$, $\gamma \geq 0$, (*Hamacher t-conorms*)
- $a\boldsymbol{t}_{F,\lambda}b = \log_{\lambda}\left(1 + \dfrac{(\lambda^a - 1)(\lambda^b - 1)}{\lambda - 1}\right)$, (*Frank t-norms*)

  $a\boldsymbol{s}_{F,\lambda}b = 1 - \log_{\lambda}\left(1 + \dfrac{(\lambda^{1-a} - 1)(\lambda^{1-b} - 1)}{\lambda - 1}\right)$, $1 \neq \lambda > 0$, (*Frank t-conorms*)
- $a\boldsymbol{t}_{W,\lambda}b = 0 \vee \dfrac{a+b-1+\lambda ab}{1+\lambda}$, (*Weber t-norms*)

  $a\boldsymbol{s}_{W,\lambda}b = 1 \wedge \dfrac{(1+\lambda)(a+b)-\lambda ab}{1+\lambda}$, $\lambda > -1$. (*Weber t-conorms*)

□

Worth noticing is that $\wedge$ and $\vee$ are not only associated. They are ν-dual for each strong negation ν as

$$a \wedge^{\nu} b = a \vee b \quad \text{and} \quad a \vee^{\nu} b = a \wedge b. \tag{2.77}$$

As one sees, the families of all t-norms and t-conorms are infinite. It is clear that from the viewpoint of applications continuous t-operations are the focus of attention. This is why all t-operations presented up to now are continuous, excluding $\boldsymbol{t}_\mathrm{d}$ and $\boldsymbol{s}_\mathrm{d}$. However, what plays a key role in practice are the t-norms $\wedge$, $\boldsymbol{t}_\mathrm{a}$ and $\boldsymbol{t}_\text{Ł}$ with special emphasis on $\wedge$, and the associated t-conorms $\vee$, $\boldsymbol{s}_\mathrm{a}$ and $\boldsymbol{s}_\text{Ł}$ with special emphasis on $\vee$. The reason is their simplicity and that their values can be easily interpreted. A purely mathematical justification for the special importance of the t-norms $\wedge$, $\boldsymbol{t}_\mathrm{a}$, $\boldsymbol{t}_\text{Ł}$ and t-conorms $\vee$, $\boldsymbol{s}_\mathrm{a}$, $\boldsymbol{s}_\text{Ł}$ will be given in Subsection 3.2.2. For these *three basic t-norms* and *three basic t-conorms*, one has (see (2.75))

$$\boldsymbol{t}_\text{Ł} \le \boldsymbol{t}_\mathrm{a} \le \wedge \le \vee \le \boldsymbol{s}_\mathrm{a} \le \boldsymbol{s}_\text{Ł}.$$

Looking at Example 2.6, we see that

$$\boldsymbol{t}_{\mathrm{H},1} = \boldsymbol{t}_\mathrm{a}, \quad \boldsymbol{s}_{\mathrm{H},1} = \boldsymbol{s}_\mathrm{a},$$
$$\boldsymbol{t}_{\mathrm{S},1} = \boldsymbol{t}_{\mathrm{Y},1} = \boldsymbol{t}_{\mathrm{W},0} = \boldsymbol{t}_\text{Ł},$$
$$\boldsymbol{s}_{\mathrm{S},1} = \boldsymbol{s}_{\mathrm{Y},1} = \boldsymbol{s}_{\mathrm{W},0} = \boldsymbol{s}_\text{Ł}.$$

The t-operations $\boldsymbol{t}_{\mathrm{H},2}$ and $\boldsymbol{s}_{\mathrm{H},2}$, respectively, are known as the *Einstein t-norm* and *t-conorm*, respectively.

Referring to the basic t-norms and t-conorms, let us emphasize exceptional limit properties of Frank t-operations, namely

$$a\boldsymbol{t}_{\mathrm{F},\lambda}b \to a \wedge b \text{ and } a\boldsymbol{s}_{\mathrm{F},\lambda}b \to a \vee b \text{ if } \lambda \to 0,$$
$$a\boldsymbol{t}_{\mathrm{F},\lambda}b \to a\boldsymbol{t}_\mathrm{a}b \text{ and } a\boldsymbol{s}_{\mathrm{F},\lambda}b \to a\boldsymbol{s}_\mathrm{a}b \text{ if } \lambda \to 1,$$
$$a\boldsymbol{t}_{\mathrm{F},\lambda}b \to a\boldsymbol{t}_\text{Ł}b \text{ and } a\boldsymbol{s}_{\mathrm{F},\lambda}b \to a\boldsymbol{s}_\text{Ł}b \text{ if } \lambda \to \infty.$$

Consequently, one defines

$$\boldsymbol{t}_{\mathrm{F},0} = \wedge, \quad \boldsymbol{t}_{\mathrm{F},1} = \boldsymbol{t}_\mathrm{a}, \quad \boldsymbol{t}_{\mathrm{F},\infty} = \boldsymbol{t}_\text{Ł},$$
$$\boldsymbol{s}_{\mathrm{F},0} = \vee, \quad \boldsymbol{s}_{\mathrm{F},1} = \boldsymbol{s}_\mathrm{a}, \quad \boldsymbol{s}_{\mathrm{F},\infty} = \boldsymbol{s}_\text{Ł}.$$

The extended families $(\boldsymbol{t}_{\mathrm{F},\lambda})_{\lambda\in[0,\infty]}$ and $(\boldsymbol{s}_{\mathrm{F},\lambda})_{\lambda\in[0,\infty]}$ are then called the *Frank families* of t-norms and t-conorms, respectively.

Finally, some more advanced aspects of t-norms and t-conorms will be discussed in Section 3.2. The reader is also referred to BELIAKOV *et al.* (2007), GOTTWALD (2001), KLEMENT *et al.* (2000, 2004a, b, c) and LOWEN (1996) for further properties and references.

2.4.4 Operations Based on Triangular Norms and Conorms

Triangular operations are a suitable tool for defining basic operations on fuzzy sets in a flexible way. Let A, B: $\boldsymbol{U} \to [0, 1]$, and let $\boldsymbol{t}$ and $\boldsymbol{s}$, respectively, denote a t-norm and a t-conorm, respectively.

The *intersection* $A \cap_t B$ *of* A *and* B *induced by* $\boldsymbol{t}$ is a fuzzy set such that

$$(A \cap_t B)(x) = A(x)\,\boldsymbol{t}\,B(x) \quad \text{for each } x \in \boldsymbol{U}.$$

The *sum of* A *and* B *induced by* $\boldsymbol{s}$ is a fuzzy set $A \cup_s B$ with

$$(A \cup_s B)(x) = A(x)\,\boldsymbol{s}\,B(x) \quad \text{for each } x \in \boldsymbol{U}.$$

A commonly used choice is then $\boldsymbol{s} = \boldsymbol{t}^*$. Finally, the *cartesian product of* A: $\boldsymbol{U} \to [0, 1]$ *and* B: $\boldsymbol{V} \to [0, 1]$ *induced by* $\boldsymbol{t}$ is a fuzzy set $A \times_t B$ in $\boldsymbol{U} \times \boldsymbol{V}$ with

$$(A \times_t B)(x, y) = A(x)\,\boldsymbol{t}\,B(y) \quad \text{for each } (x, y) \in \boldsymbol{U} \times \boldsymbol{V}.$$

Using the induced operations, $A \subset B$ and $A = B$ remain defined in the way presented in Subsection 2.2.1.

The standard operations from Subsection 2.2.1 are thus particular cases of the induced ones:

$$A \cap B = A \cap_\wedge B, \quad A \cup B = A \cup_\vee B, \quad A \times B = A \times_\wedge B.$$

If A and B collapse to sets, then $\cap_t$, $\cup_s$ and $\times_t$ lead to correct results of classical set algebra, which follows from (2.67) and (2.68).

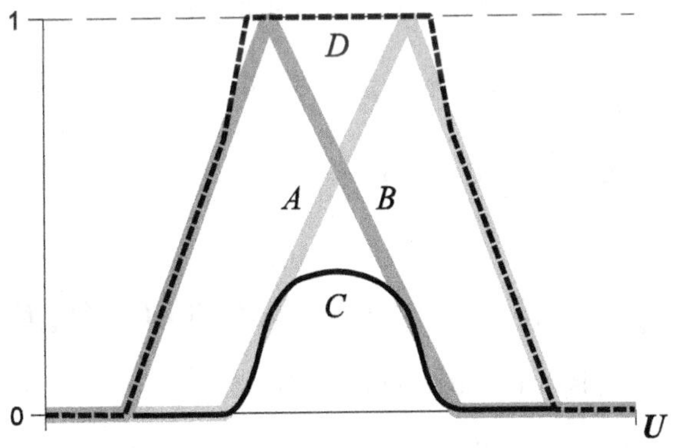

Fig. 2.29 Graphical representation of $C = \boldsymbol{A} \cap_{\boldsymbol{t}_{\mathbf{a}}} \boldsymbol{B}$ and $D = \boldsymbol{A} \cup_{\boldsymbol{s}_{\mathbf{L}}} \boldsymbol{B}$

Example 2.7. Let us present a numerical example continuing Example 2.1(b) from Subsection 2.3.2. For

$$A = 0.3/2 + 0.8/3 + 1/4 + 0.7/6$$

and

$$B = 0.5/2 + 1/3 + 0.9/4 + 1/5 + 0.2/6 + 0.3/7$$

in $\boldsymbol{U} = \{0, 1, ..., 8\}$, one gets

$$\boldsymbol{A} \cap_{t_a} \boldsymbol{B} = 0.15/2 + 0.8/3 + 0.9/4 + 0.14/6,$$

$$\boldsymbol{A} \cap_{t_L} \boldsymbol{B} = 0.8/3 + 0.9/4$$

and

$$\boldsymbol{A} \cup_{s_a} \boldsymbol{B} = 0.65/2 + 1/3 + 1/4 + 1/5 + 0.76/6 + 0.3/7,$$

$$\boldsymbol{A} \cup_{s_L} \boldsymbol{B} = 0.8/2 + 1/3 + 1/4 + 1/5 + 0.9/6 + 0.3/7.$$

For $C = 0.8/5 + 0.9/7$ and $D = 0.4/3 + 1/5 + 0.7/6$, we obtain

$$\boldsymbol{C} \times_{t_a} \boldsymbol{D} = 0.32/(5, 3) + 0.8/(5, 5) + 0.56/(5, 6) + 0.36/(7, 3) + 0.9/(7, 5) + 0.63/(7, 6),$$

$$\boldsymbol{C} \times_{t_L} \boldsymbol{D} = 0.2/(5, 3) + 0.8/(5, 5) + 0.5/(5, 6) + 0.3/(7, 3) + 0.9/(7, 5) + 0.6/(7, 6).$$

□

By virtue of (T1)-(T5), we generally have:

commutativity,

$$A \cup_s B = B \cup_s A, \quad A \cap_t B = B \cap_t A, \tag{2.78}$$

associativity,

$$A \cup_s (B \cup_s C) = (A \cup_s B) \cup_s C, \quad A \cap_t (B \cap_t C) = (A \cap_t B) \cap_t C, \tag{2.79}$$

monotonicity,

$$A \cap_t B \subset C \cap_t D \text{ and } A \cup_s B \subset C \cup_s D \text{ whenever } A \subset C \text{ and } B \subset D, \tag{2.80}$$

neutral elements:

$$A \cup_s 1_\varnothing = A, \quad A \cap_t 1_U = A. \tag{2.81}$$

On account of (2.69), (2.70) and (2.72),

$$A \cap_t A \subset A \subset A \cup_s A, \tag{2.82}$$

$$\boldsymbol{A} \cap_{t_d} \boldsymbol{B} \subset A \cap_t B \subset A \cap B \subset A, B \subset A \cup B \subset A \cup_s B \subset \boldsymbol{A} \cup_{s_d} \boldsymbol{B}, \tag{2.83}$$

and

$$(\cup_s \text{ is distributive with respect to } \cap_t) \Leftrightarrow \boldsymbol{t} = \wedge$$
$$(\cap_t \text{ is distributive with respect to } \cup_s) \Leftrightarrow \boldsymbol{s} = \vee. \tag{2.84}$$

Finally, if $\boldsymbol{t}$ and $\boldsymbol{s}$ are ν-dual, then *De Morgan laws* hold true:

$$(A \cup_s B)^\nu = A^\nu \cap_t B^\nu \text{ and } (A \cap_t B)^\nu = A^\nu \cup_s B^\nu. \tag{2.85}$$

For $\nu = \nu_Ł$, they can be rewritten as

$$(A \cup_{t^*} B)' = A' \cap_t B' \quad \text{and} \quad (A \cap_t B)' = A' \cup_{t^*} B'. \tag{2.86}$$

The loss of idempotency and distributivity in (2.82) and (2.84) causes that we leave the world of De Morgan algebras. Let us look at a few examples of the variety of properties we then deal with (see also (3.20)):

- $A \cap_t A' \neq 1_\varnothing$, $A \cup_s A' \neq 1_U$, $A \cap_t A = A \cup_s A = A$ for $\boldsymbol{t} = \wedge$, $\boldsymbol{s} = \vee$, $\nu = \nu_Ł$;
- $A \cap_t A' \neq 1_\varnothing$, $A \cup_s A' \neq 1_U$, $A \cap_t A \subset A \subset A \cup_s A$ for $\boldsymbol{t} = \boldsymbol{t}_a$, $\boldsymbol{s} = \boldsymbol{s}_a$, $\nu = \nu_Ł$;
- $A \cap_t A' = 1_\varnothing$, $A \cup_s A' = 1_U$, $A \cap_t A \subset A \subset A \cup_s A$ for $\boldsymbol{t} = \boldsymbol{t}_Ł$, $\boldsymbol{s} = \boldsymbol{s}_Ł$, $\nu = \nu_Ł$;
- $A \cap_t A^\nu = 1_\varnothing$, $A \cup_s A^\nu \neq 1_U$, $A \cap_t A \subset A \subset A \cup_s A$ for $\boldsymbol{t} = \boldsymbol{t}_a$, $\boldsymbol{s} = \boldsymbol{s}_a$, $\nu = \nu_*$.

2.4.5 Implication Operators

Besides negations and t-operations, another type of operators playing an important role when dealing with fuzzy sets are implication operators. They are used as truth functions for implication connectives occurring, say, in approximate reasoning and knowledge representation (see (2.7) and Chapter 5).

We will say that $\rightarrow: [0, 1] \times [0, 1] \rightarrow [0, 1]$ is an *implication operator* if it satisfies the following conditions for each $a, b, c, d \in [0, 1]$:

(I1) $a \rightarrow b \geq c \rightarrow b$ for $a \leq c$, (*nonincreasingness in the first argument*)

(I2) $a \rightarrow b \leq a \rightarrow d$ for $b \leq d$, (*nondecreasingness in the second argument*)

(I3) $0 \rightarrow b = 1$,

(I4) $a \rightarrow 1 = 1$, (*boundary conditions*)

(I5) $1 \rightarrow 0 = 0$.

Let us mention two important sorts of implication operators defined by (I1)-(I5):

$$a \rightarrow_t b = \bigvee \{c \in [0, 1] \colon a \boldsymbol{t} c \leq b\} \tag{2.87}$$

and

$$a \rightarrow_{s,\nu} b = \nu(a) \boldsymbol{s} b, \tag{2.88}$$

where $\boldsymbol{t}$, $\boldsymbol{s}$ and ν, respectively, denote a t-norm, t-conorm and strong negation, respectively. $\rightarrow_t$ is then called the *implication operator induced by* $\boldsymbol{t}$ (*R-implication*, in short), whereas $\rightarrow_{s,\nu}$ is known as the *implication operator induced by* $\boldsymbol{s}$ *and* ν (*S-implication*, in short). Obviously, if $\boldsymbol{t}$ is continuous, $\bigvee$ in (2.87) collapses to max. For the basic t-norms and t-conorms with the standard negation, we thus get:

- $a \to_{\wedge} b = \begin{cases} 1, & \text{if } a \le b, \\ b, & \text{otherwise,} \end{cases}$ (*Gődel implication operator*)

- $a \to_{t_a} b = \begin{cases} 1 \wedge \dfrac{b}{a}, & \text{if } a \ne 0, \\ 1, & \text{otherwise,} \end{cases}$ (*Goguen implication operator*)

- $a \to_{t_Ł} b = a \to_{s_Ł, \nu_Ł} b = 1 \wedge (1 - a + b),$ (*Łukasiewicz implication operator* $\to_Ł$)

- $a \to_{\vee, \nu_Ł} b = (1 - a) \vee b,$ (*Kleene-Dienes implication operator*)

- $a \to_{s_a, \nu_Ł} b = 1 - a + ab.$ (*Reichenbach implication operator*)

An immediate consequence of (2.87) and (T4) is that

$$a \le b \Rightarrow a \to_t b = 1, \quad 1 \to_t b = b,$$

while

$$a \to_t b = 1 \Leftrightarrow a \le b \quad \text{for continuous t-norms.}$$

In contrast to $\to_t$, one generally has $a \le b \not\Rightarrow a \to_{s,\nu} b = 1$, although $1 \to_{s,\nu} b = b$.

R-implications can be used to create negations. Indeed, for each $\boldsymbol{t}$, ν_t with

$$\nu_t(a) = a \to_t 0 = \bigvee\{c \in [0,1]\colon a\boldsymbol{t}c = 0\} \tag{2.89}$$

is a negation known as the *negation induced by* $\boldsymbol{t}$ (cf. (2.4)). In particular, we get (see (2.60) and (2.61)):

- $\nu_t = \nu_*$ for $\boldsymbol{t} = \wedge$ and $\boldsymbol{t} = \boldsymbol{t}_a$,
- $\nu_t = \nu_Ł$ for $\boldsymbol{t} = \boldsymbol{t}_Ł$,
- $\nu_t = \nu_{Y,p}$ for $\boldsymbol{t} = \boldsymbol{t}_{S,p}$,
- $\nu_t = \nu_{S,\lambda}$ for $\boldsymbol{t} = \boldsymbol{t}_{W,\lambda}$.

More facts concerning the induced negations will be presented in Subsection 3.2.4.

Although the system of axioms (I1)-(I5) looks reasonable, it is not fully satisfactory and does not cover all types of operators used as implication operators in the theory of fuzzy sets. An instance is $\to_{t,s,\nu}$, the *implication operator induced by a t-norm* $\boldsymbol{t}$, *t-conorm* $\boldsymbol{s}$, *and strong negation* ν (*QL-implication*, in short), where

$$a \to_{t,s,\nu} b = \nu(a)\, \boldsymbol{s}\, (a \boldsymbol{t} b). \tag{2.90}$$

Then

$$a \to_{\wedge,\vee,\nu_Ł} b = (1 - a) \vee (a \wedge b),$$

which is called the *Zadeh implication operator*. We see that $\to_{t,s,\nu}$ does not fulfil (I1) and (I4).

The interested reader is referred to BACZYŃSKI/JAYARAM (2008), the most comprehensive source of information about theory and applications of implication operators; see also FODOR/ROUBENS (1994) and GOTTWALD (2001).

2.4.6 Logical Background

One should realize that using the induced intersections, sums and complements we change the logical foundation of fuzzy sets. It is no longer $Ł_\infty$, but *many-valued logic with triangular norms* in which a negation ν, t-norm $\boldsymbol{t}$, t-conorm $\boldsymbol{s}$, and implication operator $\to$ become truth functions for basic logical connectives. More precisely, we then define (cf. (2.4)-(2.10)):

$$[\![\neg_{\mathrm{m}} \boldsymbol{p}]\!] = \nu([\![\boldsymbol{p}]\!]), \tag{2.91}$$

$$[\![\boldsymbol{p} \,\&_{\mathrm{m}}\, \boldsymbol{q}]\!] = [\![\boldsymbol{p}]\!] \,\boldsymbol{t}\, [\![\boldsymbol{q}]\!], \tag{2.92}$$

$$[\![\boldsymbol{p} \,\#_{\mathrm{m}}\, \boldsymbol{q}]\!] = [\![\boldsymbol{p}]\!] \,\boldsymbol{s}\, [\![\boldsymbol{q}]\!], \tag{2.93}$$

$$[\![\boldsymbol{p} \Rightarrow_{\mathrm{m}} \boldsymbol{q}]\!] = [\![\boldsymbol{p}]\!] \to [\![\boldsymbol{q}]\!], \tag{2.94}$$

$$[\![\boldsymbol{p} \Leftrightarrow_{\mathrm{m}} \boldsymbol{q}]\!] = [\![\boldsymbol{p} \Rightarrow_{\mathrm{m}} \boldsymbol{q} \,\&_{\mathrm{m}}\, \boldsymbol{q} \Rightarrow_{\mathrm{m}} \boldsymbol{p}]\!], \tag{2.95}$$

$$[\![\forall_{\mathrm{m}} x \in \boldsymbol{D}\colon \boldsymbol{r}(x)]\!] = \bigwedge_{a \in \boldsymbol{D}} [\![\boldsymbol{r}(x \mid a)]\!], \tag{2.96}$$

$$[\![\exists_{\mathrm{m}} x \in \boldsymbol{D}\colon \boldsymbol{r}(x)]\!] = \bigvee_{a \in \boldsymbol{D}} [\![\boldsymbol{r}(x \mid a)]\!]. \tag{2.97}$$

If $\boldsymbol{D}$ is finite, say, $\boldsymbol{D} = \{a_1, a_2, \dots, a_k\}$ with $k \geq 1$, then direct generalizations of formulae (2.92) and (2.93) can be used instead of (2.96) and (2.97) as alternative rules of numerical interpretation for quantified sentences:

$$[\![\forall_{\mathrm{m}} x \in \boldsymbol{D}\colon \boldsymbol{r}(x)]\!] = [\![\boldsymbol{r}(x \mid a_1)]\!] \,\boldsymbol{t}\, [\![\boldsymbol{r}(x \mid a_2)]\!] \,\boldsymbol{t} \dots \boldsymbol{t}\, [\![\boldsymbol{r}(x \mid a_k)]\!] \tag{2.98}$$

and

$$[\![\exists_{\mathrm{m}} x \in \boldsymbol{D}\colon \boldsymbol{r}(x)]\!] = [\![\boldsymbol{r}(x \mid a_1)]\!] \,\boldsymbol{s}\, [\![\boldsymbol{r}(x \mid a_2)]\!] \,\boldsymbol{s} \dots \boldsymbol{s}\, [\![\boldsymbol{r}(x \mid a_k)]\!]. \tag{2.99}$$

The definitions of induced operations are connected with (2.91)-(2.93) in exactly the same way as the standard definitions from Subsection 2.2.1 are connected with $Ł_\infty$.

2.5 Fuzzy Numbers and Linguistic Variables

The subject of this section are three extremely important notions of fuzzy logic in both the wide sense and the narrow one (see Section 1.1). We mean the idea of fuzzy numbers, the extension principle making it possible to operate on fuzzy numbers, and the concept of a linguistic variable.

2.5.1 Fuzzy Numbers and Their Types

Let NUM denote a subset of real numbers, NUM $\subset \mathbb{R}$. By a *fuzzy number* one means a fuzzy set in NUM, i.e. a function NUM $\to$ [0, 1]. Usually, but not always, that fuzzy set is assumed to be normal and convex, whereas $\mathbb{R}$, $\mathbb{N}$, the set of all integers or an interval in one of these three sets is taken as NUM. Notice that each real number x and each interval $[x, y]$ can be viewed as fuzzy numbers $1_{\{x\}}$ and $1_{[x,y]}$, respectively.

Fuzzy numbers, speaking generally, are a tool for modeling imprecise numerical data. Some kinds of fuzzy numbers seem to be especially important in practice. They are illustrated and briefly discussed below.

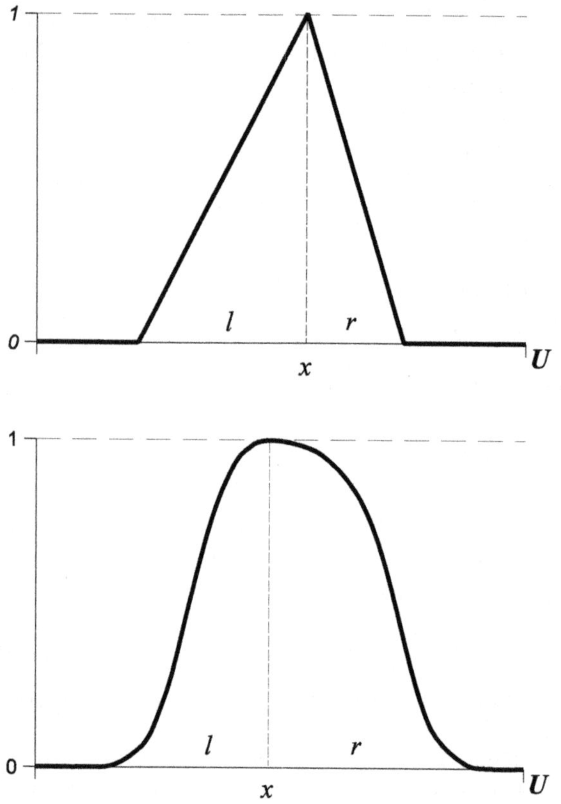

Fig. 2.30 Examples (top) of a triangular and (bottom) a bell-shaped fuzzy number

Triangular fuzzy numbers, exemplified in Figure 2.30(top), together with *bell-shaped fuzzy numbers* from Figure 2.30(bottom) are used to model imprecise numerical data such as “about x” or “approximately x”, e.g. “the temperature is about 25 °C”. The point x is called the *center* or *core* of the fuzzy number. Of special interest seem to be

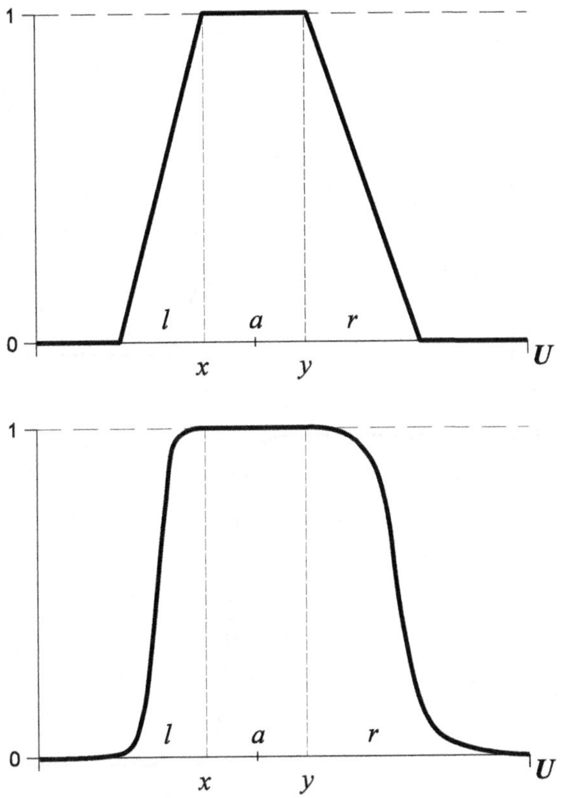

Fig. 2.31 Two forms of trapezoidal fuzzy numbers

symmetrical forms. The *left spread*, l, is then equal to the *right spread*, r (see Figures 2.4(top) and 2.5 in Subsection 2.1.2).

As to *trapezoidal fuzzy numbers* in Figure 2.31, they represent imprecise numerical data like "from about x to about y", "approximately between x and y", or "medium", i.e. fuzzy intervals, intervals with fuzzy endpoints. An instance is the statement "The system will be under repair about 4–5 hours". The ordinary interval $[x, y]$ is now the *core* or *center* of the fuzzy number. Again, symmetrical forms with identical spreads, $l = r$, are of special importance (see e.g. Figure 2.4(bottom)). If there is a full tolerance to "small" deviations from a, trapezoidal fuzzy numbers can also be used to model imprecise data "about a".

The next figure presents two forms of *z-shaped fuzzy numbers* which model imprecise data such as "at most about x", e.g. "The cost will not be greater than approximately 10000 €". We thus accept "small" violations of the condition "$\leq x$". The deeper the violation, the lower the degree of our satisfaction. This degree collapses to zero whenever the violation is greater than or equal to the right spread, r.

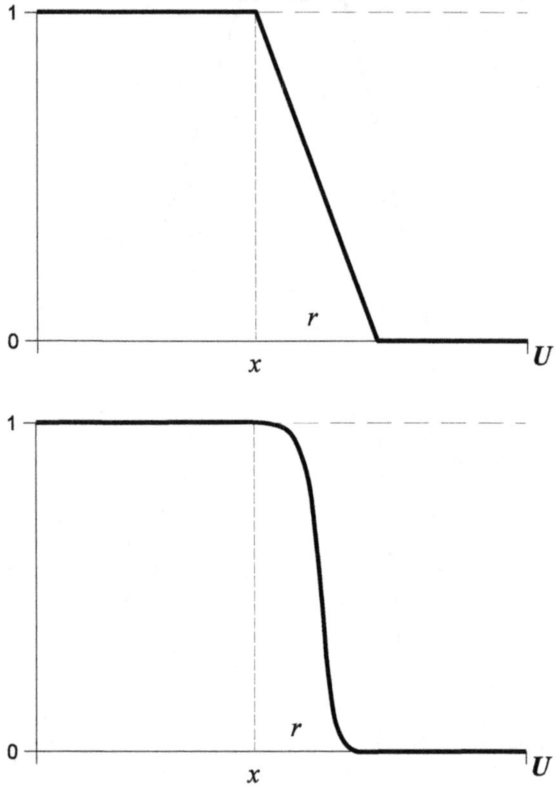

Fig. 2.32 z-shaped fuzzy numbers

A natural area of applications of z-shaped fuzzy numbers is thus the modeling of terms like "young", "short", "small", etc.

Finally, Figure 2.33 shows *s-shaped fuzzy numbers*. They are used to model imprecise numerical data of the form "at least about x", e.g. "Adam is at least about 35 years old". Slight violations of the condition "$\geq x$" are now accepted to a degree. Applications of s-shaped fuzzy numbers thus encompass the modeling of notions such as "long", "tall", "large", etc.

One should stress that the most frequently used fuzzy numbers are the piecewise linear top forms in Figures 2.30-2.33. The main reason is their simplicity. They can be uniquely encoded in a straightforward way as

(x, l, r) (see Figure 2.30(top)),

(x, y, l, r) (see Figure 2.31(top)),

$(x, r)_z$ (see Figure 2.32(top))

and $(x, l)_s$ (see Figure 2.33(top)).

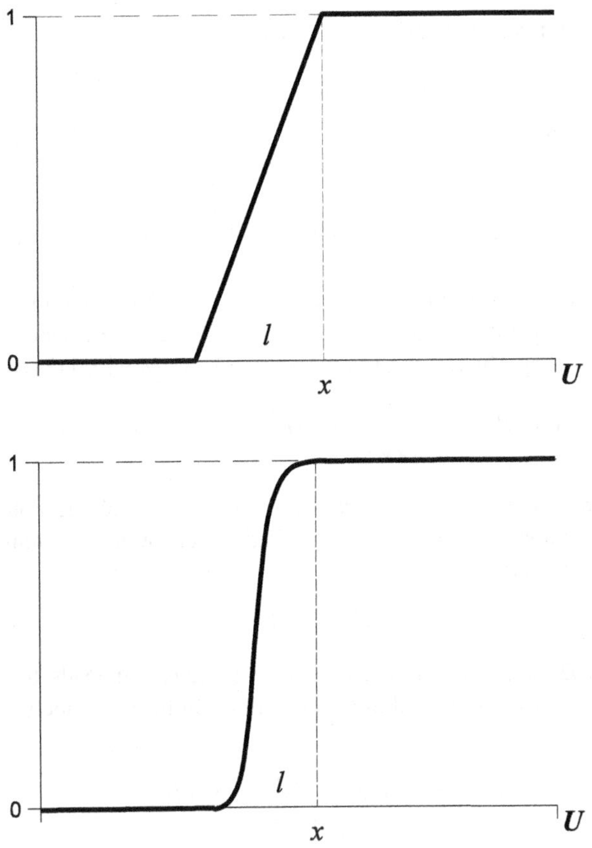

Fig. 2.33 s-shaped fuzzy numbers

For instance, the fuzzy number in Figure 2.3 can be encoded as $(185, 15)_s$, whereas (4, 1, 1) and (6, 8, 2, 2) represent the triangular fuzzy number and the trapezoidal one, respectively, shown in Figure 2.4.

2.5.2 The Extension Principle and Operations on Fuzzy Numbers

Let us move on to basic arithmetic aspects of fuzzy numbers. One says that a fuzzy number A: NUM → [0, 1] is

- *positive* if $\text{supp}(A) \subset (0,\infty)$,
- *negative* if $\text{supp}(A) \subset (-\infty, 0)$,
- *a zero fuzzy number* if $0 \in \text{supp}(A)$.

The *opposite of A* is a fuzzy number $-A$ such that

$$-A(x) = A(-x) \quad \text{for each } x \in \text{NUM}. \tag{2.100}$$

Further, A^{-1} with

$$A^{-1}(x) = \begin{cases} A\left(\frac{1}{x}\right), & \text{if } x \neq 0, \\ 0, & \text{otherwise,} \end{cases} \tag{2.101}$$

is called the *inverse of A* provided that A is a non-zero fuzzy number.

Each binary operation $*$ on numbers from NUM can be extended to fuzzy sets A, B: NUM $\to [0, 1]$. The resulting fuzzy set $A * B$: NUM $\to [0, 1]$ is defined as

$$(A * B)(z) = \bigvee_{x * y = z} A(x) \, t \, B(y) \quad \text{for each } z \in \text{NUM} \tag{2.102}$$

with a t-norm t and "$x * y = z$" as an abbreviated form of the notation "$\{(x, y)$: $x * y = z\}$". To emphasize which t-norm is used, we can write $A *_t B$ instead of $A * B$. By (2.73), one thus has

$$A *_t B \subset A *_\wedge B.$$

Formula (2.102) is known as the *extension principle.* It leads to the following definitions of the four basic arithmetic operations on fuzzy numbers A and B:

- *addition*

$$(A + B)(z) = \bigvee_{x + y = z} A(x) \, t \, B(y), \tag{2.103}$$

- *subtraction*

$$(A - B)(z) = \bigvee_{x - y = z} A(x) \, t \, B(y), \tag{2.104}$$

- *multiplication*

$$(A \cdot B)(z) = \bigvee_{x \cdot y = z} A(x) \, t \, B(y), \tag{2.105}$$

- *division*

$$(A : B)(z) = \bigvee_{x : y = z} A(x) \, t \, B(y) \tag{2.106}$$

for each $z \in$ NUM. For the division, one assumes that B is a non-zero fuzzy number.

Example 2.8. Let NUM $= \mathbb{N}$, $t = \wedge$, and

$$A = 0.5/1 + 1/2 + 0.8/3, \quad B = 0.3/2 + 1/3 + 0.7/4.$$

By (2.103) and (2.105), for each $k \in \mathbb{N}$, we have

$$(A + B)(k) = \bigvee_{i + j = k} A(i) \wedge B(j) \quad \text{and} \quad (A \cdot B)(k) = \bigvee_{i \cdot j = k} A(i) \wedge B(j).$$

Thus, for instance,

$$(A+B)(3) = (A(0) \wedge B(3)) \vee (A(1) \wedge B(2)) \vee (A(2) \wedge B(1)) \vee (A(3) \wedge B(0)) = 0.3,$$
$$(A \cdot B)(4) = (A(1) \wedge B(4)) \vee (A(2) \wedge B(2)) \vee (A(4) \wedge B(1)) = 0.5.$$

Finally,

$$A+B = 0.3/3 + 0.5/4 + 1/5 + 0.8/6 + 0.7/7$$

and

$$A \cdot B = 0.3/2 + 0.5/3 + 0.5/4 + 1/6 + 0.7/8 + 0.8/9 + 0.7/12. \qquad \square$$

$A * B$ with $* \in \{+, -, \cdot, :\}$ is always normal. The above example suggests, however, that $A \cdot B$ is not generally convex in contrast to $A+B$.

Both the addition and multiplication of fuzzy numbers are commutative and associative, and have neutral elements:

$$A + 1_{\{0\}} = A \quad \text{and} \quad A \cdot 1_{\{1\}} = A.$$

On the other hand, the distributivity law $A \cdot (B+C) = (A \cdot B) + (A \cdot C)$ holds true only if B and C are of the same sign. The most essential difference between the arithmetic of fuzzy numbers and the classical arithmetic, however, is that both the equations

$$A + X = 1_{\{0\}}, \quad A \cdot X = 1_{\{1\}}$$

have no solutions in general. In particular,

$$A + (-A) \neq 1_{\{0\}}, \quad A \cdot A^{-1} \neq 1_{\{1\}}. \tag{2.107}$$

The opposite as well as the inverse element do not generally exist in the case of fuzzy numbers, in other words.

One sees that computing with fuzzy numbers by means of the very extension principle (2.102) is rather arduous. Worth mentioning is therefore a simplification in the important case when NUM $= \mathbb{R}$ and $\boldsymbol{t} = \wedge$. For the four basic arithmetic operations, if all t-cuts A_t and B_t of fuzzy numbers A, B: $\mathbb{R} \to [0, 1]$ are closed intervals, we then have (see NGUYEN (1978) and NGUYEN/WALKER (2005))

$$(A * B)_t = A_t * B_t \tag{2.108}$$

for each $t \in [0, 1]$. Referring to interval arithmetic, the interval operations on the right-hand side of (2.108) are defined as

$$[a, b] + [c, d] = [a+c, b+d], \tag{2.109}$$
$$[a, b] - [c, d] = [a-d, b-c], \tag{2.110}$$
$$[a, b] \cdot [c, d] = [\min(ac, ad, bc, bd), \max(ac, ad, bc, bd)], \tag{2.111}$$
$$[a, b] : [c, d] = [\min(a:c, a:d, b:c, b:d), \max(a:c, a:d, b:c, b:d)] \tag{2.112}$$

with $c, d \neq 0$ in (2.112). If $a, b, c, d \geq 0$, then (2.111) and (2.112) collapse to

$$[a, b] \cdot [c, d] = [ac, bd], \tag{2.113}$$

$$[a, b] : [c, d] = [a : d, b : c]. \tag{2.114}$$

The decomposition property thus gives

$$A * B = \bigcup_{t \in (0,1]} t\, 1_{(A*B)_t} = \bigcup_{t \in (0,1]} t\, 1_{A_t * B_t}. \tag{2.115}$$

Especially convenient formulae can be used for the addition and subtraction of triangular fuzzy numbers and (piecewise linear) trapezoidal fuzzy numbers. Let $A = (x_A, l_A, r_A)$ and $B = (x_B, l_B, r_B)$. Then

$$A + B = (x_A + x_B, l_A + l_B, r_A + r_B), \quad -B = (-x_B, r_B, l_B)$$

and

$$A - B = A + (-B) = (x_A - x_B, l_A + r_B, r_A + l_B). \tag{2.116}$$

Using trapezoidal fuzzy numbers $A = (x_A, y_A, l_A, r_A)$ and $B = (x_B, y_B, l_B, r_B)$, one has

$$A + B = (x_A + x_B, y_A + y_B, l_A + l_B, r_A + r_B), \quad -B = (-y_B, -x_B, r_B, l_B)$$

and

$$A - B = (x_A - y_B, y_A - x_B, l_A + r_B, r_A + l_B). \tag{2.117}$$

As to the multiplication, only some approximate formulae of type (2.116)-(2.117) are available in that case.

2.5.3 Comparisons of Fuzzy Numbers

The last issue related to fuzzy numbers we like to outline is the question of their comparisons. That issue is neither obvious nor simple, and there exist many reasonable approaches to it (see e.g. CHANG/LEE (1994)). Let us present two of them which seem to be extreme in a way.

The classical intervals of real numbers are partially ordered by means of a relation defined as

$$[a, b] \leq [c, d] \Leftrightarrow a \leq c \;\&\; b \leq d. \tag{2.118}$$

For fuzzy numbers A and B whose t-cuts are closed intervals of reals, one then defines

$$A \leq B \Leftrightarrow \forall t \in (0, 1]: A_t \leq B_t. \tag{2.119}$$

This way of comparing is mathematically elegant and easy to interpret. However, it is (too) restrictive because too many fuzzy numbers are then incomparable. An illustrative example is given below.

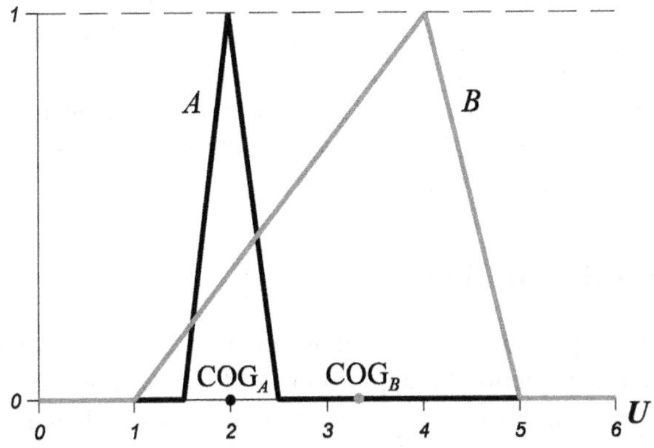

Fig. 2.34 Two triangular fuzzy numbers incomparable with respect to (2.119)

The fuzzy numbers in Figure 2.34 are incomparable if (2.119) is used. On the other hand, our intuition suggests that $A < B$. This is an impulse to introduce an extremely less restrictive method in which

$$A < B \Leftrightarrow \mathrm{COG}_A < \mathrm{COG}_B$$

and

$$A = B \Leftrightarrow \mathrm{COG}_A = \mathrm{COG}_B, \tag{2.120}$$

where COG_E denotes the first coordinate of the *center of gravity* (*COG*) of the figure determined by the horizontal axis and the membership function E. Thus, for fuzzy sets in $\mathbb{R}$, one has

$$\mathrm{COG}_E = \frac{\int_{\mathbb{R}} xE(x)\,dx}{\int_{\mathbb{R}} E(x)\,dx}. \tag{2.121}$$

If E is (or is viewed as) a finite fuzzy set with $\mathrm{supp}(E) = \{x_1, x_2, \dots, x_k\}$, then

$$\mathrm{COG}_E = \frac{\sum_{i=1}^{k} x_i E(x_i)}{\sum_{i=1}^{k} E(x_i)}. \tag{2.122}$$

As to the fuzzy numbers in Figure 2.34, it is obvious that $\mathrm{COG}_A = 2$. Applying (2.122) with $k = 5$ and $x_i = i$ for $i = 1, \dots, 5$, we get $\mathrm{COG}_B = 3\frac{1}{3}$. By (2.120), we conclude that $A < B$.

We finish the presentation of fuzzy numbers. It has been restricted to those notions, properties, and methods which are more or less relevant for our discussion in this book. For further topics, details and references, the reader is referred e.g. to DUBOIS/PRADE (1978, 1987b), KLEMENT *et al.* (2000), KLIR/YUAN (1995), NGUYEN/ WALKER (2005) and PEDRYCZ/GOMIDE (2007).

2.5.4 Linguistic Variables

The concept of a linguistic variable, introduced in ZADEH (1975a, b, c), seems to be one of the most crucial notion for applications of fuzzy sets. A *linguistic variable* is a variable attaining linguistic values. More precisely, those values are expressions of natural language and, semantically, one identifies them with some fuzzy sets. In a simple formal formulation, a linguistic variable can be viewed as a quadruple

$$(\mu, \boldsymbol{U}, \boldsymbol{V}, \boldsymbol{I})$$

in which μ is the name of the linguistic variable, $\boldsymbol{V}$ denotes a set (list) of its values, and $\boldsymbol{I}$ is a set of interpretations of those values. Each interpretation forms a fuzzy set in $\boldsymbol{U}$, a function $\boldsymbol{U} \to [0, 1]$. The universe $\boldsymbol{U}$ is usually a set of numbers and the interpretations become fuzzy numbers. Familiar examples of linguistic variables are *age, height, distance, color, salary, population density, speed.*

Example 2.9. Let us look a bit closer at the linguistic variable *speed* (of a car). It can be specified as

$$(\textit{speed}, [0, 300], \boldsymbol{V}, \boldsymbol{I})$$

with

$\boldsymbol{V}$ = { *about k kph, approximately between k1 and k2 kph, at most about k, at least about k, very low, low, medium, high, very high, ...*}.

The set $\boldsymbol{I}$ is composed of fuzzy numbers assigned to and interpreting the terms from $\boldsymbol{V}$. For instance (see Subsection 2.5.1),

"about *k* kph" can be interpreted as $(k, 5, 5)$,
"approximately between *k1* and *k2* kph" as $(k1, k2, 5, 5)$,
"at least about *k*" as $(k, 10)_s$,
"very low" as $(10, 10)_z$,
"high" as $(120, 140, 10, 10)$.

If imprecision is only an option, $\boldsymbol{V}$ should be supplemented by such conventional values as "*k* kph", "from *k1* to *k2* kph", "at most *k*" and "at least *k*" interpreted by means of fuzzy numbers $1_{\{k\}}$, $1_{[k1, k2]}$, $1_{[0, k]}$ and $1_{[k, 300]}$, respectively. □

A linguistic value can be a modification of another value by adding a *modifier* such as *very*, *rather*, etc. This is the case, say, of "very low". On the other hand, it happens that a linguistic value is made of other ones by using logical connectives, e.g. "not low and not high" = "medium". What we deal with are thus *primary*, *atomic* linguistic values (*young*, *low*, ...) and *complex* ones (*very low*, *medium*, ...). Using some simple rules, interpretations of complex values can be generated automatically from given ("handmade") interpretations of primary ones. Indeed, assume that A, B: $\boldsymbol{U} \to [0, 1]$, A interprets a linguistic value ρ, and B interprets σ. As previously, let $\boldsymbol{t}$, $\boldsymbol{s}$, and ν, respectively, denote a t-norm, t-conorm, and negation, respectively. Moreover, let A^p with $p > 0$ be a fuzzy set defined as $A^p(x) = (A(x))^p$. Typical forms of the generating rules are then:

$$\text{"not } \rho\text{", e.g. "not small", is interpreted by } A^{\nu}, \tag{2.123}$$

$$\text{"very } \rho\text{", e.g. "very young", is interpreted by } A^{2}, \tag{2.124}$$

$$\text{"rather } \rho\text{", e.g. "rather tall", is interpreted by } A^{0.5}, \tag{2.125}$$

$$\text{"}\rho \text{ and } \sigma\text{" by } A \cap_t B, \tag{2.126}$$

$$\text{"}\rho \text{ or } \sigma\text{" by } A \cup_s B. \tag{2.127}$$

Example 2.10. Assume "short" and "tall" are primary values of linguistic variable "height". We like to treat "medium" as a complex value defined as "not short and not tall". Further, A and B, respectively, are given fuzzy sets interpreting "short" and "tall", respectively (see Figure 2.35). For simplicity, let $\nu = \nu_L$, $\boldsymbol{t} = \wedge$ and $\boldsymbol{s} = \vee$.

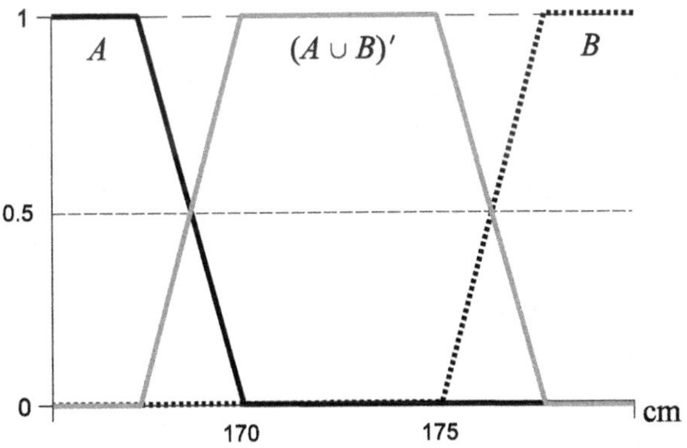

Fig. 2.35 Interpretation of "medium" generated from interpretations of "short" and "tall"

By (2.123), (2.126) and De Morgan laws (2.31), "medium" is then interpreted by means of $A' \cap B' = (A \cup B)'$. □

If ***V*** is a large set, its description via a list of elements is a troublesome task. A more effective approach then seems to be the following:

- fix a small set ***PV*** of primary linguistic values,
- construct a formal grammar ***G*** offering some syntactic rules of producing permissible complex linguistic values from the primary values collected in ***PV***.

Consequently, a linguistic variable can be understood in a more advanced manner as

$$(\mu, \boldsymbol{U}, \boldsymbol{G}, \boldsymbol{S}),$$

where ***S*** is a set of semantical rules exemplified by (2.123)-(2.127) and assigning to each produced complex value its interpretation.

Chapter 3

Further Aspects of Triangular Norms – A Study Inspired by Flexible Querying

The main aim of this chapter is a study of further aspects of triangular norms. However, we like to do it in a varied and a bit non-standard way growing out of applications and practice. We therefore begin with a presentation of the basics of flexible querying in databases, one of classical areas of applications of fuzzy sets. Some questions arising from flexible querying combined with the use of t-norms are then treated in Section 3.2 as a natural pretext and inspiration for analyzing a few more advanced issues in t-norms. We mean classification and generators of t-norms and t-conorms, induced negations, and complementary t-norms and t-conorms.

3.1 Flexible Querying in Databases

Database querying belongs to already classical areas of applications of fuzzy sets and their methodology. That methodology enables the user to formulate *imprecisely, fuzzily specified queries*, and to get satisfactory answers to them, which is known as *flexible* (*soft, fuzzy*) *querying*. More precisely, we mean queries of the form

"Which objects in the database are ***p***?",

where ***p*** denotes an arbitrary, generally imprecise property expressed in a natural language. Examples of such queries are:

(FQ1) "Which *affordable* hotels are situated in the *vicinity* of the center of Paris?",

(FQ2) "Which males in the database are *rather tall*, *about* 30 years old, and live in the *vicinity* of Warsaw?",

(FQ3) "In which samples of water *most* contamination indicators *considerably* exceed the limits?".

Speaking more generally, what we then deal with is the problem of information retrieval in relational databases, in the Internet, etc. However, we allow for imprecise values of attributes and imprecise relationships between those values. Conventional database management systems require a precise specification of both the values and relationships as in the following sharp counterparts of (FQ1) and (FQ2):

M. Wygralak: *Intelligent Counting Under Information Imprecision*, STUDFUZZ 292, pp. 71–91.
DOI: 10.1007/978-3-642-34685-9_3

(SQ1) "Which hotels at 70 €/night at the most are situated no more than 15 min by metro from the center of Paris?",

(SQ2) "Which males in the database are 175-180 cm tall, 28-32 years old, and live at most 50 km from Warsaw?".

One sees that flexible querying makes it possible to formulate queries to the database in a human-consistent way. In comparison with conventional querying languages, the class of acceptable queries becomes much wider and interesting from the viewpoint of practice. One more advantage of flexible querying is that a precise formulation of a query is sometimes difficult or even impossible. This may be the case, say, of (FQ2) if the imprecise description therein is the only one the police have on the basis of eyewitness account.

Referring to (FQ1) and its sharp counterpart (SQ1), we like to present a procedure of constructing the answer to a flexible query. We will concentrate on those basic aspects which are relevant for our discussion skipping over many details and variants of queries, e.g. queries like (FQ3) involving a linguistic quantifier (see Section 10.3). For a comprehensive treatment of the issue, the reader is referred to BOSC/KACPRZYK (1995), BOSC *et al.* (2005), BOSC/PIVERT (1991, 1994a, b), KACPRZYK/ZADROŻNY (1997, 2001b, 2010b), ZADROŻNY/KACPRZYK (1996); see also CHRISTIANSEN *et al.* (2004), LARSEN *et al.* (2001).

3.1.1 Constructing the Answer to a Flexible Query

The answer to a conventional query to the database is the set of all objects satisfying the properties (conditions) included in the query. So, the answer to (SQ1) is the intersection $\boldsymbol{A} \cap \boldsymbol{B}$ of two sets:

$\boldsymbol{A}$ - set of all hotels at $\leq$ 70 €/night,

$\boldsymbol{B}$ - set of all hotels situated $\leq$ 15 min by metro from the center of Paris.

The conditions in (FQ1) are imprecise instead, and the answer is now the intersection $A \cap_t B$ of two fuzzy sets with a chosen t-norm $\boldsymbol{t}$:

A - fuzzy set of *affordable* hotels,

B - fuzzy set of hotels in the *vicinity* of the center of Paris.

Clearly, the answer to a query containing $n \geq 1$ elementary conditions $A_1, A_2, \dots, A_n$ connected by AND is a fuzzy set

$$A_1 \cap_t A_2 \cap_t \dots \cap_t A_n. \tag{3.1}$$

We understand that $\cap_t$ should be replaced with $\cup_s$ using a t-conorm $\boldsymbol{s}$ whenever OR connects two elementary conditions.

The answer to a fuzzy query, speaking generally, is thus a fuzzy set being an aggregation of fuzzy sets representing the elementary conditions contained in the query (see Chapter 4). As we will see, such fuzzy, imprecise answers are not worse at all than answers to conventional queries. On the contrary, they lead to better choices and better decisions. A general procedure of constructing the answer to a flexible query looks as follows:

- for each object x in the database, the database system computes the degree of fulfilment of each elementary condition in the query;
- these partial evaluations are then aggregated into one final evaluation of x which forms a total membership degree $Ans(x)$ of x in the answer;
- for practical reasons, the answer may be restricted to those pairs $(x, Ans(x))$ in which $Ans(x)$ is sufficiently "large".

Let us come back to the case of (FQ1). For each hotel in the database, we assume that exact data about room price and distance from the center of Paris (time of journey by metro) are given. Moreover, for simplicity, we accept that all rooms in a hotel cost the same. Consider the following short list of hotels stored in the database.

Table 3.1 Example of data records about hotels in Paris

Id	Hotel name	Room price/night (€)	Distance from the city center (min)
1	*Savoy*	115	5
2	*Napoleon*	95	25
3	*Mont Blanc*	63	18
4	*Rose*	71	8
5	*Valois*	66	15
6	*Simenon*	85	20
7	*Orlean*	45	35

Notice that hotels *Mont Blanc* and *Rose*, respectively, show only slight violations of the distance condition "$\leq$ 15 min" and price condition "$\leq$ 70 €" in (SQ1).

The above data are sufficient to answer the conventional query (SQ1). Trying to answer (FQ1), the database has to be equipped with two linguistic variables:

$$(room\_price,\ \boldsymbol{U}_1,\ \{\ \ldots,\ affordable,\ \ldots\},\ \boldsymbol{I}_1\},$$
$$(distance\_from\_the\_center,\ \boldsymbol{U}_2,\ \{\ \ldots,\ in\_the\_vicinity\_of\_the\_center,\ \ldots\},\ \boldsymbol{I}_2\}.$$

Assume the linguistic values *affordable* and *in_the_vicinity_of_the_center* are interpreted by means of z-shaped fuzzy numbers $(60, 30)_z$ and $(10, 20)_z$, respectively (see below).

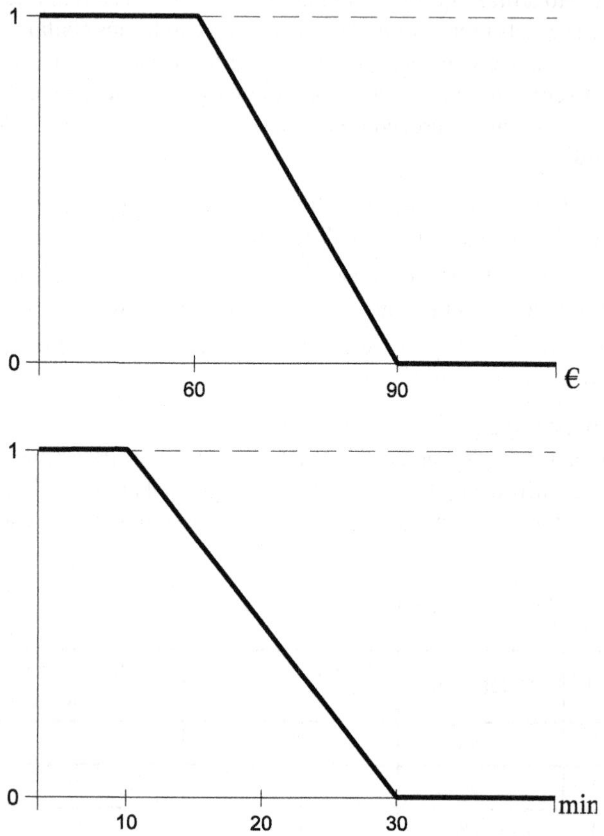

Fig. 3.1 Interpretation (top) of linguistic value *affordable* and (bottom) of *in_the_vicinity_of_the_center*

According to the general procedure, the answer to (FQ1) will be constructed in the following three steps.

Step 1. On the basis of the above interpretations, two partial evaluations will be assigned to each hotel (see Table 3.2):

- $A(i)$, a degree of membership of the ith hotel in fuzzy set A, i.e. an evaluation of the ith hotel with respect to condition *affordable*;
- $B(i)$, a membership degree of the ith hotel in fuzzy set B, i.e. an evaluation with respect to condition *in_the_vicinity_of_the_center*.

Table 3.2 List of partial evaluations of the hotels

i	$A(i)$	$B(i)$
1	0	1
2	0	0.25
3	0.90	0.60
4	0.63	1
5	0.80	0.75
6	0.17	0.50
7	1	0

These evaluations can be expressed in the language of truth degrees as

$A(i) = [\![\text{the } i\text{th hotel is affordable}]\!]$,
$B(i) = [\![\text{the } i\text{th hotel lies in the vicinity of the center of Paris}]\!]$.

Step 2. Each pair $(A(i), B(i))$ of partial evaluations (each n-tuple, in general) is then aggregated into a final evaluation

$$A(i)\, \boldsymbol{t}\, B(i) = (A \cap_t B)(i).$$

In other words, for each hotel, the database system generates a degree of simultaneous fulfilment of both the conditions. Table 3.3 presents the resulting final evaluations involving the three basic t-norms.

Table 3.3 Final evaluations

i	$(A \cap_t B)(i)$		
	$t = \wedge$	$t = t_a$	$t = t_Ł$
1	0	0	0
2	0	0	0
3	0.60	0.54	0.50
4	0.63	**0.63**	**0.63**
5	**0.75**	0.60	0.55
6	0.17	0.09	0
7	0	0	0

We see that the choice of t generally influences the ranking of hotels. For $t = \wedge$, hotel no. 5 gets the best final evaluation. If t_a or $t_Ł$ is used, it is hotel no. 4. Since atb is always less than or equal to $a \wedge b$, $\wedge$ produces more optimistic final evaluations than other t-norms.

Step 3. The database is now ready to present the answer, $A \cap_t B$. Speaking practically, that answer is a list of hotels together with and sorted according to their final evaluations. For instance, using $t = \wedge$, one obtains:

Valois	66€	15 min	0.75
Rose	71€	8 min	0.63
Mont Blanc	63€	18 min	0.60
Simenon	85€	20 min	0.17

This list does not contain those hotels whose final evaluation is 0 as the user's interest is evidently restricted to supp($A \cap_t B$). Moreover, one can establish a quality threshold $t \in (0, 1]$ and, then, the answer would be limited to the t-cut $(A \cap_t B)_t$. Thus, say, $t = 0.5$ eliminates Hotel Simenon from the list.

Let us emphasize that the answer to the conventional query (SQ1) is 1-element, namely

Valois	66€	15 min	0.75

If this hotel were situated a bit farther from the center of Paris, the answer would be empty at all.

Concluding, the fuzzy answer to (FQ1) seems to be more complete and more useful for the user (e.g. a tourist) than the 1-element answer to (SQ1). It allows to make better decisions. In the answer to (SQ1) one loses two hotels, *Rose* and *Mont Blanc*, which only slightly violate the constraints on price and distance and, therefore, are also of interest to the user.

One should also notice that *affordable* and *in_the_vicinity_of_the_center* are competitive conditions in principle. Usually, the shorter the distance from the city center, the higher the room price. The user is aware of this relationship and, thus, his/her true intention is in essence to find a satisfactory compromise between price and distance, more generally: between competitive conditions like price and quality, price and size, etc. In contrast to conventional querying, flexible querying and the methodology of fuzzy sets make it possible to find such compromises.

3.1.2 Unequally Important Elementary Conditions

The discussion in the previous subsection was based on an implicit assumption that all elementary conditions in the query are equally important. It happens, however, that some conditions are in reality more important for the user than the other ones. For instance, in (FQ1), affordable room price may be more essential than distance from the center of Paris. This is why we like to consider the following more general case:

- the query contains a conjunction of $n \geq 1$ conditions modeled by means of fuzzy sets $A_1, A_2, \ldots, A_n$,
- to each condition A_i a weight $w_i > 0$ has been assigned by the user, $i = 1, 2, \ldots, n$.

The answer is then

$$A_1^{w_1} \cap_t A_2^{w_2} \cap_t \ldots \cap_t A_n^{w_n}. \tag{3.2}$$

So, $w_1 = \ldots = w_n = 1$ gives (3.1). As to upper bounds and interpretations of the weights, at least two approaches are possible.

(a) $w_i \in (0, 1]$. $w_i = 1$ then signifies that the *i*th condition is fully important. This importance is only partial whenever $w_i < 1$. Returning to the fuzzy query (FQ1) from Subsection 3.1.1, let us assume that the price condition is fully important, whereas the importance of the distance condition is average, say, equal to 0.5. The answer is now $A \cap_t B^{0.5}$ with a t-norm $\boldsymbol{t}$. The final evaluations for the three basic t-norms are collected in the following.

Table 3.4 Final evaluations for $w_1 = 1$ and $w_2 = 0.5$

i	$(A \cap_t B^{0.5})(i) = A(i)\, \boldsymbol{t}\, (B(i))^{0.5}$		
	$\boldsymbol{t} = \wedge$	$\boldsymbol{t} = \boldsymbol{t}_a$	$\boldsymbol{t} = \boldsymbol{t}_Ł$
1	0	0	0
2	0	0	0
3	0.77	**0.70**	**0.675**
4	0.63	0.63	0.63
5	**0.80**	0.69	0.666
6	0.17	0.12	0
7	0	0	0

The evaluations are now generally higher than in Table 3.3. For $\boldsymbol{t} = \wedge$, *Valois* is still the most suitable hotel, whereas *Mont Blanc* moves to the second place. If $\boldsymbol{t}_a$ or $\boldsymbol{t}_Ł$ is used, *Mont Blanc* becomes the best choice (lower price decides, longer distance is less important).

(b) $w_i \in (0, \infty)$. This time $w_i = 1$ means that the importance of the *i*th condition is "typical", "standard". If $w_i < 1$ ($w_i > 1$, respectively), that importance is less (greater, respectively) than a typical one. In practice, it usually suffices to use weights from (0, 10] and, then, $w_i = 10$ means that the *i*th condition is of "supreme", "absolute" importance. The reason for that restriction is that, say, $0.9^{10} \approx 0.35$ and $0.8^{10} \approx 0.11$. Even a small deviation from 1 of the fulfilment degree of an absolutely important elementary condition thus leads to a low final evaluation. For instance,

$$\left(A_1^{w_1} \cap_t A_2^{w_2} \cap_t \dots \cap_t A_n^{w_n}\right)(x) = (A_1(x))^{w_1}\, t\, (A_2(x))^{w_2}\, t \dots t\, (A_n(x))^{w_n} \leq 0.11$$

whenever $w_i = 10$ and $A_i(x) = 0.8$ for some $i \leq n$ (see (2.70)).

As to (FQ1), suppose that the importance of the price condition is equal to 2 ("very important", see (2.124)), whereas the weight assigned to the distance condition is 0.5. So, the answer collapses to $A^2 \cap_t B^{0.5}$.

Table 3.5 Final evaluations for $w_1 = 2$ and $w_2 = 0.5$

i	$(A^2 \cap_t B^{0.5})(i) = (A(i))^2\, t\, (B(i))^{0.5}$		
	$t = \wedge$	$t = t_a$	$t = t_L$
1	0	0	0
2	0	0	0
3	**0.77**	**0.63**	**0.58**
4	0.40	0.40	0.40
5	0.64	0.55	0.51
6	0.03	0.02	0
7	0	0	0

The best option is now definitely Hotel Mont Blanc as the price criterion dominates.

3.2 The Case of Hotel Simenon – More Advanced Aspects of Triangular Norms

Looking at Tables 3.3-3.5 in the previous section, one sees that the final evaluation of Hotel Simenon involving the Łukasiewicz t-norm is always equal to zero although the corresponding partial evaluations as well as the final evaluations with t-norms $\wedge$ and t_a are positive. We like to explain a theoretical background of that phenomenon. This will be a good pretext for presenting some more advanced aspects of t-norms. For compactness, we will omit most proofs, especially those longer ones, and purely technical details. The interested reader can find them in GOTTWALD (2001) and KLEMENT *et al.* (2000); see also LOWEN (1996) and WEBER (1983).

3.2.1 Classes of Triangular Norms

The families of all t-norms and t-conorms are large. Our discussion will be focused on continuous t-operations (see remarks in Subsection 2.4.3). We will understand this continuity in a simple way as continuity in each variable. Similarly, that a t-operation $\boldsymbol{u}$ is strictly increasing will be understood as its strict increasingness with respect to each variable, i.e. $a\,\boldsymbol{u}\,b < a\,\boldsymbol{u}\,c$ whenever $a \in (0, 1)$ and $b < c$.

For a t-norm $\boldsymbol{t}$, t-conorm $\boldsymbol{s}$, and $k \in \mathbb{N}$, let us define

$$a_t^{(k)} = a\,\boldsymbol{t}\,a\,\boldsymbol{t} \ldots \boldsymbol{t}\,a \quad \text{and} \quad a_s^{(k)} = a\,\boldsymbol{s}\,a\,\boldsymbol{s} \ldots \boldsymbol{s}\,a$$

with k arguments a on the right hand sides, and with the understanding that

$$a_t^{(0)} = 1, \quad a_t^{(1)} = a, \quad a_s^{(0)} = 0, \quad a_s^{(1)} = a.$$

Definition 3.1. Let $\boldsymbol{t}$ and $\boldsymbol{s}$ be continuous.

(a) One says that $\boldsymbol{t}$ is an *Archimedean t-norm* if

$$a_t^{(2)} = a\,\boldsymbol{t}\,a < a \quad \text{for each } a \in (0, 1).$$

$\boldsymbol{s}$ is called an *Archimedean t-conorm* whenever

$$a_s^{(2)} = a\,\boldsymbol{s}\,a > a \quad \text{for each } a \in (0, 1).$$

(b) $\boldsymbol{t}$ and $\boldsymbol{s}$ are said to be *strict* if they are a strictly increasing.

(c) We will say that $\boldsymbol{t}$ is a *nilpotent t-norm* if, for each $a \in (0, 1)$, there exists $k \geq 2$ such that

$$a_t^{(k)} = 0.$$

Finally, $\boldsymbol{s}$ is called a *nilpotent t-conorm* whenever, for each $a \in (0, 1)$, there exists $k \geq 2$ giving

$$a_s^{(k)} = 1.$$

Being a strict t-operation thus means to be continuous and strictly increasing. There are simple relationships between the subclasses introduced in Definition 3.1.

- Each strict t-operation is Archimedean since $a\,\boldsymbol{t}\,a < a\,\boldsymbol{t}\,1 = a$ and $a\,\boldsymbol{s}\,a > a\,\boldsymbol{s}\,0 = a$ whenever $\boldsymbol{t}$ and $\boldsymbol{s}$ are strict, and $a \in (0, 1)$.
- Nilpotent t-operation are Archimedean, too. For instance, suppose a t-norm $\boldsymbol{t}$ is continuous and not Archimedean. Then $a\,\boldsymbol{t}\,a = a$ for some $a \in (0, 1)$ (see (2.69)). This implies $a_t^{(3)} = a_t^{(2)}\,\boldsymbol{t}\,a = a\,\boldsymbol{t}\,a = a$ and, hence, $a_t^{(k)} = a > 0$ for each $k \geq 2$, i.e. $\boldsymbol{t}$ is not nilpotent.
- A t-operation is both Archimedean and nonstrict iff it is nilpotent.

Archimedean t-norms thus split into two disjoint subclasses: each Archimedean t-norm is either strict or nilpotent. The same holds true for Archimedean t-conorms.

Example 3.2. (a) Clearly, the drastic t-operations are discontinuous and, hence, not Archimedean. $\wedge$ and $\vee$ exemplify continuous and non-Archimedean t-operations as $a \wedge a = a \vee a = a$.

(b) The algebraic and Łukasiewicz t-norms and t-conorms as well as those listed in Example 2.6 are all Archimedean. In particular, $\boldsymbol{t}_{\mathrm{a}}$, $\boldsymbol{s}_{\mathrm{a}}$, $\boldsymbol{t}_{\mathrm{H},\gamma}$, $\boldsymbol{s}_{\mathrm{H},\gamma}$, $\boldsymbol{t}_{\mathrm{F},\lambda}$ and $\boldsymbol{s}_{\mathrm{F},\lambda}$ are strict, whereas $\boldsymbol{t}_{Ł}$, $\boldsymbol{s}_{Ł}$, $\boldsymbol{t}_{\mathrm{S},p}$, $\boldsymbol{s}_{\mathrm{S},p}$, $\boldsymbol{t}_{\mathrm{Y},p}$, $\boldsymbol{s}_{\mathrm{Y},p}$, $\boldsymbol{t}_{\mathrm{W},\lambda}$ and $\boldsymbol{s}_{\mathrm{W},\lambda}$ are nonstrict, i.e. nilpotent. □

In connection with the above example, let us point out that if $\boldsymbol{t}$ and $\boldsymbol{s}$ are ν-dual with a strong negation ν, $\boldsymbol{s} = \boldsymbol{t}^{\nu}$, then the following equivalences hold:

$$\boldsymbol{t} \text{ continuous} \Leftrightarrow \boldsymbol{s} \text{ continuous},$$

$$\boldsymbol{t} \text{ Archimedean} \Leftrightarrow \boldsymbol{s} \text{ Archimedean},$$

$$\boldsymbol{t} \text{ strict} \Leftrightarrow \boldsymbol{s} \text{ strict}$$

and, consequently,

$$\boldsymbol{t} \text{ nilpotent} \Leftrightarrow \boldsymbol{s} \text{ nilpotent}.$$

Worth noticing is also that each strictly increasing and, thus, each strict t-norm

- satisfies the *cancellation law*, i.e.

$$(a\boldsymbol{t}b = a\boldsymbol{t}c \;\&\; a>0) \Rightarrow b = c, \tag{3.3}$$

- *does not have zero divisors*, i.e.

$$a, b>0 \Rightarrow a\boldsymbol{t}b>0. \tag{3.4}$$

Indeed, $a\boldsymbol{t}b = a\boldsymbol{t}c$ with $a > 0$ and $b \neq c$ as well as $a\boldsymbol{t}b = 0$ with $a, b > 0$ would contradict the strict monotonicity of $\boldsymbol{t}$. It is trivial that $\wedge$ does not have zero divisors, too. On the other hand, nilpotent t-norms *have zero divisors*, namely

$$a\boldsymbol{t}b = 0 \quad \text{for some } a, b > 0,$$

e.g. $0.3\, \boldsymbol{t}_{Ł}\, 0.4 = 0$.

Thus, each Archimedean t-norm is either strict (and, then, does not have zero divisors) or nilpotent (i.e. has zero divisors). That nilpotent t-norms possess zero divisors has a nice counterpart in the world of nilpotent t-conorms. If $\boldsymbol{s}$ is nilpotent, then

$$a\boldsymbol{s}b = 1 \quad \text{for some } a, b < 1.$$

An instance is $0.7\, \boldsymbol{s}_{Ł}\, 0.6 = 1$.

Let us come back to the evaluation process in Subsections 3.1.1-3.1.2. The final evaluation of Hotel Simenon involving the Łukasiewicz t-norm was always equal to zero despite positive partial evaluations with respect to price and distance. As we now see, this is a consequence of the fact that $\boldsymbol{t}_{Ł}$ is nilpotent and, thus, has zero divisors. One can say that the t-norms having zero divisors show some "inertia" in attaining positive values: a positive argument from (0, 1) is then still treated as zero whenever the other one is not sufficiently "large". For instance, $0.3\, \boldsymbol{t}_{Ł}\, 0.5 = 0$ and $0.3\, \boldsymbol{t}_{Ł}\, 0.8 > 0$; see also (3.23) and related comments. This property is very useful in applications. It makes it possible to ignore objects which, admittedly, fulfil given elementary conditions to positive degrees, but those degrees (treated *en bloc*) are too "small" to take them into consideration. And just this is the case of Hotel Simenon.

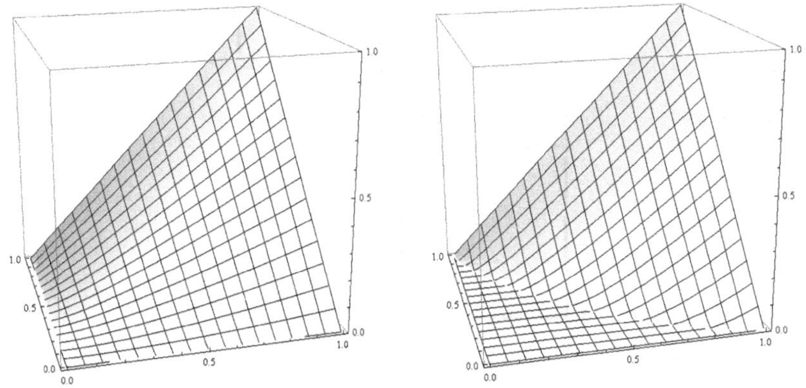

Fig. 3.2 3D plots of $t_{S,0.2}$ (left) and $t_{S,0.7}$ (right)

Using a parameterized family of nilpotent t-norms, say, the family of Schweizer t-norms $t_{S,p}$, one can establish a desired level of inertia by choosing a suitable value of the parameter. Figure 3.2 presents an example. $p = 0.2$ gives a small-inertia t-norm, whereas $p = 0.7$ generates a t-norm showing a much higher inertia level.

3.2.2 Continuous and Archimedean Triangular Norms

The class of continuous t-norms has an interesting structure. It is worth presenting and suggests that Archimedean t-norms play a key role in that class. Let us begin with defining a notion which is necessary for our further discussion.

Definition 3.3. Let J denote a nonempty and at most countable set of indices. Assume that $(t_i)_{i\in J}$ is a family of t-norms, and $((a_i, b_i))_{i\in J}$ forms a family of nonempty, pairwise disjoint open subintervals of [0, 1]. The *ordinal sum* of the family $((t_i, [a_i, b_i]))_{i\in J}$ of the t-norms t_i assigned to the intervals $[a_i, b_i]$ is a binary operation t in [0, 1] with

$$a\,t\,b = \begin{cases} a_i + (b_i - a_i)\left(\dfrac{a - a_i}{b_i - a_i}\; t_i\; \dfrac{b - a_i}{b_i - a_i}\right), & \text{if } a, b \in [a_i, b_i], \\ a \wedge b, & \text{otherwise.} \end{cases}$$

One proves that ordinal sums are always t-norms. The concept of an ordinal sum is thus a convenient tool for constructing new, heterogeneously specified t-norms from families of given t-norms.

Example 3.4. Take $J = \{1.2\}$. Construct the ordinal sum of the family $((t_a, [0, 0.5]), (t_Ł, [0.75, 1]))$ illustrated below.

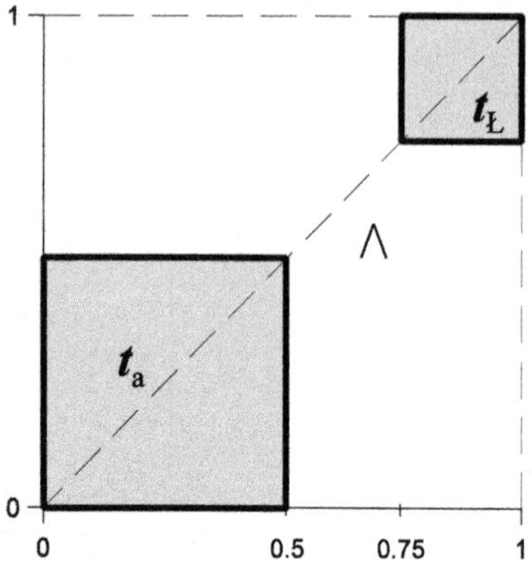

Fig. 3.3 Graphical representation of $((t_a, [0, 0.5]), (t_Ł, [0.75, 1]))$

By Definition 3.3, for $a, b \in [a_1, b_1] = [0, 0.5]$, we get

$$a\,t\,b = 0.5\left(\frac{a}{0.5}\,t_a\,\frac{b}{0.5}\right) = 2ab.$$

If $a, b \in [a_2, b_2] = [0.75, 1]$, then

$$a\,t\,b = 0.75 + 0.25\left(\frac{a-0.75}{0.25}\,t_Ł\,\frac{b-0.75}{0.25}\right)$$
$$= 0.75 + 0 \vee (a + b - 1.75).$$

The ordinal sum is thus equal to

$$a\,t\,b = \begin{cases} 2ab, & \text{if } a, b \in [0, 0.5], \\ 0.75 + 0 \vee (a + b - 1.75), & \text{if } a, b \in [0.75, 1], \\ a \wedge b, & \text{otherwise.} \end{cases}$$

□

The following theorem shows the structure of the family of continuous t-norms and, simultaneously, underlines the importance of Archimedean t-norms and ordinal sums. Its proof can be found in GOTTWALD (2001).

Theorem 3.5. *Each continuous t-norm either is equal to* $\wedge$ *or is Archimedean, or is the ordinal sum of a family of Archimedean t-norms.*

We like to present one more property involving ordinal sums and coming from FRANK (1979). It forms another argument for exceptional features of the Frank families of t-norms and t-conorms from Subsection 2.4.3.

Theorem 3.6 (*Frank theorem*). *A continuous t-norm **t** and a continuous t-conorm **s** satisfy the functional equation*

$$a\,\boldsymbol{t}\,b + a\,\boldsymbol{s}\,b = a + b \quad \text{for each } a, b \in [0,1]$$

iff $\boldsymbol{t} = \boldsymbol{t}_{\mathrm{F},\lambda}$ *and* $\boldsymbol{s} = \boldsymbol{s}_{\mathrm{F},\lambda}$ *for some* $\lambda \in [0, \infty]$ *or **t** is the ordinal sum of a family* $((\boldsymbol{t}_{\mathrm{F},\lambda}, [a_\lambda, b_\lambda]))_{\lambda \in J \subset (0,\infty]}$, *whereas **s** is determined via the above equation.*

Let us recollect that an *automorphism of the unit interval* [0, 1] is understood as a continuous and strictly increasing mapping of [0, 1] onto itself.

Theorem 3.7. *Let **t** denote an Archimedean t-norm.*

(a) ***t** is strict iff there exists an automorphism* φ *of* $[0,1]$ *such that*

$$a\,\boldsymbol{t}\,b = \varphi^{-1}(\varphi(a)\cdot\varphi(b)) = \varphi^{-1}(\varphi(a)\,\boldsymbol{t}_{\mathrm{a}}\,\varphi(b)) \quad \text{for each } a, b \in [0,1].$$

(b) ***t** is nilpotent iff there exists an automorphism* φ *of* $[0,1]$ *with*

$$a\,\boldsymbol{t}\,b = \varphi^{-1}(0 \vee (\varphi(a) + \varphi(b) - 1)) = \varphi^{-1}(\varphi(a)\,\boldsymbol{t}_{\text{Ł}}\,\varphi(b)) \quad \text{for each } a, b \in [0,1].$$

Although the family of Archimedean t-norms is infinite and numerically varied, this variety is thus only seeming. The algebraic t-norm and the Łukasiewicz one, respectively, are prototypes, models of all strict t-norms and all nilpotent ones, respectively. $\varphi = id$ gives $\boldsymbol{t}_{\mathrm{a}}$ and $\boldsymbol{t}_{\text{Ł}}$. In other words, all strict t-norms are isomorphic (similar) to $\boldsymbol{t}_{\mathrm{a}}$, whereas all nilpotent ones are isomorphic to $\boldsymbol{t}_{\text{Ł}}$. Theorems 3.5 and 3.7 can be viewed as a mathematical justification for calling $\wedge$, $\boldsymbol{t}_{\mathrm{a}}$ and $\boldsymbol{t}_{\text{Ł}}$ the three basic t-norms, which was initiated in Subsection 2.4.3.

Finally, let us mention that there exists a counterpart of ordinal sums for t-conorms as well as an analogue of Theorem 3.5 for continuous t-conorms. Clearly, $\wedge$ is then replaced with $\vee$. A counterpart of Theorem 3.7 for Archimedean t-conorms is also true. It says that $\boldsymbol{s}_{\mathrm{a}}$ and $\boldsymbol{s}_{\text{Ł}}$, respectively, are prototypes of strict t-conorms and nilpotent ones, respectively.

3.2.3 Generators

What Theorem 3.7 and its analogue for t-conorms offer us can also be viewed as characterizations of Archimedean t-operations. We like to recollect another, but closely related characterization proposed in LING (1965). It is a powerful tool for constructing arbitrary Archimedean t-norms and t-conorms.

Theorem 3.8 (*Ling characterization theorem*). *Let* $\mathbf{t}, \mathbf{s}$: $[0, 1] \times [0, 1] \to [0, 1]$.

(a) $\mathbf{t}$ *is an Archimedean t-norm iff there exists its generator, i.e. a strictly decreasing and continuous function* g: $[0, 1] \to [0, \infty]$ *such that* $g(1) = 0$ *and*

$$a\mathbf{t}b = g^{-1}(g(0) \wedge (g(a) + g(b))) \quad \textit{for each } a, b \in [0, 1].$$

Moreover, $\mathbf{t}$ *is strict iff* $g(0) = \infty$.

(b) $\mathbf{s}$ *is an Archimedean t-conorm iff there exists its generator, i.e. a strictly increasing and continuous function h*: $[0, 1] \to [0, \infty]$ *with* $h(0) = 0$ *and such that*

$$a\mathbf{s}b = h^{-1}(h(1) \wedge (h(a) + h(b))) \quad \textit{for each } a, b \in [0, 1].$$

Finally, $\mathbf{s}$ *is strict iff* $h(1) = \infty$.

Generators of Archimedean t-norms and t-conorms are uniquely determined up to a positive constant factor. In the subject literature, they are also called *additive generators*.

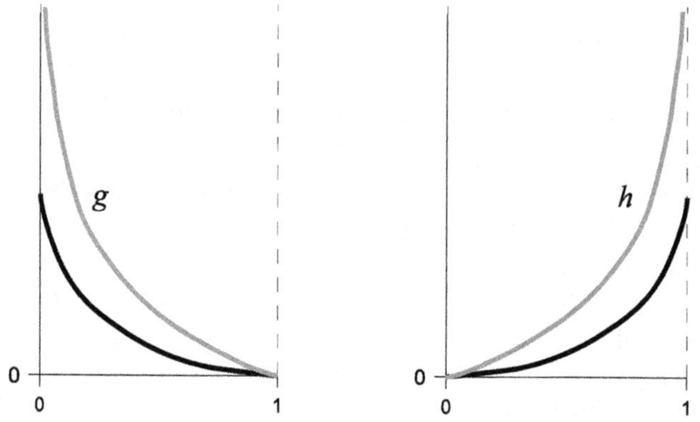

Fig. 3.4 Generators of Archimedean t-norms (left) and t-conorms (right)

By Theorem 3.8, the upper functions in Figure 3.4 are generators of a strict t-norm (left) and strict t-conorm (right). The lower ones generate nilpotent t-operations.

Generator g of a nilpotent t-norm is called *normed* if $g(0) = 1$. Similarly, generator h of a nilpotent t-conorm is said to be *normed* whenever $h(1) = 1$. It is clear that each generator of a nilpotent t-norm or t-conorm can always be normed by multiplying it by $1/g(0)$ or $1/h(1)$, respectively. Without a loss of generality, any discussion involving generators of nilpotent t-operations can thus be restricted to normed generators.

It is worth noticing that there exists a simple relationship between generators of ν-dual Archimedean t-norms and t-conorms.

Theorem 3.9. *Let* ν *denote a strong negation. If* ***t*** *is an Archimedean t-norm with generator g and* ***s*** *is the* ν*-dual t-conorm with generator h, then*

$$h(a) = g(\nu(a)) \quad \textit{for each } a \in [0,1].$$

Indeed, by (2.74) and Theorem 3.8(a),

$$g(\nu(a\,\mathbf{s}\,b)) = g(0) \wedge (g(\nu(a)) + g(\nu(b))).$$

We see that h with $h(x) = g(\nu(x))$ generates a t-conorm and, by Theorem 3.8(b), that t-conorm is just $\mathbf{t}^{\nu}$ as one has $h(a\,\mathbf{s}\,b) = h(1) \wedge (h(a) + h(b))$, which completes the proof.

If ***t*** and ***s*** are associated, $\mathbf{s} = \mathbf{t}^*$, the above theorem leads to the following relationship between their generators:

$$h(a) = g(1 - a) \quad \text{for each } a \in [0,1]. \tag{3.5}$$

We are now ready to give concrete examples of generators of Archimedean t-norms and t-conorms.

Example 3.10. Let us present a list of generators of t-norms from Example 2.6. Those of nilpotent t-norms are given in the normed form. The generators of the associated t-conorms are created by means of (3.5).

- $\mathbf{t}_{\mathrm{a}}$: $g(a) = -\ln a$, $\mathbf{s}_{\mathrm{a}}$: $h(a) = -\ln(1-a)$,
- $\mathbf{t}_{\text{Ł}}$: $g(a) = 1 - a$, $\mathbf{s}_{\text{Ł}}$: $h(a) = a$,
- $\mathbf{t}_{\mathrm{S},p}$: $g(a) = 1 - a^p$, $\mathbf{s}_{\mathrm{S},p}$: $h(a) = 1 - (1-a)^p$,
- $\mathbf{t}_{\mathrm{Y},p}$: $g(a) = (1-a)^p$, $\mathbf{s}_{\mathrm{Y},p}$: $h(a) = a^p$,
- $\mathbf{t}_{\mathrm{H},\gamma}$: $g(a) = \begin{cases} \ln\frac{\gamma + (1-\gamma)a}{a}, & \text{if } \gamma > 0, \\ \frac{1-a}{a}, & \text{if } \gamma = 0, \end{cases}$

  $\mathbf{s}_{\mathrm{H},\gamma}$: $h(a) = \begin{cases} \ln\frac{\gamma + (1-\gamma)(1-a)}{1-a}, & \text{if } \gamma > 0, \\ \frac{a}{1-a}, & \text{if } \gamma = 0, \end{cases}$

- $t_{F,\lambda}$: $g(a) = \log_\lambda \frac{\lambda - 1}{\lambda^a - 1}$, $s_{F,\lambda}$: $h(a) = \log_\lambda \frac{\lambda - 1}{\lambda^{1-a} - 1}$,

- $t_{W,\lambda}$: $g(a) = \begin{cases} 1 - \frac{\ln(1+\lambda a)}{\ln(1+\lambda)}, & \text{if } \lambda \neq 0, \\ 1 - a, & \text{if } \lambda = 0, \end{cases}$

$s_{W,\lambda}$: $h(a) = \begin{cases} 1 - \frac{\ln(1+\lambda(1-a))}{\ln(1+\lambda)}, & \text{if } \lambda \neq 0, \\ a, & \text{if } \lambda = 0. \end{cases}$ □

The Ling characterization theorem makes it possible to construct in a convenient way n-ary extensions of t-operations (see also (3.21)). By mathematical induction, we get

$$a_1 \, t \, a_2 \, t \ldots t \, a_n = g^{-1}\left(g(0) \wedge \sum_{i=1}^{n} g(a_i) \right) \quad \text{for } n \geq 2, \tag{3.6}$$

where t is an Archimedean t-norm with generator g, and $a_1, a_2, \ldots, a_n \in [0, 1]$. An analogous formula can be easily created for Archimedean t-conorms. By (3.6), one obtains

$$a_1 \, t \, a_2 \, t \ldots t \, a_n > 0 \Leftrightarrow \sum_{i=1}^{n} g(a_i) < g(0) \tag{3.7}$$

and, moreover,

$$a_1 \, t \, a_2 \, t \ldots t \, a_n \geq t \Leftrightarrow \sum_{i=1}^{n} g(a_i) \leq g(t) \tag{3.8}$$

for each $t \in (0, 1]$. Thus, $a_1 \, t \, a_2 \, t \ldots t \, a_n > 0 \Leftrightarrow a_1, a_2, \ldots, a_n > 0$ whenever t is strict. Clearly, this cannot be extended to nilpotent t-norms. For instance, $t_Ł$ is generated by $g(a) = 1 - a$ and, hence,

$$a_1 \, t_Ł \, a_2 \, t_Ł \ldots t_Ł \, a_n > 0 \Leftrightarrow \sum_{i=1}^{n} a_i > n - 1 \tag{3.9}$$

and

$$a_1 \, t_Ł \, a_2 \, t_Ł \ldots t_Ł \, a_n \geq t \Leftrightarrow \sum_{i=1}^{n} a_i \geq n - (1 - t). \tag{3.10}$$

Notice that any strong negation can play the role of the normed generator of a nilpotent t-norm. In particular, that negation may be a broken negation $\nu_{=c}$ with $c \in (0, 1)$ (see (2.63)). Examples of the resulting t-norms are presented below.

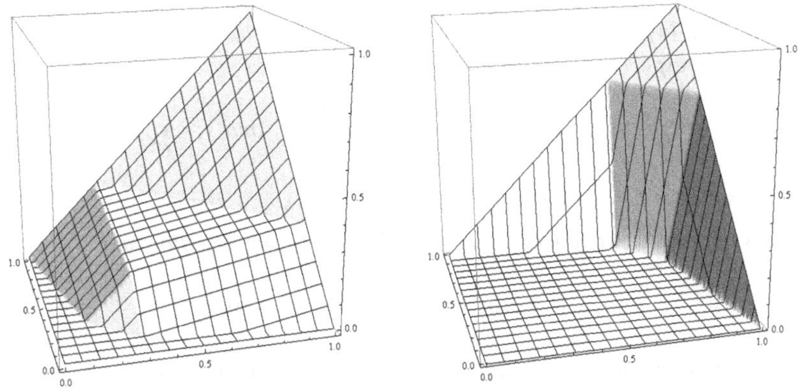

Fig. 3.5 3D plots of $a\,t\,b = \nu(1 \wedge (\nu(a) + \nu(b)))$ for $\nu = \nu_{=c}$ with $c = 0.3$ (left) and $c = 0.7$ (right)

Negations in combination with t-norms are generally wort discussing a bit deeper. And this issue will be the subject of the next subsection.

3.2.4 Induced Negations and Complementarity

First of all, as pointed out in FODOR (1993) and TRILLAS (1979), there exists a prototypical form of all strong and even strict negations in exactly the same way as there are prototypes of Archimedean t-norms and t-conorms. And, again, that prototype is the simplest strong negation: the standard negation $\nu_Ł$.

Theorem 3.11. *A function* $\nu: [0,1] \to [0,1]$ *is*
(a) *a strong negation iff there exists an automorphism* φ *of* $[0,1]$ *with*

$$\nu(a) = \varphi^{-1}(1 - \varphi(a)) \quad \textit{for each } a \in [0,1];$$

(b) *a strict negation iff there exist automorphisms* φ *and* ψ *of* $[0,1]$ *such that*

$$\nu(a) = \psi(1 - \varphi(a)) \quad \textit{for each } a \in [0,1].$$

Similarly to generators of t-operations, the above automorphisms are not uniquely determined. It is obvious that $\varphi = \psi = id$ gives $\nu_Ł$.

Worth recollecting and developing is the concept of negations induced by t-norms, defined in (2.89). For a t-norm $\boldsymbol{t}$ and a t-conorm $\boldsymbol{s}$, let us introduce the following unary operations $\nu_t, \nu_s: [0,1] \to [0,1]$:

$$\nu_t(a) = \bigvee\{c \in [0,1]: a\,t\,c = 0\}, \quad \nu_s(a) = \bigwedge\{c \in [0,1]: a\,s\,c = 1\}. \tag{3.11}$$

Theorem 3.12. (a) *For each* $\boldsymbol{t}$ *and* $\boldsymbol{s}$, ν_t *and* ν_s *are negations. These negations are antitonic with respect to* $\boldsymbol{t}$ *and* $\boldsymbol{s}$, *i.e.*

$$\nu_{t_2} \le \nu_{t_1} \ \textit{and}\ \nu_{s_2} \le \nu_{s_1} \ \textit{whenever}\ t_1 \le t_2 \ \textit{and}\ s_1 \le s_2.$$

(b) *If* $\boldsymbol{t}$ *is strict or* $\boldsymbol{t} = \wedge$, *then* $\nu_t = \nu_*$. *If* $\boldsymbol{s}$ *is strict or* $\boldsymbol{s} = \vee$, *then* $\nu_s = \nu^*$.

(c) *If* $\boldsymbol{t}$ *is nilpotent and generated by* g, *then* ν_t *is a strong negation and*

$$\nu_t(a) = g^{-1}(g(0) - g(a)) \ \textit{for each}\ a \in [0, 1].$$

If $\boldsymbol{s}$ *is nilpotent and* h *is its generator, then* ν_s *is strong, too, and*

$$\nu_s(a) = h^{-1}(h(1) - h(a)) \ \textit{for each}\ a \in [0, 1].$$

It is a routine matter to show (a) and (b). (c) follows from Theorem 3.8 and its consequences. For instance, the formula in the first part of (c) results from (3.7):

$$a\boldsymbol{t}c = 0 \Leftrightarrow g(0) \le g(a) + g(c) \Leftrightarrow c \le g^{-1}(g(0) - g(a)).$$

Similarly to ν_t, ν_s will be called the *negation induced by* $\boldsymbol{s}$. A combination of Theorems 3.11(a) and 3.12(c) suggests that each strong negation ν is nothing else than a negation ν_s induced by a nilpotent t-conorm $\boldsymbol{s}$. Just the normed generator of $\boldsymbol{s}$ can be used to create that strong negation.

Example 3.13. Thanks to Theorem 3.12, the instances of induced negations ν_t from Subsection 2.4.5 can be easily verified without referring to definition (2.89). Applying Example 3.10 and Theorem 3.12, let us present a few examples of negations induced by t-conorms.

- $\boldsymbol{s} = \vee$, $\boldsymbol{s} = \boldsymbol{s}_{\mathrm{a}}$: $\nu_s = \nu^*$,
- $\boldsymbol{s} = \boldsymbol{s}_{Ł}$: $\nu_s = \nu_{Ł}$,
- $\boldsymbol{s} = \boldsymbol{s}_{\mathrm{S},p}$: $\nu_s(a) = 1 - (1 - (1 - a)^p)^{1/p}$,
- $\boldsymbol{s} = \boldsymbol{s}_{\mathrm{Y},p}$: $\nu_s(a) = (1 - a^p)^{1/p}$ (see (2.61)),
- $\boldsymbol{s} = \boldsymbol{s}_{\mathrm{W},\lambda}$: $\nu_s(a) = 1 - \frac{a}{1 + \lambda(1 - a)}$. □

Let us look for a moment at strong negations in the context of their equilibrium points (see Subsection 2.4.1). By Theorem 3.12(c), if ν is strong, i.e. forms a negation ν_s induced by a nilpotent t-conorm $\boldsymbol{s}$ with normed generator h, then

$$e(\nu) = h^{-1}(0.5). \tag{3.12}$$

This generalizes the fact that the equilibrium point of the Łukasiewicz negation is 0.5.

In particular, we get (see Example 3.13))

$$e(\nu) = 1 - 0.5^{1/p} \quad \text{for } \nu = \nu_s \text{ with } \boldsymbol{s} = \boldsymbol{s}_{\mathrm{S},p}. \tag{3.13}$$

Returning to (2.62), it says that

$$e(\nu) = 0.5^{1/p} \quad \text{for } \nu = \nu_s \text{ with } \boldsymbol{s} = \boldsymbol{s}_{\mathrm{Y},p}. \tag{3.14}$$

By choosing a suitable value $p>0$ in (3.13) and (3.14), $e(\nu)$ can thus lie as near to 0 or 1 as one likes. Finally, let us mention an extreme property of equilibrium points. If $\boldsymbol{t}$ is a nilpotent t-norm and $\nu = \nu_t$ is used, then

$$a\,\boldsymbol{t}\,a = 0 \Leftrightarrow a \le e(\nu), \tag{3.15}$$

which follows from (3.11).

Negations induced by nilpotent t-operations are strong. As such, they can thus be applied to establish via (2.74) a correspondence between nilpotent t-norms and t-conorms. Let

$$a\,\boldsymbol{t}^\circ\, b = \nu_t(\nu_t(a)\,\boldsymbol{t}\,\nu_t(b)) \quad \text{and} \quad a\,\boldsymbol{s}^\circ\, b = \nu_s(\nu_s(a)\,\boldsymbol{s}\,\nu_s(b)) \tag{3.16}$$

for $a, b \in [0, 1]$, where $\boldsymbol{t}$ and $\boldsymbol{s}$ are a nilpotent t-norm and nilpotent t-conorm, respectively. The nilpotent t-conorm $\boldsymbol{t}^\circ$ is thus $\boldsymbol{t}^\nu$ with $\nu = \nu_t$, whereas the nilpotent t-norm $\boldsymbol{s}^\circ$ is nothing else than $\boldsymbol{s}^\nu$ with $\nu = \nu_s$. We will say that $\boldsymbol{t}$ and $\boldsymbol{s}$ are *complementary* if $\boldsymbol{s} = \boldsymbol{t}^\circ$ or, equivalently, $\boldsymbol{t} = \boldsymbol{s}^\circ$. For instance, for each $p > 0$, $(\boldsymbol{t}_{\mathrm{Y},p}, \boldsymbol{s}_{\mathrm{S},p})$ and $(\boldsymbol{t}_{\mathrm{S},p}, \boldsymbol{s}_{\mathrm{Y},p})$ are pairs of complementary t-operations.

Assume that $\boldsymbol{t}$ with normed generator g and $\boldsymbol{s}$ with normed generator h are complementary. A corollary from Theorems 3.9 and 3.12(c) is the relationship

$$h(a) = 1 - g(a) \quad \text{for each } a \in [0, 1]. \tag{3.17}$$

Indeed, $h(a) = g(\nu_t(a)) = g(0) - g(a) = 1 - g(a)$. Another easily verifiable consequence of Theorem 3.12(c) and (3.17) is that

$$\nu_t = \nu_{t^\circ} \quad \text{and} \quad \nu_s = \nu_{s^\circ} \tag{3.18}$$

for each nilpotent $\boldsymbol{t}$ and $\boldsymbol{s}$.

Complementarity thus offers a correspondence between t-norms and t-conorms that generally differs from being associated. $\boldsymbol{t}_{\mathrm{L}}$ and $\boldsymbol{s}_{\mathrm{L}}$, however, are both associated and complementary. As we see, complementarity forms a "dynamic" correspondence between nilpotent t-norms and t-conorms. One assigns a t-conorm to a t-norm $\boldsymbol{t}$ by means of a $\boldsymbol{t}$-dependent negation, ν_t. The relationship of being associated is always based on the same negation, ν_{L}.

It is worth noticing that complementarity leads to a bit surprising, boolean properties of operations on fuzzy sets. By routine transformations involving Theorems 3.8 and 3.12(c), we get

$$a\,t\,\nu_t(a) = 0 \quad \text{and} \quad a\,t^{\circ}\,\nu_t(a) = 1 \tag{3.19}$$

for each nilpotent t-norm t and each $a \in [0, 1]$. Hence (cf. (2.32))

$$A \cap_t A^{\nu_t} = 1_{\varnothing} \quad \text{and} \quad A \cup_{t^{\circ}} A^{\nu_t} = 1_U \tag{3.20}$$

for each fuzzy set A.

Finally, we like to presents a useful consequence of Theorem 3.8(a) and (3.17) (BELLUCE *et al.* (1991)). Its proof can be found also in WYGRALAK (2003a).

Theorem 3.14. *If t is a nilpotent t-norm and h denotes the normed generator of t°, then*

$$a\,t\,b = h^{-1}(0 \vee (h(a) + h(b) - 1)) \quad \textit{for each } a, b \in [0, 1].$$

Worth discussing are some corollaries from this theorem. It suggests that the normed generator of a nilpotent t-conorm can be used in Theorem 3.7(b) as an automorphism of [0, 1] giving the complementary t-norm. Let us point to more practical consequences. By mathematical induction, we get

$$a_1\,t\,a_2\,t \ldots t\,a_n = h^{-1}\left(0 \vee \left(\sum_{i=1}^{n} h(a_i) - (n-1)\right)\right) \tag{3.21}$$

for each $a_1, a_2, \ldots, a_n \in [0, 1]$, $n \geq 2$. First, this formula forms an alternative tool for creating n-ary extensions of nilpotent t-norms (cf. (3.6)). Second, taking a bit closer look at (3.21) with $a_1\,t\,a_2\,t \ldots t\,a_n \in (0, 1)$, we see that $h(a_1\,t\,a_2\,t \ldots t\,a_n)$ then forms the decimal part of the real number $\Sigma_{i=1,n}\, h(a_i)$, whereas the integer part is $n-1$ (see WYGRALAK (2009)). This can be informally written as

$$\sum_{i=1}^{n} h(a_i) = (n-1).h(a_1\,t\,a_2\,t \ldots t\,a_n). \tag{3.22}$$

In particular, if t is the Łukasiewicz t-norm, then $h = id$ and, thus, $a_1\,t_Ł\,a_2\,t_Ł \ldots t_Ł\,a_n$ is just the decimal part of $\Sigma_{i=1,n}\, a_i$.

Furthermore, (3.21) implies the following equivalences:

$$a_1\,t\,a_2\,t \ldots t\,a_n > 0 \Leftrightarrow \sum_{i=1}^{n} (1 - h(a_i)) < 1 \tag{3.23}$$

and

$$a_1\,t\,a_2\,t \ldots t\,a_n \geq t \Leftrightarrow \sum_{i=1}^{n} (1 - h(a_i)) \leq 1 - h(t) \tag{3.24}$$

for each $t \in (0, 1]$. By the way, (3.23)-(3.24) as well as (3.21) can also be derived from (3.6)-(3.8) by putting $g(a) = 1 - h(a)$ (see (3.17)). What (3.23) says is in essence that $a_1 \, t \, a_2 \, t \ldots t \, a_n > 0$ iff the sequence $h(a_1), h(a_2), \ldots, h(a_n)$ is sufficiently "similar" to the sequence 1, 1, ... , 1 of n ones. Speaking more precisely, the sum of deviations of the $h(a_i)$'s from 1 – just the Hamming distance between those two sequences – must be "small", less than 1, if one likes to have $a_1 \, t \, a_2 \, t \ldots t \, a_n > 0$. In particular, $a_1 \, t_Ł \, a_2 \, t_Ł \ldots t_Ł \, a_n > 0$ iff the sum of deviations of the a_i's from 1 is less than 1, i.e. iff the sequence $a_1, a_2, \ldots, a_n$ is sufficiently "similar" to the sequence of n ones. An analogous way of interpreting can be applied to (3.24).

It seems that the above observations give a better insight into the phenomenon of zero divisors of nilpotent t-norms by showing its numerical background.

Chapter 4

Aggregation of Information and Aggregation Operators

Huge amount of information one meets nowadays in a lot of areas of human activity makes information aggregation an extremely important and challenging issue. This chapter presents aggregation operators, a formal tool for aggregation of numerical information. Main types and properties of those operators will be discussed. Moreover, we will show applications of aggregation operators in the Bellman-Zadeh model of decision making in a fuzzy environment.

4.1 The Issue of Information Aggregation

Aggregation of information (data) is understood as any process in which different pieces of information - possibly provided by several sources - are gathered and simultaneously used with the aim of integrating them into one piece of information which represents and summarizes in a way all the gathered pieces. The objects of aggregation can be numerical data as well as, say, texts or multimedia data. In any case, however, if the aggregation process is performed by means of a computer, it requires and, in essence, collapses to aggregation of numbers. Our further discussion will thus be focused on this fundamental task of numerical data aggregation. It can be formally expressed as follows:

$$\text{assign a number } a \in [0, 1] \text{ to a given system of numbers } a_1, \ldots, a_n \in [0, 1],\ n \geq 1.$$

The a is then viewed as a representative of the whole system $a_1, \ldots, a_n$, a synthesis of information brought by it.

Notice that the restriction to the unit interval [0, 1] in the aggregation task is quite sufficient from the viewpoint of theory and practice. Indeed, we can assume that the data to be aggregated are from a closed interval $[b, c]$ of real numbers. At worst this interval can be understood as that of numbers which are representable in the memory of our computer. Each interval $[b, c]$, on the other hand, is isomorphic with and can be transformed into [0, 1].

The main objectives of aggregation are:

M. Wygralak: *Intelligent Counting Under Information Imprecision*, STUDFUZZ 292, pp. 93–109.
DOI: 10.1007/978-3-642-34685-9_4

- to enable comparisons between groups, vectors, or multidimensional arrays of data through comparisons between their representatives; their results are then a basis for coming to a conclusion or decision;
- to reduce the amount of information to be processed further by replacing many pieces of information with a single one;
- to get more – or more advanced – information about particular groups of some objects, e.g. groups of consumers, market segments, etc.;
- to reduce the influence of untypical data resulting from measurement errors, etc.

It is clear that the idea of aggregation can be extended element by element to fuzzy sets. For instance, the intersection $A_1 \cap_t \dots \cap_t A_n$ and sum $A_1 \cup_s \dots \cup_s A_n$ of fuzzy sets $A_1, \dots, A_n$ are then results of aggregation of these fuzzy sets. Each membership degree $(A_1 \cap_t \dots \cap_t A_n)(x) = A_1(x)\, \boldsymbol{t} \dots \boldsymbol{t}\, A_n(x)$ and $(A_1 \cup_s \dots \cup_s A_n)(x) = A_1(x)\, \boldsymbol{s} \dots \boldsymbol{s}\, A_n(x)$ is obtained through aggregation of $A_1(x), \dots, A_n(x)$ done by means of a t-norm $\boldsymbol{t}$ or t-conorm $\boldsymbol{s}$.

Aggregation techniques have multiple applications. Standard instances are data mining, decision support, control, multicriteria decision making (MCDM), economics and management. Worth emphasizing is that economic and stock market indices, a basis for most decisions in economics and finance, are usually results of aggregation of some data.

As to MCDM, each decision alternative is then evaluated with respect to two or more criteria. The resulting partial evaluations of each individual alternative can be aggregated into one final evaluation. Comparing all the final evaluations, the decision maker decides on the winning alternative. We see that many everyday human activities like shopping, recruitment, financial and investment decisions, choosing a holiday or educational offer are in essence MCDM processes involving multicriteria evaluation procedures and aggregation. An example is also the hotel evaluation task from Section 3.1.

4.2 Aggregation Operators

Aggregation operators are a formal tool for information aggregation. The variety of related applications requires an appropriately general and universal definition of such operators. A suitable solution seems to be an axiomatic formulation (see below). Another approach offering the concept of linguistic quantifier driven aggregation will be outlined in Section 10.3.

Definition 4.1. An *aggregation operator* is a function

$$Aggr\colon \bigcup_{n \geq 1} [0, 1]^n \to [0, 1]$$

satisfying the following conditions:

(A1) $Aggr(a_1, \ldots, a_n) \leq Aggr(b_1, \ldots, b_n)$ whenever $a_i \leq b_i$ for $i = 1, \ldots, n$,

(A2) $Aggr(a) = a$ for each $a \in [0, 1]$,

(A3) $Aggr(\underbrace{0, \ldots, 0}_{n}) = 0$ and $Aggr(\underbrace{1, \ldots, 1}_{n}) = 1$.

The number $Aggr(a_1, \ldots, a_n)$ is then called the *aggregation of* $a_1, \ldots, a_n$.

To each n-tuple of numbers from [0, 1], an aggregation operator $Aggr$ thus assigns a single number from the closed unit interval. This assignment is monotonic, (A1), satisfies the identity condition (A2) as well as the boundary condition (A3). One should emphasize that $Aggr$ does not have to be commutative, associative, or idempotent, i.e. an aggregation of n a's from (0, 1) does not have to be equal to a, $n \geq 2$. In Sections 4.3-4.5, we like to present and comment on the main types and instances of aggregation operators.

4.3 Aggregation Operators Involving Triangular Norms and Conorms

By (T1)-(T5) and (2.67) placed in Subsection 2.4.3, all t-norms and t-conorms are commutative and associative aggregation operators. They are not idempotent, except for $\wedge$ and $\vee$. An aggregation operator is also $Aggr(a_1, \ldots, a_n) = \boldsymbol{a_1}^{\boldsymbol{w_1}} \boldsymbol{t} \ldots \boldsymbol{t}\, \boldsymbol{a_n}^{\boldsymbol{w_n}}$ from (3.2), where $w_i > 0$ is a weight assigned to the ith argument, $i = 1, \ldots, n$, and $w_1 = 1$ whenever $n = 1$ (see (A2)).

4.3.1 Compensatory Operators

For a t-norm $\boldsymbol{t}$, t-conorm $\boldsymbol{s}$, and $\lambda \in [0, 1]$, let us define

$$E_{t,s}(a_1, \ldots, a_n; \lambda) = (a_1\, \boldsymbol{t} \ldots \boldsymbol{t}\, a_n)^{1-\lambda}\, (a_1\, \boldsymbol{s} \ldots \boldsymbol{s}\, a_n)^{\lambda}$$

and

$$L_{t,s}(a_1, \ldots, a_n; \lambda) = (1-\lambda)(a_1\, \boldsymbol{t} \ldots \boldsymbol{t}\, a_n) + \lambda(a_1\, \boldsymbol{s} \ldots \boldsymbol{s}\, a_n). \tag{4.1}$$

These two commutative aggregation operators, known as *compensatory operators* and ranging in the interval $[a_1\, \boldsymbol{t} \ldots \boldsymbol{t}\, a_n,\ a_1\, \boldsymbol{s} \ldots \boldsymbol{s}\, a_n]$, are thus a logarithmic convex combination and a linear convex combination of $a_1\, \boldsymbol{t} \ldots \boldsymbol{t}\, a_n$ and $a_1\, \boldsymbol{s} \ldots \boldsymbol{s}\, a_n$. In general, they are not associative and not idempotent. Let us look at two very particular examples of compensatory operators, namely

$$E_{\wedge,\vee}(a_1, a_2; \lambda) = (a_1 \wedge a_2)^{1-\lambda}\, (a_1 \vee a_2)^{\lambda}$$

and

$$L_{\wedge,\vee}(a_1, a_2; \lambda) = (1-\lambda)(a_1 \wedge a_2) + \lambda(a_1 \vee a_2).$$

They collapse to a weighted geometric mean and weighted arithmetic mean of a_1 and a_2 (see Section 4.4).

4.3.2 Soft Triangular Norms and Conorms

As previously, choose a t-norm $\boldsymbol{t}$, t-conorm $\boldsymbol{s}$, and $\lambda \in [0, 1]$. A *soft t-norm* $\underline{\boldsymbol{t}}$ and *soft t-conorm* $\underline{\boldsymbol{s}}$ are then defined as

$$\underline{\boldsymbol{t}}(a_1, \ldots, a_n; \lambda) = (1 - \lambda)\frac{a_1 + \ldots + a_n}{n} + \lambda(a_1\, \boldsymbol{t} \ldots \boldsymbol{t}\, a_n)$$

and (4.2)

$$\underline{\boldsymbol{s}}(a_1, \ldots, a_n; \lambda) = (1 - \lambda)\frac{a_1 + \ldots + a_n}{n} + \lambda(a_1\, \boldsymbol{s} \ldots \boldsymbol{s}\, a_n).$$

For instance, we have

$$\underline{\wedge}(a_1, a_2; \lambda) = (1 - \lambda)\frac{a_1 + a_2}{2} + \lambda(a_1 \wedge a_2),$$

$$\underline{\vee}(a_1, a_2; \lambda) = (1 - \lambda)\frac{a_1 + a_2}{2} + \lambda(a_1 \vee a_2),$$

$$\underline{\boldsymbol{t}}(a_1, a_2; \lambda) = (1 - \lambda)\frac{a_1 + a_2}{2} + \lambda a_1 a_2 \;\text{ for } \boldsymbol{t} = \boldsymbol{t}_a.$$

Soft t-norms and t-conorms are commutative aggregation operators. However, they are neither associative nor idempotent in general.

4.4 Averaging Operators

By an *averaging operator* one means an aggregation operators *Aggr* such that

$$Aggr(a_1, \ldots, a_n) \in [a_1 \wedge \ldots \wedge a_n, a_1 \vee \ldots \vee a_n]$$

for each n-tuple $(a_1, \ldots, a_n)$ and $n > 1$.We will present three groups of averaging operators: means, weighted means, and OWA operators.

4.4.1 Means

Their basic, well-known and most widely used examples are:

$$AR(a_1, \ldots, a_n) = \frac{1}{n}\sum_{i=1}^{n} a_i, \qquad (\textit{arithmetic mean})$$

$$GM(a_1, \ldots, a_n) = \left(\prod_{i=1}^{n} a_i\right)^{\frac{1}{n}}, \qquad (\textit{geometric mean})$$

$$HR(a_1, \ldots, a_n) = \frac{n}{\sum_{i=1}^{n} \frac{1}{a_i}}, \qquad \text{(harmonic mean)}$$

$$QD(a_1, \ldots, a_n) = \left(\frac{1}{n} \sum_{i=1}^{n} a_i^2 \right)^{\frac{1}{2}}. \qquad \text{(quadratic mean)}$$

Each of these operators is commutative and idempotent. None of them is associative. Moreover, one has

$$a_1 \wedge \ldots \wedge a_n \leq HR(a_1, \ldots, a_n) \leq GM(a_1, \ldots, a_n) \leq$$
$$AR(a_1, \ldots, a_n) \leq QD(a_1, \ldots, a_n) \leq a_1 \vee \ldots \vee a_n.$$

A common generalization of the above means is the idea of a *power mean* or *r-mean* $M_r(a_1, \ldots, a_n)$ with $r \in \mathbb{R}$ and

$$M_r(a_1, \ldots, a_n) = \left(\frac{1}{n} \sum_{i=1}^{n} a_i^r \right)^{\frac{1}{r}}. \tag{4.3}$$

So, $M_1 = AR$, $M_{-1} = HR$, $M_2 = QD$ and, moreover, we have

$$M_r(a_1, \ldots, a_n) \to GM(a_1, \ldots, a_n) \text{ if } r \to 0,$$
$$M_r(a_1, \ldots, a_n) \to a_1 \wedge \ldots \wedge a_n \text{ if } r \to -\infty,$$
$$M_r(a_1, \ldots, a_n) \to a_1 \vee \ldots \vee a_n \text{ if } r \to \infty.$$

Power means are commutative and idempotent, but are not associative.

Finally, let us mention an even more general family of aggregation operators which are defined by

$$M_\mu(a_1, \ldots, a_n) = \mu^{-1}\left(\frac{1}{n} \sum_{i=1}^{n} \mu(a_i) \right), \tag{4.4}$$

where μ: $[0, 1] \to [-\infty, \infty]$ is a continuous and strictly monotonic function such that $\{\mu(0), \mu(1)\} \neq \{-\infty, \infty\}$. $M_\mu(a_1, \ldots, a_n)$ is called a *quasi-arithmetic mean* of $a_1, \ldots, a_n$. Particular cases are $M_{id} = AR$, $M_\mu = GM$ for $\mu(x) = -\ln x$, $M_\mu = HR$ for $\mu(x) = 1/x$, and - more generally - $M_\mu = M_r$ for $\mu(x) = x^r$.

4.4.2 Weighted Means

Again, let us begin with recollecting their basic, well-known examples:

$$AR(a_1, \dots, a_n; p_1, \dots, p_n) = \frac{1}{p}\sum_{i=1}^{n} p_i a_i, \qquad (\textit{weighted arithmetic mean})$$

$$GM(a_1, \dots, a_n; p_1, \dots, p_n) = \left(\prod_{i=1}^{n} a_i^{p_i}\right)^{\frac{1}{p}}, \qquad (\textit{weighted geometric mean})$$

$$HR(a_1, \dots, a_n; p_1, \dots, p_n) = \frac{p}{\sum_{i=1}^{n} \frac{p_i}{a_i}}, \qquad (\textit{weighted harmonic mean})$$

$$QD(a_1, \dots, a_n; p_1, \dots, p_n) = \left(\frac{1}{p}\sum_{i=1}^{n} p_i a_i^2\right)^{\frac{1}{2}}, \qquad (\textit{weighted quadratic mean})$$

where $p_1, \dots, p_n \geq 0$ is a system of weights with p_i assigned to the ith argument of aggregation, and $p = \Sigma_{i=1,n} p_i > 0$. Especially important is the case when the system of weights is normalized, $p = 1$, which can always be done. Trivially, $p_1 = \dots = p_n = 1$ and $p_1 = \dots = p_n = 1/n$ ($p = 1$) give the usual means. The chain of inequalities between the four basic means in Subsection 4.4.1 is still valid for their weighted counterparts.

Finally, the above weighted means have generalizations similar to those offered in (4.3)-(4.4). They are of the form of a *weighted power mean* or *weighted r-mean* $M_r(a_1, \dots, a_n; p_1, \dots, p_n)$ and *weighted quasi-arithmetic mean* $M_\mu(a_1, \dots, a_n; p_1, \dots, p_n)$, where

$$M_r(a_1, \dots, a_n; p_1, \dots, p_n) = \left(\frac{1}{p}\sum_{i=1}^{n} p_i a_i^r\right)^{\frac{1}{r}}$$

and (4.5)

$$M_\mu(a_1, \dots, a_n; p_1, \dots, p_n) = \mu^{-1}\left(\frac{1}{p}\sum_{i=1}^{n} p_i \mu(a_i)\right).$$

These aggregation operators are still idempotent, but are neither commutative nor associative.

4.4.3 OWA Operators

They were introduced in YAGER (1988). Let $d_1, \dots, d_n$ be a sequence obtained from $a_1, \dots, a_n$ by arranging its terms in nonincreasing order, i.e. $d_1 \geq \dots \geq d_n$. Moreover, let

$w = (w_1, \dots, w_n)$ be a (normalized) weighting vector with $w_i \geq 0$ for $i = 1, \dots, n$, and $\Sigma_{i=1,n} w_i = 1$. An *OWA (ordered weighted averaging) operator* is defined as

$$OWA(a_1, \dots, a_n; w) = \sum_{i=1}^{n} w_i d_i. \tag{4.6}$$

One then says that $(d_1, \dots, d_n)$ is the *ordered argument of OWA*. This operator is commutative and idempotent, but not associative in general.

Comparing OWA operators with weighted arithmetic means in the context of multicriteria evaluation, the novelty seems to be that w_i is now a weight assigned to d_i, the ith largest element in $(a_1, \dots, a_n)$. This simple modification enables us to distinguish, positively or negatively, partial evaluations of a specific rank. Notice that this way of doing is widely used, say, in the refereeing process in many sports like ski jumping, diving, and gymnastics. For instance, the style of a ski jump is judged separately by five referees. Two extreme results are ignored, whereas the remaining three are added up and that sum forms the final result. As we see, this is nothing else than $3 \cdot OWA(a_1, \dots, a_5; w)$ with $w = (0, 1/3, 1/3, 1/3, 0)$ after a prior transformation of partial results from the interval [0, 20] used in ski jumping to [0, 1].

Let us present other examples of the weighting vectors together with the resulting OWA aggregations. First, assume that w^k with $1 \leq k \leq n$ is a vector defined as

$$w_i^k = \begin{cases} 1, & \text{if } i = k, \\ 0, & \text{otherwise.} \end{cases}$$

The importance is now concentrated on the kth largest element of $(a_1, \dots, a_n)$, and

$$OWA(a_1, \dots, a_n; w^k) = d_k. \tag{4.7}$$

In particular,

$$OWA(a_1, \dots, a_n; w^1) = a_1 \vee \dots \vee a_n \tag{4.8}$$

and

$$OWA(a_1, \dots, a_n; w^n) = a_1 \wedge \dots \wedge a_n. \tag{4.9}$$

If the importance is uniformly distributed, i.e. the weighting vector collapses to w^c with

$$w_i^c = 1/n \quad \text{for } i = 1, \dots, n,$$

we get

$$OWA(a_1, \dots, a_n; w^c) = \frac{1}{n}\sum_{i=1}^{n} d_i = \frac{1}{n}\sum_{i=1}^{n} a_i. \tag{4.10}$$

The OWA aggregation thus becomes the usual arithmetic mean of $a_1, \dots, a_n$.

4.5 Conclusions and Systematization

Although the overview of aggregation operators in Sections 4.3-4.4 is far from being exhaustive, it gives an image of their richness as to forms and properties. Sections 4.6 and 4.7, respectively, will present an application of aggregation operators in decision making and some theoretical aspects. A comprehensive study of aggregation operators and detailed references the reader can find in BELIAKOV *et al.* (2007); see also CALVO *et al.* (2002), DETYNIECKI (2000), FODOR/ROUBENS (1994), FODOR/YAGER (2000), KLEMENT *et al.* (2000), and WAGENKNECHT/KALININA (2006).

The choice of a suitable aggregation operator is always problem-dependent and must be thought over by the user. In this context, worth taking a look at are ranges of various aggregation operators.

$$a_1 \boldsymbol{t}_\mathrm{d} \ldots \boldsymbol{t}_\mathrm{d}\, a_n \le a_1 \boldsymbol{t} \ldots \boldsymbol{t}\, a_n \le a_1 \wedge \ldots \wedge a_n \le AR(a_1, \ldots, a_n) \le a_1 \vee \ldots \vee a_n \le a_1 \boldsymbol{s} \ldots \boldsymbol{s}\, a_n \le a_1 \boldsymbol{s}_\mathrm{d} \ldots \boldsymbol{s}_\mathrm{d}\, a_n$$

t-norms — averaging operators — t-conorms

soft t-norms — soft t-conorms

compensatory operators

Fig. 4.1 Ranges of selected aggregation operators

This suggests a natural division of aggregation operators *Aggr* into four classes.

- *Averaging operators*, i.e. operators such that

$$Aggr(a_1, \ldots, a_n) \in [a_1 \wedge \ldots \wedge a_n, a_1 \vee \ldots \vee a_n]$$

  for each n-tuple $(a_1, \ldots, a_n)$ and $n \ge 2$. Their main examples were presented in the previous section.

- *Conjunctive operators* satisfying the condition

$$Aggr(a_1, \ldots, a_n) \le a_1 \wedge \ldots \wedge a_n$$

  for each $(a_1, \ldots, a_n)$, and exemplified by t-norms.

- *Disjunctive operators* such that

$$Aggr(a_1, \ldots, a_n) \ge a_1 \vee \ldots \vee a_n$$

  for each $(a_1, \ldots, a_n) \in [0, 1]^n$ and $n \ge 2$. Their typical examples are t-conorms.

- *Mixed operators*, i.e. those which do not belong to any of the above three classes. Instances are soft t-norms, soft t-conorms, and compensatory operators. The reader is referred to BELIAKOV *et al.* (2007) for further examples like uninorms and nullnorms.

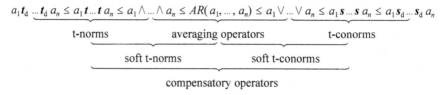

Let us look at these four groups of operators in the context multicriteria evaluation of some objects.

First, conjunctive and disjunctive operators with special reference to t-norms and t-conorms lead to aggregations having a strong "logical" flavor. Recollect that t-operations are used as truth functions of many-valued conjunction and disjunction connectives (see (2.92) and (2.93)). Aggregations involving a t-norm $\boldsymbol{t}$ give "pessimistic" results. Indeed, since $a_1 \,\boldsymbol{t} \ldots \boldsymbol{t}\, a_n \leq a_1 \wedge \ldots \wedge a_n$, the final evaluation $a_1 \,\boldsymbol{t} \ldots \boldsymbol{t}\, a_n$ becomes low whenever even one of the partial evaluations $a_1, \ldots, a_n$ is low. If $a_i = 0$ for some i, then $a_1 \,\boldsymbol{t} \ldots \boldsymbol{t}\, a_n = 0$. For nilpotent t-norms, $a_1 \,\boldsymbol{t} \ldots \boldsymbol{t}\, a_n = 0$ is possible even if $a_1, \ldots, a_n > 0$. The use of a t-conorm $\boldsymbol{s}$, on the other hand, leads to "optimistic" results of aggregation. If even one partial evaluation a_i is high, so is the final evaluation $a_1 \,\boldsymbol{s} \ldots \boldsymbol{s}\, a_n$ as $a_1 \,\boldsymbol{s} \ldots \boldsymbol{s}\, a_n \geq a_1 \vee \ldots \vee a_n$. In particular, $a_1 \,\boldsymbol{s} \ldots \boldsymbol{s}\, a_n = 1$ whenever $a_i = 1$ for some $1 \leq i \leq n$. For nilpotent t-conorms, $a_1 \,\boldsymbol{s} \ldots \boldsymbol{s}\, a_n = 1$ is possible if $a_1, \ldots, a_n < 1$.

Second, the behavior of averaging operators is quite different. They offer a compromise: the final evaluation $Aggr(a_1, \ldots, a_n)$ lies somewhere between $a_1 \wedge \ldots \wedge a_n$ and $a_1 \vee \ldots \vee a_n$, i.e. lies between the lowest partial evaluation and the highest one.

Notice that the worlds of conjunctive and averaging operators are not disjoint as $\wedge$ is both conjunctive and averaging. Analogously, the worlds of disjunctive and averaging aggregation operators are connected by $\vee$. This seems to be an additional justification for the popularity of $\wedge$ and $\vee$ in practice. Finally, the behavior of mixed operators is heterogeneous and varies on different parts of the domain.

Concluding, aggregation is a complex and advanced operation forming a common generalization of very different types of behavior like averaging, computing truth degrees, etc. Choosing a suitable aggregation operator, the user has to decide which of the three basic behaviors (averaging, conjunctive or disjunctive) or which type of a heterogeneous behavior is concerned.

4.6 Applications to Decision Making in a Fuzzy Environment

Decision making is inextricably linked with the notion of goals, which should be reached, and constraints, say, financial ones. However, a characteristic feature of human decision processes is imprecision (fuzziness, flexibility) of goals and constraints. It can be exemplified by a familiar situation when someone wants to rent a flat and defines his/her

- goal: one-bedroom flat with a *large* kitchen area,
- constraints: price not greater than *approximately* 800 €/month,
  quiet neighborhood, *convenient* commute.

This flexibility has been incorporated into the *Bellman-Zadeh model of decision making in a fuzzy environment* introduced in BELLMAN/ZADEH (1970) (see also KACPRZYK (1997)). The fuzzy environment in which a decision making process proceeds is then understood as consisting just of fuzzy goals, fuzzy constraints, and (!) a fuzzy decision. We like to outline that famous and powerful model. Its advantage is simplicity, intuitiveness and adequacy.

4.6.1 Bellman-Zadeh Model

The universe $\boldsymbol{U}$ is now a given set of possible decision alternatives: options, variants, choices, etc. Let us begin with the simplest case of one fuzzy goal and one fuzzy constraint. The *fuzzy goal*, G, and *fuzzy constraint*, C, are modeled as fuzzy sets in $\boldsymbol{U}$, i.e. $G, C: \boldsymbol{U} \to [0, 1]$. It is convenient to treat the membership degrees $G(x)$ and $C(x)$ as *satisfaction levels* of the decision maker accompanying the choice of x as decision. Speaking more precisely, they are levels of satisfaction with the realization of the goal and the constraint, respectively, by choosing x. The *fuzzy decision*, D, is defined as a fuzzy set

$$D = G * C, \tag{4.11}$$

where $*$ denotes an aggregation operator extended to fuzzy sets, i.e.

$$D(x) = Aggr(G(x), C(x)).$$

$D(x)$ thus forms an aggregated level of satisfaction accompanying the choice of decision alternative x and involving both $G(x)$ and $C(x)$.

It is clear that if the decision maker formulates many goals ($G_1, \dots, G_j$) and many constraints ($C_1, \dots, C_k$), the fuzzy decision is of the form

$$D = G_1 * \dots * G_j * C_1 * \dots * C_k. \tag{4.12}$$

In each case, however, the fuzzy decision as a fuzzy set in $\boldsymbol{U}$ is not feasible. An additional step becomes necessary, namely a *defuzzification* (*sharpening*) of D into a crisp decision $d^* \in \boldsymbol{U}$. Let us present two basic defuzzification techniques.

- We take as d^* a *maximizing decision*: d^* is such that $D(d^*) = \max\{D(x): x \in \boldsymbol{U}\}$. In other words,

  $$d^* \in \operatorname*{argmax}_x D(x) \tag{4.13}$$

  and we implicitly assume that a satisfaction-maximizing decision d^* exists at all. This d^* is not generally unique, on the other hand.
- Using (2.121) and (2.122), let us put

  $$d^* = \mathrm{COG}_D. \tag{4.14}$$

  This sharp decision can be viewed as a kind of compromise taking into account different possible choices with different satisfaction levels.

Finally, we like to refer to Section 3.1 and to the task of constructing the answer to the flexible query (FQ1): "Which *affordable* hotels are situated in the *vicinity* of the center of Paris?". The reader sees that it can be reformulated in the language of decision making in a fuzzy environment as

- goal: *affordable* hotel,
- constraint: in the *vicinity* of the center of Paris

or, more or less equivalently,

- goal: hotel in the *vicinity* of the center of Paris,
- constraint: *affordable* price.

For instance, the answer $A \cap B$ in Subsection 3.1.1 is now nothing else than the fuzzy decision with $* = \cap$, whereas Hotel Valois becomes the maximizing decision with satisfaction level 0.75. We also refer the reader to Section 10.4 presenting a modification of the Bellman-Zadeh model based on counting.

4.6.2 Computational Examples

Let us show applications of the Bellman-Zadeh model in two decision problems in which maximizing decisions are of interest.

First, assume we want to buy a TV set. Looking at four TV models, $\boldsymbol{U} = \{m_1, m_2, m_3, m_4\}$, we think about which one to choose. Our goal and constraints are

G: *reliable* TV set,
C_1: *high* picture and sound quality,
C_2: *energy-saving*,
C_3: *good* guarantee conditions.

The following assessments have been formulated by experts:

$$G = 0.7/m_1 + 0.3/m_2 + 0.7/m_3 + 0.9/m_4,$$
$$C_1 = 0.9/m_1 + 0.7/m_2 + 0.6/m_3 + 0.3/m_4,$$
$$C_2 = 0.7/m_1 + 0.5/m_2 + 0.9/m_3 + 0.7/m_4,$$
$$C_4 = 0.3/m_1 + 0.4/m_2 + 0.6/m_3 + 0.9/m_4.$$

By (4.12), the fuzzy decision is generally equal to

$$D = G * C_1 * C_2 * C_3,$$

i.e.

$$D(m_i) = Aggr(G(m_i), C_1(m_i), C_2(m_i), C_3(m_i))$$

for $i = 1, 2, 3, 4$. Let us look at D obtained for different aggregation operators $Aggr$.

- *Aggr* = *AR*. So, the arithmetic mean is used, and gives

$$D = 0.65/m_1 + 0.475/m_2 + 0.7/m_3 + 0.7/m_4.$$

TV models m_3 and m_4 are two maximizing decisions.

- *Aggr* = $\wedge$. For each model m_i, we then take into account its lowest evaluation from $\{G(m_i), C_1(m_i), C_2(m_i), C_3(m_i)\}$. This way of doing leads to

$$D = G \cap C_1 \cap C_2 \cap C_3 = 0.3/m_1 + 0.3/m_2 + 0.6/m_3 + 0.3/m_4$$

with m_3 as a unique maximizing decision.

- *Aggr* = *OWA* with weighting vector $w = (0.5, 0, 0, 0.5)$. So,

$$D(m_i) = OWA(G(m_i), C_1(m_i), C_2(m_i), C_3(m_i); 0.5, 0, 0, 0.5).$$

Our attention is now focused on the spread determined by the highest and the lowest evaluations from $\{G(m_i), C_1(m_i), C_2(m_i), C_3(m_i)\}$. The ordered argument of OWA operator is of the form

$$\begin{aligned} &(0.9, 0.7, 0.7, 0.3) \quad \text{for } i = 1, \\ &(0.7, 0.5, 0.4, 0.3) \quad \text{for } i = 2, \\ &(0.9, 0.7, 0.6, 0.6) \quad \text{for } i = 3, \\ &(0.9, 0.9, 0.7, 0.3) \quad \text{for } i = 4 \end{aligned}$$

and, by (4.6), we get

$$\begin{aligned} D(m_1) &= 0.5 \cdot 0.9 + 0.5 \cdot 0.3 = 0.6, \\ D(m_2) &= 0.5 \cdot 0.7 + 0.5 \cdot 0.3 = 0.5, \\ D(m_3) &= 0.5 \cdot 0.9 + 0.5 \cdot 0.6 = 0.75, \\ D(m_4) &= 0.5 \cdot 0.9 + 0.5 \cdot 0.3 = 0.6. \end{aligned}$$

Hence

$$D = 0.6/m_1 + 0.5/m_2 + 0.75/m_3 + 0.6/m_4.$$

Again, the choice of m_3 is thus recommended.

The second example is quite different and concerns the problem of pricing a new product. This task is important and fairly difficult in practice. Assume that the production cost of the product is 20 €, whereas 35 € is considered by market experts to be a perfect competitive price. The producer has defined a goal and two constraints on the price sought. It

G: must be *close* to doubled production cost,

C_1: should be *low*,

C_2: should be *close* to competitive price.

The goal and constraints has been modeled as the following fuzzy numbers in the universe of prices (see Subsection 2.5.1):

$G = (40, 5, 5)$, a triangular fuzzy number,
$C_1 = (20, 30)_z$, a z-shaped fuzzy number,
$C_2 = (35, 5, 5)$.

Use $Aggr = \wedge$. The fuzzy decision is then $D = G \cap C_1 \cap C_2$ (see Figure 4.2).

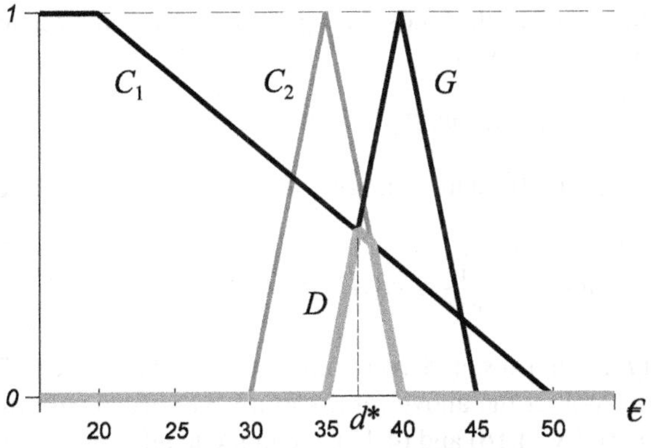

Fig. 4.2 The fuzzy goal, fuzzy constraints, fuzzy decision and maximizing decision in the pricing task

By elementary computations, a unique maximizing decision is $d^* = 37.14$ €. It forms a price suggested by the Bellman-Zadeh model.

4.7 Two Issues Related to Aggregation

We like to come back to some theoretical aspects of aggregation operators. Our subject will be mean values of aggregations and, second, a relationship between quasi-arithmetic means and averaging with respect to t-norms and t-conorms.

4.7.1 Mean Values of Aggregations

Let us begin with the general case of arbitrary n-ary aggregations $Aggr(a_1, \dots, a_n)$ from Definition 4.1, $n \geq 2$. Then

$$m(Aggr; n) = \int_{[0,1]^n} Aggr(a_1, \dots, a_n)\, da_1 \dots da_n \tag{4.15}$$

is the *mean value* of $Aggr$ on $[0, 1]^n$. Routine computations lead to the following particular results (cf. KOLESÁROVÁ (2006)):

$$m(AR;n)=\frac{1}{2},\quad m(GM;n)=\left(\frac{n}{n+1}\right)^n,$$

$$m(\wedge;n)=\frac{1}{n+1},\quad m(\vee;n)=\frac{n}{n+1},\tag{4.16}$$

$$m(t_a;n)=\frac{1}{2^n},\quad m(t_Ł;n)=\frac{1}{(n+1)!},\quad m(t_d;n)=0.$$

The arithmetic mean is thus the "most average" mathematical mean and aggregation operator in general. Notice that

$$m(t^*;n)=1-m(t;n)\tag{4.17}$$

for each t-norm t. By (4.16) and monotonicity of $m(Aggr;n)$ with respect to $Aggr$, we obtain

$$0\le m(t;n)\le\frac{1}{n+1}\le m(AVOP;n)\le\frac{n}{n+1}\le m(s;n)\le 1\tag{4.18}$$

for each t-norm t, t-conorm s, and averaging operator $AVOP$ (see (2.70)).

Finally, let us look at the above results in a very special case of binary aggregations $Aggr(a,b)$. By (4.16) and (4.17), we immediately get

$$m(AR;2)=\int_0^1\!\!\int_0^1\frac{a+b}{2}\,da\,db=\frac{1}{2},\quad m(GM;2)=\int_0^1\!\!\int_0^1(ab)^{1/2}\,da\,db=\frac{4}{9},$$

$$m(\wedge;2)=\int_0^1\!\!\int_0^1 a\wedge b\,da\,db=\frac{1}{3},\quad m(\vee;2)=\frac{2}{3},\tag{4.19}$$

$$m(t_a;2)=\frac{1}{4},\quad m(t_Ł;2)=\frac{1}{6},\quad m(s_a;2)=\frac{3}{4},\quad m(s_Ł;2)=\frac{5}{6},$$

whereas (4.18) implies

$$0\le m(t;2)\le\frac{1}{3}\le m(AVOP;2)\le\frac{2}{3}\le m(s;2)\le 1\tag{4.20}$$

for each t, s, and $AVOP$.

4.7.2 Averaging with Respect to Triangular Norms and Conorms

Referring to the essence, say, of arithmetic and geometric means, let us formulate the following definition.

Definition 4.2. For given real numbers $x_1,\dots,x_n$, $n\ge 2$, and a binary operation $\circledast$ on them, if there exists a unique number $c\in\mathbb{R}$ such that

$$x_1 \circledast \ldots \circledast x_n = \underbrace{c \circledast \ldots \circledast c}_{n},$$

then we say that c is the *mean of* $x_1, \ldots, x_n$ *with respect to* $\circledast$ or, in brief, is the $\circledast$-*mean* of these numbers. We will denote it by $M^{\circledast}(x_1, \ldots, x_n)$.

It is quite obvious that

$$M^{+}(x_1, \ldots, x_n) = AR(x_1, \ldots, x_n), \quad M^{\cdot}(x_1, \ldots, x_n) = GM(x_1, \ldots, x_n)$$

and, generally,

$$M^{\circledast}(x_1, \ldots, x_n) = M_r(x_1, \ldots, x_n)$$

whenever $x_1 \circledast \ldots \circledast x_n = \boldsymbol{x_1^r + \ldots + x_n^r}$, $0 \neq r \in \mathbb{R}$ (see (4.3)). AR, GM and M_r are here extended to arguments from $\mathbb{R}$, which is not confusing.

Our further discussion will be focused on $\boldsymbol{t}$-means and $\boldsymbol{s}$-means with a t-norm $\boldsymbol{t}$ and a t-conorm $\boldsymbol{s}$. We thus like to ask if there is a unique $c \in [0, 1]$ such that

$$a_1 \boldsymbol{t} \ldots \boldsymbol{t} a_n = c \boldsymbol{t} \ldots \boldsymbol{t} c \quad \text{or} \quad a_1 \boldsymbol{s} \ldots \boldsymbol{s} a_n = c \boldsymbol{s} \ldots \boldsymbol{s} c$$

for $a_1, \ldots, a_n \in [0, 1]$, where the number of the c's is equal to n in both the cases. One immediately gets

$$M^{\wedge}(a_1, \ldots, a_n) = a_1 \wedge \ldots \wedge a_n \quad \text{and} \quad M^{\vee}(a_1, \ldots, a_n) = a_1 \vee \ldots \vee a_n. \tag{4.21}$$

However, we should be careful if discontinuous or non-strictly increasing t-operations are involved. For instance,

$$\boldsymbol{M}^{\boldsymbol{t}_{\mathrm{d}}}(0.7, 1), \quad \boldsymbol{M}^{\boldsymbol{s}_{\mathrm{d}}}(0.4, 0), \quad \boldsymbol{M}^{\boldsymbol{t}_{Ł}}(0.3, 0.4), \quad \boldsymbol{M}^{\boldsymbol{s}_{Ł}}(0.7, 0.6)$$

do not exist as

$$0.7\, \boldsymbol{t}_{\mathrm{d}}\, 1 = 0.7 \quad \text{and} \quad c\, \boldsymbol{t}_{\mathrm{d}}\, c \neq 0.7 \quad \text{for each } c \in [0, 1],$$
$$0.4\, \boldsymbol{s}_{\mathrm{d}}\, 0 = 0.4 \quad \text{and} \quad c\, \boldsymbol{s}_{\mathrm{d}}\, c \neq 0.4 \quad \text{for each } c \in [0, 1],$$

whereas (non-uniqueness)

$$0.3\, \boldsymbol{t}_{Ł}\, 0.4 = 0 \quad \text{and} \quad c\, \boldsymbol{t}_{Ł}\, c = 0 \quad \text{for each } c \in [0, 0.5],$$
$$0.7\, \boldsymbol{s}_{\mathrm{d}}\, 0.6 = 1 \quad \text{and} \quad c\, \boldsymbol{s}_{Ł}\, c = 1 \quad \text{for each } c \in [0.5, 1].$$

The existence of $\boldsymbol{t}$-means and $\boldsymbol{s}$-means is guaranteed whenever $\boldsymbol{t}$ and $\boldsymbol{s}$ are strict. This is also the case of nilpotent t-norms and t-conorms provided that $a_1 \boldsymbol{t} \ldots \boldsymbol{t} a_n > 0$ and $a_1 \boldsymbol{s} \ldots \boldsymbol{s} a_n < 1$, respectively (see below).

Theorem 4.3. *Let* $\boldsymbol{t}$ *be an Archimedean t-norm with generator g, normed whenever* $\boldsymbol{t}$ *is nilpotent. Then the* $\boldsymbol{t}$*-mean* $M^{\boldsymbol{t}}(a_1, \ldots, a_n)$ *of numbers* $a_1, \ldots, a_n \in [0, 1]$, $n > 1$, *exists and is equal to their quasi-arithmetic mean* $M_\mu(a_1, \ldots, a_n)$ *with* $\mu = g$, *i.e.*

$$M^{\boldsymbol{t}}(a_1, \ldots, a_n) = M_g(a_1, \ldots, a_n) = g^{-1}\left(\frac{1}{n}\sum_{i=1}^{n} g(a_i)\right),$$

unless $\boldsymbol{t}$ *is nilpotent and* $a_1 \boldsymbol{t} \ldots \boldsymbol{t} a_n = 0$. *In that case, each* $c \leq g^{-1}(1/n)$ *satisfies the condition* $c\boldsymbol{t} \ldots \boldsymbol{t} c = 0$ *with n factors c, i.e.* $M^{\boldsymbol{t}}(a_1, \ldots, a_n)$ *does not exist, and*

$$M_g(a_1, \ldots, a_n) \leq g^{-1}\left(\frac{1}{n}\right).$$

Indeed, by (3.6) and Theorem 3.8(a), if $\boldsymbol{t}$ is strict and g is its generator, the condition $a_1 \boldsymbol{t} \ldots \boldsymbol{t} a_n = c\boldsymbol{t} \ldots \boldsymbol{t} c$ with $n > 1$ arguments c is equivalent to

$$g^{-1}\left(\sum_{i=1}^{n} g(a_i)\right) = g^{-1}(ng(c))$$

and, hence,

$$c = g^{-1}\left(\frac{1}{n}\sum_{i=1}^{n} g(a_i)\right) = M_g(a_1, \ldots, a_n) = M^{\boldsymbol{t}}(a_1, \ldots, a_n).$$

On the other hand, that condition collapses to

$$1 \wedge \sum_{i=1}^{n} g(a_i) = 1 \wedge ng(c)$$

whenever $\boldsymbol{t}$ is nilpotent and g forms its normed generator. If $a_1 \boldsymbol{t} \ldots \boldsymbol{t} a_n > 0$, this resolves itself into

$$\sum_{i=1}^{n} g(a_i) = ng(c) < 1$$

and, consequently, $c = M^{\boldsymbol{t}}(a_1, \ldots, a_n) = M_g(a_1, \ldots, a_n)$. Finally, for $a_1 \boldsymbol{t} \ldots \boldsymbol{t} a_n = 0$, we get

$$\sum_{i=1}^{n} g(a_i) \geq 1, \quad \text{i.e. } M_g(a_1, \ldots, a_n) \leq g^{-1}\left(\frac{1}{n}\right).$$

$M^{\boldsymbol{t}}(a_1, \ldots, a_n)$ does not exist in this case as the equality $c\boldsymbol{t} \ldots \boldsymbol{t} c = 0$ with n arguments c means that $g^{-1}(1 \wedge ng(c)) = 0$, which is satisfied by each $c \leq g^{-1}(1/n)$.

One can thus say that $\boldsymbol{t}$-means with an Archimedean $\boldsymbol{t}$ are some quasi-arithmetic means and, on the other hand, t-norms become operations with respect to which those quasi-arithmetic means are created (cf. KLEMENT *et al.* (2000)). In particular, we have

$$M^{t}(a_1, \ldots, a_n) = GM(a_1, \ldots, a_n) \quad \text{for } \boldsymbol{t} = \boldsymbol{t}_{\mathrm{a}}$$

and

$$M^{t}(a_1, \ldots, a_n) = HR(a_1, \ldots, a_n) \quad \text{for } \boldsymbol{t} = \boldsymbol{t}_{\mathrm{H},0}.$$

If $a_1 \boldsymbol{t} \ldots \boldsymbol{t} a_n > 0$, one obtains

$$M^{t}(a_1, \ldots, a_n) = M_p(a_1, \ldots, a_n) \quad \text{for } \boldsymbol{t} = \boldsymbol{t}_{\mathrm{S},p},$$

i.e. $\boldsymbol{t}_{\mathrm{S},p}$-means collapse to power means. So,

$$M^{t}(a_1, \ldots, a_n) = AR(a_1, \ldots, a_n) \quad \text{for } \boldsymbol{t} = \boldsymbol{t}_{\mathrm{Ł}}$$

and

$$M^{t}(a_1, \ldots, a_n) = QD(a_1, \ldots, a_n) \quad \text{for } \boldsymbol{t} = \boldsymbol{t}_{\mathrm{S},2}.$$

Let us formulate a counterpart of Theorem 4.3 for Archimedean t-conorms and $\boldsymbol{s}$-means. Its proof is quite analogous.

Theorem 4.4. *Assume* $\boldsymbol{s}$ *is an Archimedean t-conorm with generator h, normed whenever* $\boldsymbol{s}$ *is nilpotent. Then the* $\boldsymbol{s}$*-mean* $M^{s}(a_1, \ldots, a_n)$ *of* $a_1, \ldots, a_n \in [0, 1]$, $n > 1$, *exists and is equal to the quasi-arithmetic mean* $M_h(a_1, \ldots, a_n)$, *i.e.*

$$M^{s}(a_1, \ldots, a_n) = M_h(a_1, \ldots, a_n) = \boldsymbol{h}^{-1}\left(\frac{1}{n}\sum_{i=1}^{n} \boldsymbol{h}(a_i)\right),$$

unless $\boldsymbol{s}$ *is nilpotent and* $a_1 \boldsymbol{s} \ldots \boldsymbol{s} a_n = 1$. *In this case, each* $c \geq h^{-1}(1/n)$ *fulfils the condition* $c\,\boldsymbol{s} \ldots \boldsymbol{s}\, c = 1$ *with n components c and, thus,* $M^{s}(a_1, \ldots, a_n)$ *does not exist. Moreover,*

$$M_h(a_1, \ldots, a_n) \geq h^{-1}\left(\frac{1}{n}\right).$$

It is easy to check that (see (3.5))

$$M^{s}(a_1, \ldots, a_n) = 1 - M^{t}(1 - a_1, \ldots, 1 - a_n) \qquad (4.22)$$

for associated and Archimedean $\boldsymbol{t}$ and $\boldsymbol{s}$.

Chapter 5

Fuzzy Relations, Approximate Reasoning, Fuzzy Rule-Based Systems

This chapter presents a chain of notions and issues whose importance for fuzzy logic and, especially, for fuzzy logic in the narrow sense is absolutely fundamental. We begin with fuzzy relations and their types with special reference to similarity measures. The next subject will be elements of approximate reasoning, including fuzzy conditional statements and the compositional rule of inference. Finally, we will show the idea of fuzzy control and fuzzy rule-based systems. Questions of rule bases, fuzzification, inference, and defuzzification of results will be discussed and illustrated by simple examples.

5.1 Fuzzy Relations

The notion of a relation belongs to the most elementary and fundamental notions of mathematics. More or less consciously, it is commonly used in practically all areas of human knowledge and activity whenever there is a need for expressing sharp, precisely specified associations between some objects. However, a lot of real associations, e.g. similarity, cannot be adequately modeled in the conventional $\{0, 1\}$-valued way in which objects are either related or not. A feature of those associations is that they can be more or less intensive. The concept of a fuzzy relation makes it possible to model them.

Fuzzy relations are a useful tool in information retrieval, classification, control, knowledge representation, decision making, etc. We will concentrate around selected aspects of fuzzy relations which are important for our further discussion in this book. A comprehensive presentation of fuzzy relations and detailed references can be found in BOIXANDER *et al.* (2000), GOTTWALD (2001) and OVCHINNIKOV (2000); see also NGUYEN/WALKER (2005) and PEDRYCZ/GOMIDE (2007).

5.1.1 The Concept of a Fuzzy Relation

An n-ary relation is formally a subset of the cartesian product of n sets. It is a set of ordered n-tuples, in other words. This definition has a natural generalization.

M. Wygralak: *Intelligent Counting Under Information Imprecision*, STUDFUZZ 292, pp. 111–137.
DOI: 10.1007/978-3-642-34685-9_5

Definition 5.1. Each fuzzy subset R of the cartesian product $\boldsymbol{U}_1 \times \ldots \times \boldsymbol{U}_n$ of $n \geq 2$ sets will be called an *n-ary fuzzy relation on* $\boldsymbol{U}_1, \ldots, \boldsymbol{U}_n$. If $\boldsymbol{U}_i = \boldsymbol{U}$ for $i = 1, \ldots, n$, then R is said to be an *n-ary fuzzy relation in* $\boldsymbol{U}$.

An n-ary fuzzy relation is thus a function

$$R\colon \boldsymbol{U}_1 \times \ldots \times \boldsymbol{U}_n \to [0, 1]. \tag{5.1}$$

$R(x_1, \ldots, x_n) \in [0, 1]$ is then the *degree to which* $x_1, \ldots, x_n$ *are related*, the *degree of saturation* or *intensity of R* for the n-tuple $(x_1, \ldots, x_n)$. Ordinary n-ary relations as functions $\boldsymbol{U}_1 \times \ldots \times \boldsymbol{U}_n \to \{0, 1\}$ become a very particular case of fuzzy relations.

The concept of a fuzzy relation has deep motivations and makes it possible to model in an adequate way real relations which can occur with different intensities. Their most familiar example is the relation of similarity of two persons, notions, objects, etc. Different intensities are then expressed by humans by means of terms such as "very similar", "pretty similar", "totally similar". Other instances are the relations of cooperation ("full", "casual", ...), preference, and correlation. Let us look at a binary fuzzy relation $R\colon \boldsymbol{U} \times \boldsymbol{U} \to [0, 1]$ describing similarity of persons from population $\boldsymbol{U}$ with respect to some attribute (e.g. age, appearance, qualifications, ...). Assume that $R(x_1, x_3) = 0.2$, $R(x_4, x_7) = 1$, and $R(x_5, x_9) = 0.85$. So, x_1 and x_3 are similar to degree 0.2 (are slightly similar, in other words), x_4 and x_7 are similar to degree 1 (absolutely similar, even identical), whereas x_5 and x_9 are similar to degree 0.85 (very similar).

We will focus attention just on binary fuzzy relations as they seem to play a key role in practice. Let $R\colon \boldsymbol{U} \times \boldsymbol{V} \to [0, 1]$. If $\boldsymbol{U}$ and $\boldsymbol{V}$ are small sets, R can be equivalently presented in a convenient matrix form, called a *fuzzy relation matrix*. For instance, let $\boldsymbol{U} = \{x_1, x_2, x_3\}$, $\boldsymbol{V} = \{y_1, y_2, y_3, y_4\}$, and

$$R = \begin{bmatrix} \mathbf{0.4} & \mathbf{1} & \mathbf{0.6} & \mathbf{0.3} \\ \mathbf{1} & \mathbf{0.2} & \mathbf{0.7} & \mathbf{0.9} \\ \mathbf{1} & \mathbf{0.8} & \mathbf{0.6} & \mathbf{0} \end{bmatrix}.$$

The number placed in the ith row and jth column is then $R(x_i, y_j)$, the degree to which x_i and y_j are related; $i \in \{1, 2, 3\}$ and $j \in \{1, 2, 3, 4\}$. The above matrix notation thus means that, say, $R(x_2, x_3) = 0.7$ and $R(x_3, x_2) = 0.8$. Obviously, if R collapses to an ordinary binary relation, its matrix becomes a zero-one matrix.

Since fuzzy relations are fuzzy sets, it is clear that all notions and constructions of the language of fuzzy sets introduced in Sections 2.2-2.4 remain meaningful for fuzzy relations. For instance, if $R, S\colon \boldsymbol{U} \times \boldsymbol{V} \to [0, 1]$ are two fuzzy relations, then

$$R \subset S \Leftrightarrow \forall (x, y) \in \boldsymbol{U} \times \boldsymbol{V}\colon\ R(x, y) \leq S(x, y), \qquad (R \textit{ included in } S)$$

$$R_t = \{(x, y) \in \boldsymbol{U} \times \boldsymbol{V}\colon\ R(x, y) \geq t\}, \qquad (t\textit{-cut of } R)$$

$$\mathrm{supp}(R) = \{(x, y) \in \boldsymbol{U} \times \boldsymbol{V}\colon\ R(x, y) > 0\}, \qquad (\textit{support of } R)$$

$$(R \cup S)(x, y) = R(x, y) \vee S(x, y). \qquad (\textit{sum of } R \textit{ and } S)$$

Let us move on to some notions which are more specific for fuzzy relations. Recollect that the domain $\mathrm{dom}\,\rho$ of an ordinary relation $\rho \subset \boldsymbol{U} \times \boldsymbol{V}$ is the set of all first elements of the ordered pairs forming ρ, i.e.

$$\mathrm{dom}\,\rho = \{x \in \boldsymbol{U}\colon\ \rho(x, y) \text{ for some } y \in \boldsymbol{V}\}.$$

The set $\mathrm{ran}\,\rho$ of all second elements of those pairs is called the range of ρ:

$$\mathrm{ran}\,\rho = \{y \in \boldsymbol{V}\colon\ \rho(x, y) \text{ for some } x \in \boldsymbol{U}\}.$$

These two notions have nice generalizations to fuzzy relations. For $R\colon \boldsymbol{U} \times \boldsymbol{V} \to [0, 1]$, its *domain* is s fuzzy set $\mathrm{dom}\,R\colon \boldsymbol{U} \to [0, 1]$ with

$$\mathrm{dom}\,R(x) = \bigvee_{y \in \boldsymbol{V}} R(x, y), \tag{5.2}$$

whereas its *range*, a fuzzy set $\mathrm{ran}\,R\colon \boldsymbol{V} \to [0, 1]$, is defined as

$$\mathrm{ran}\,R(y) = \bigvee_{x \in \boldsymbol{U}} R(x, y). \tag{5.3}$$

Consider an example with $\boldsymbol{U} = \{x_1, x_2, x_3, x_4\}$, $\boldsymbol{V} = \{y_1, y_2, y_3, y_4, y_5\}$, and

$$R = \begin{bmatrix} 0.1 & 0.6 & 0.1 & 0.7 & 0.3 \\ 0.1 & 0.8 & 1 & 0.3 & 0.5 \\ 0.2 & 0 & 0.4 & 0.9 & 0.8 \\ 0.3 & 0.1 & 0.6 & 1 & 1 \end{bmatrix}.$$

By (5.2) and (5.3), we get

$$\mathrm{dom}\,R = 0.7/x_1 + 1/x_2 + 0.9/x_3 + 1/x_4,$$
$$\mathrm{ran}\,R = 0.3/y_1 + 0.8/y_2 + 1/y_3 + 1/y_4 + 1/y_5.$$

Finally, the *inverse* of a fuzzy relation $R\colon \boldsymbol{U} \times \boldsymbol{V} \to [0, 1]$ is a fuzzy relation $R^{-1}\colon \boldsymbol{V} \times \boldsymbol{U} \to [0, 1]$ such that

$$R^{-1}(y, x) = R(x, y) \quad \text{for each } (x, y) \in \boldsymbol{U} \times \boldsymbol{V}. \tag{5.4}$$

The matrix form of R^{-1} is thus created by transposing the matrix of R.

5.1.2 Composition of Fuzzy Relations

Let R: $\boldsymbol{U}\times\boldsymbol{V}\to[0,1]$ and S: $\boldsymbol{V}\times\boldsymbol{W}\to[0,1]$ be two fuzzy relations, and let $\boldsymbol{t}$ denote a t-norm. The *composition of R and S* is a fuzzy relation $R\circ_t S$: $\boldsymbol{U}\times\boldsymbol{W}\to[0,1]$ with

$$(R\circ_t S)(x,z)=\bigvee_{y\in\boldsymbol{V}} R(x,y)\,\boldsymbol{t}\,S(y,z)\quad\text{for each }(x,z)\in\boldsymbol{U}\times\boldsymbol{W}. \tag{5.5}$$

Using a more specific terminology, $R\circ_t S$ is also called the *composition of R and S with respect to* ***t*** or their *sup-****t*** *composition.* It is easy to see that $R\circ_t S$ becomes the usual composition whenever R and S collapse to ordinary relations.

Example 5.2. For computational simplicity, take $\boldsymbol{U}=\{x_1,x_2\}$, $\boldsymbol{V}=\{y_1,y_2,y_3\}$ and $\boldsymbol{W}=\{z_1,z_2\}$. Consider fuzzy relations R: $\boldsymbol{U}\times\boldsymbol{V}\to[0,1]$ and S: $\boldsymbol{V}\times\boldsymbol{W}\to[0,1]$ with

$$R=\begin{bmatrix}0.3 & 0.8 & 1\\ 0.6 & 0.9 & 0.7\end{bmatrix},\quad S=\begin{bmatrix}0.5 & 0.9\\ 0.4 & 1\\ 0.2 & 0.8\end{bmatrix}.$$

By (5.5), we obtain

$$R\circ_\wedge S=\begin{bmatrix}0.4 & 0.8\\ 0.5 & 0.9\end{bmatrix},\quad R\circ_{t_a} S=\begin{bmatrix}0.32 & 0.8\\ 0.36 & 0.9\end{bmatrix},\quad R\circ_{t_L} S=\begin{bmatrix}0.2 & 0.8\\ 0.3 & 0.9\end{bmatrix}.$$

For instance,

$$(R\circ_\wedge S)(x_1,z_1)=[R(x_1,y_1)\wedge S(y_1,z_1)]\vee[R(x_1,y_2)\wedge S(y_2,z_1)]\vee[R(x_1,y_3)\wedge S(y_3,z_1)]$$
$$=(0.3\wedge 0.5)\vee(0.8\wedge 0.4)\vee(1\wedge 0.2)=0.4,$$

$$(R\circ_{t_a} S)(x_2,z_1)=[R(x_2,y_1)\cdot S(y_1,z_1)]\vee[R(x_2,y_2)\cdot S(y_2,z_1)]\vee[R(x_2,y_3)\cdot S(y_3,z_1)]$$
$$=0.6\cdot 0.5\vee 0.9\cdot 0.4\vee 0.7\cdot 0.2=0.36.$$ □

A routine matter is to verify the following general properties of sup-$\boldsymbol{t}$ compositions:

$$(R\circ_t S)^{-1}=S^{-1}\circ_t R^{-1},$$
$$P\circ_t R\subset P\circ_t S\ \text{and}\ R\circ_t Q\subset S\circ_t Q\quad\text{whenever } R\subset S, \tag{5.6}$$
$$P\circ_t(R\circ_t S)=(P\circ_t R)\circ_t S\quad\text{whenever } \boldsymbol{t}\text{ is continuous.}$$

As previously, let us use a t-norm $\boldsymbol{t}$, and let $\mathbf{1}$ denote a 1-element set. A fuzzy set A: $\boldsymbol{U}\to[0,1]$ can be formally viewed as a fuzzy relation, namely A: $\mathbf{1}\times\boldsymbol{U}\to[0.1]$. Thus, we can construct the composition $A\circ_t R$ of that fuzzy set and a fuzzy relation R: $\boldsymbol{U}\times\boldsymbol{V}\to[0,1]$. By (5.5), $A\circ_t R$: $\mathbf{1}\times\boldsymbol{V}\to[0,1]$, i.e. this composition is a fuzzy set in $\boldsymbol{V}$ such that

$$(A \circ_t R)(y) = \bigvee_{x \in U} A(x) \, t \, R(x, y) \quad \text{for each } y \in V. \tag{5.7}$$

Similarly, a fuzzy set B: $V \to [0, 1]$ is equivalent to a fuzzy relation, B: $V \times \mathbf{1} \to [0. 1]$. We can thus construct the composition $R \circ_t B$ of a fuzzy relation R: $U \times V \to [0, 1]$ and that fuzzy set, where $R \circ_t B$: $U \times \mathbf{1} \to [0, 1]$ with

$$(R \circ_t B)(x) = \bigvee_{y \in V} R(x, y) \, t \, B(y) \quad \text{for each } x \in U, \tag{5.8}$$

i.e. $R \circ_t B$ forms a fuzzy set in U.

5.1.3 Types of Fuzzy Relations

Let us focus on binary fuzzy relations in U. Again, t denotes a t-norm. We will say that a fuzzy relation R: $U \times U \to [0, 1]$ is

- *reflexive* whenever $R(x, x) = 1$ for each $x \in U$,
- *antireflexive* whenever $R(x, x) = 0$ for each $x \in U$,
- *symmetrical* whenever $R(x, y) = R(y, x)$ for each $x, y \in U$,
- *antisymmetrical* whenever $R(x, y) > 0$ & $R(y, x) > 0 \Rightarrow x = y$ for each $x, y \in U$,
- *transitive with respect to* t or t*-transitive* whenever

$$R(x, z) \geq R(x, y) \, t \, R(y, z) \quad \text{for each } x, y, z \in U, \tag{5.9}$$

- *connected* whenever $R(x, y) > 0$ or $R(y, x) > 0$ for each $x, y \in U$.

As we see, (5.9) is equivalent to

$$R(x, z) \geq \bigvee_{y \in U} R(x, y) \, t \, R(y, z) \quad \text{for each } x, z \in U, \text{ i.e. } R \circ_t R \subset R. \tag{5.10}$$

If R: $U \times U \to \{0, 1\}$, the above definitions collapse to the well-known classical definitions of basic types of ordinary relations.

Compositions of t*-transitive* fuzzy relations do not have to be t-transitive. By (5.9), the following properties are satisfied:

- If R is t-transitive and $t_1 \leq t$, then R is t_1-transitive, too.
- Consequently, $\wedge$-transitivity implies t-transitivity for each t and

$$\wedge\text{-transitivity} \Rightarrow t_a\text{-transitivity} \Rightarrow t_Ł\text{-transitivity}. \tag{5.11}$$

- If R is t-transitive and reflexive, then

$$R(x, z) = \bigvee_{y \in U} R(x, y) \, t \, R(y, z) \quad \text{for each } x, z \in U, \text{ i.e. } R \circ_t R = R. \tag{5.12}$$

Indeed, (5.10) implies

$$R(x,z) \ge \bigvee_{y \in \boldsymbol{U}} R(x,y)\, \boldsymbol{t}\, R(y,z) \ge R(x,x)\, \boldsymbol{t}\, R(x,z) = R(x,z) \quad \text{for each } x, z \in \boldsymbol{U}.$$

Finally, for a t-norm $\boldsymbol{t}$, $k \ge 1$, and $R\colon \boldsymbol{U} \times \boldsymbol{U} \to [0,1]$, let us define

$$R^k = R \circ_t \ldots \circ_t R \tag{5.13}$$

with k factors R on the right-hand side. R^k is called the kth *power of* R. Further, let

$$\bar{R} = R \cup R^2 \cup R^3 \cup \ldots, \tag{5.14}$$

which is said to be the *transitive closure of* R. Its basic properties are given below.

- If R is $\boldsymbol{t}$-transitive, then

$$R \supset R^2 \supset R^3 \supset \ldots, \quad \text{i.e. } R = \bar{R}. \tag{5.15}$$

  Indeed, by (5.10) and (5.6), we get

$$R^2 \subset R, \; R^3 = R^2 \circ_t R \subset R \circ_t R = R^2, \text{ etc.}$$

- If $\boldsymbol{t}$ is continuous, then $\bar{R}$ is $\boldsymbol{t}$-transitive and, by (5.15), $\bar{R} = \bar{\bar{R}}$.

5.1.4 Similarity Measures and Similarity Classes

We like to introduce two more complex types of fuzzy relations.

Definition 5.3. Let $\boldsymbol{t}$ denote a t-norm, and let $R\colon \boldsymbol{U} \times \boldsymbol{U} \to [0,1]$.
(a) If R is a reflexive and symmetrical fuzzy relation, one says that R is a *weak similarity* or a *resemblance relation.*

(b) If R is reflexive, symmetrical and $\boldsymbol{t}$-transitive, it is called a *similarity relation* or, in short, a *similarity*. R is also termed a $\boldsymbol{t}$*-similarity* or $\boldsymbol{t}$*-equivalence*.

Resemblance relations together with similarity relations will be called *similarity measures* in $\boldsymbol{U}$.

As one knows, an ordinary relation $\rho \subset \boldsymbol{U} \times \boldsymbol{U}$ being reflexive, symmetrical and transitive is said to be an equivalence relation. ρ splits $\boldsymbol{U}$ into disjoint classes, *equivalence classes*, of elements which are identical with respect to an attribute. ρ thus leads to a classification of elements from $\boldsymbol{U}$, and that classification can be identified with knowledge about $\boldsymbol{U}$. For instance, if $\boldsymbol{U}$ denotes a population of persons and the attribute under consideration are qualifications, the equivalence classes consist of persons having the same qualifications, i.e. persons which are indistinguishable with respect to qualifications.

A similarity relation R splits $\boldsymbol{U}$ into *similarity classes* which are fuzzy sets of elements being *similar* with respect to a given attribute, e.g. similarity classes of persons having similar qualifications, persons at a similar age, objects in a similar shape, etc. These classes are not disjoint in general. Consequently, one says that R leads to a *soft classification* in $\boldsymbol{U}$. A similarity relation R: $\boldsymbol{U} \times \boldsymbol{U} \to \{0, 1\}$ collapses to an ordinary equivalence relation. The concept of similarity classes can be formalized as follows (ZADEH (1971)).

Definition 5.4. Let R: $\boldsymbol{U} \times \boldsymbol{U} \to [0, 1]$ be a similarity relation. The *similarity class* of an element $x \in \boldsymbol{U}$ is a fuzzy set $[x]$: $\boldsymbol{U} \to [0, 1]$ with

$$[x](y) = R(x, y) \quad \text{for each } y \in \boldsymbol{U}.$$

The similarity class $[x]$ is thus a fuzzy set of elements from $\boldsymbol{U}$ which are more or less similar to x.

Example 5.5. Assume $\boldsymbol{U} = \{x_1, x_2, x_3, x_4, x_5, x_6\}$ and

$$R = \begin{bmatrix} 1 & 0.2 & 1 & 0.6 & 0.2 & 0.6 \\ 0.2 & 1 & 0.2 & 0.2 & 0.9 & 0.2 \\ 1 & 0.2 & 1 & 0.6 & 0.2 & 0.6 \\ 0.6 & 0.2 & 0.6 & 1 & 0.2 & 0.9 \\ 0.2 & 0.9 & 0.2 & 0.2 & 1 & 0.2 \\ 0.6 & 0.2 & 0.6 & 0.9 & 0.2 & 1 \end{bmatrix}.$$

Obviously, R is both reflexive and symmetrical. R is also $\wedge$-transitive, i.e. forms a $\wedge$-similarity. $R(x_i, x_j)$ expresses the degree to which x_i and x_j are similar. We get

$$\begin{aligned}
&[x_1] = [x_3] = 1/x_1 + 0.2/x_2 + 1/x_3 + 0.6/x_4 + 0.2/x_5 + 0.6/x_6, \\
&[x_2] = 0.2/x_1 + 1/x_2 + 0.2/x_3 + 0.2/x_4 + 0.9/x_5 + 0.2/x_6, \\
&[x_4] = 0.6/x_1 + 0.2/x_2 + 0.6/x_3 + 1/x_4 + 0.2/x_5 + 0.9/x_6, \\
&[x_5] = 0.2/x_1 + 0.9/x_2 + 0.2/x_3 + 0.2/x_4 + 1/x_5 + 0.2/x_6, \\
&[x_6] = 0.6/x_1 + 0.2/x_2 + 0.6/x_3 + 0.9/x_4 + 0.2/x_5 + 1/x_6.
\end{aligned}$$

These similarity classes are not disjoint, e.g. x_1 belongs to $[x_1]$ and $[x_3]$ to degree 1, belongs to $[x_2]$ and $[x_5]$ to degree 0.2, and – finally – belongs to $[x_4]$ and $[x_6]$ to degree 0.6. As one sees, the membership degrees in $[x_i]$ are just the numbers from the ith row of matrix R. □

An immediate consequence of the properties of t-transitivity mentioned in Subsection 5.1.3 is that $\wedge$-similarity is the strongest form of similarity. More precisely, we have the following:

- if R is a $\boldsymbol{t}$-similarity and $\boldsymbol{t}_1 \le \boldsymbol{t}$, then R is also a $\boldsymbol{t}_1$-similarity;
- thus, $\wedge$-similarity implies $\boldsymbol{t}$-similarity for each t-norm $\boldsymbol{t}$ and, in particular,

$$\wedge\text{-similarity} \Rightarrow \boldsymbol{t}_{\mathrm{a}}\text{-similarity} \Rightarrow \boldsymbol{t}_{\mathrm{Ł}}\text{-similarity}.$$

A routine task is to verify

Theorem 5.6. R: $\boldsymbol{U} \times \boldsymbol{U} \to [0, 1]$ is a $\wedge$-similarity iff R_t with each $t \in (0, 1]$ is an equivalence relation in $\boldsymbol{U}$.

So, each t-cut of a $\wedge$-similarity collapses to an ordinary equivalence relation. One shows that $\wedge$-similarities are equivalent to partition trees in $\boldsymbol{U}$.

Example 5.7. Consider the $\wedge$-similarity from Example 5.5. By Theorem 5.6, each $R_t = \{(x, y) \in \boldsymbol{U} \times \boldsymbol{U}\text{: } R(x, y) \ge t\}$ is then an equivalence relation. Its equivalence classes are composed of elements being mutually similar to degree $\ge t$:

t	equivalence classes of R_t
$t \in (0, 0.2]$	$\{x_1, x_2, x_3, x_4, x_5, x_6\}$
$t \in (0.2, 0.6]$	$\{x_1, x_3, x_4, x_6\}$ $\{x_2, x_5\}$
$t \in (0.6, 0.9]$	$\{x_1, x_3\}$ $\{x_4, x_6\}$ $\{x_2, x_5\}$
$t \in (0.9, 1]$	$\{x_1, x_3\}$ $\{x_4\}$ $\{x_6\}$ $\{x_2\}$ $\{x_5\}$

If the threshold t increases, the existing equivalence classes split up into finer classes and, consequently, the classification of elements from $\boldsymbol{U}$ becomes more and more exact and detailed. □

The last part of this subsection will be devoted to methods of construction of similarity relations. Our intuition suggests that similarity can be viewed as a concept closely connected with distance: it can be modeled as a decreasing function of distance. Basic construction methods of similarity measures are therefore based on distance and assume a linear, hyperbolic, or exponential relationship between similarity and distance. They are presented below. Let R: $\boldsymbol{U} \times \boldsymbol{U} \to [0, 1]$.

Method 1. Using a normed metric d in $\boldsymbol{U}$, one defines

$$R(x, y) = 1 - d(x, y). \tag{5.16}$$

Main properties of this fuzzy relation R are listed below.

- $R(x, y) \in [0, 1]$.
- R is a $\boldsymbol{t}_{\mathrm{Ł}}$-similarity. Indeed, we have $R(x, x) = 1 - d(x, x) = 1$, $R(x, y) = R(y, x)$, and $R(x, z) \ge R(x, y)\, \boldsymbol{t}_{\mathrm{Ł}}\, R(y, z)$ as

$$d(x, z) \le d(x, y) + d(y, z) \Leftrightarrow R(x, z) \ge R(x, y) + R(y, z) - 1.$$

- Conversely, if R is a t_{L}-similarity and $R(x, y) = 1 \Leftrightarrow x = y$, then d with

$$d(x, y) = 1 - R(x, y)$$

becomes a normed metric in $\boldsymbol{U}$.

Method 2. For an arbitrary metric d in $\boldsymbol{U}$, let

$$R(x, y) = \frac{1}{1 + d(x, y)}. \tag{5.17}$$

Then $R(x, y) \in (0, 1]$ and R is a t_{a}-similarity.

Method 3. Using again an arbitrary metric d, we define

$$R(x, y) = e^{-d(x, y)}. \tag{5.18}$$

As previously, R is a t_{a}-similarity and $R(x, y) \in (0, 1]$.

Our attention will be focused on (5.16). For instance, applying the normed Minkowski distance d_p from (2.54) and putting

$$R_p(A, B) = 1 - d_p(A, B), \tag{5.19}$$

we get a t_{L}-similarity relation for fuzzy sets A and B in $\boldsymbol{U}$; $p \geq 1$. In particular,

$$R_1(A, B) = 1 - \frac{1}{|\boldsymbol{U}|} \sum_{x \in \boldsymbol{U}} |A(x) - B(x)| \tag{5.20}$$

and

$$R_2(A, B) = 1 - \left(\frac{1}{|\boldsymbol{U}|} \sum_{x \in \boldsymbol{U}} (A(x) - B(x))^2 \right)^{1/2} \tag{5.21}$$

whenever $\boldsymbol{U}$ is finite.

5.1.5 Cardinality-Based Similarity Measures for Sets

That two given data sets differ, say, in hundreds of elements and that they differ in only one or two elements are quite different situations in practice. This forms a motivation and justification for introducing and using similarity measures of ordinary sets. Examples of their applications are data analysis, data mining, databases, and decision making. We like to present three cardinality-based similarity measures. Throughout this subsection, one assumes that $\boldsymbol{U}$ is finite, $\boldsymbol{A}, \boldsymbol{B} \subset \boldsymbol{U}$, and one puts $\frac{0}{0} = 1$.

(A) Jaccard coefficient. It is also called the *compatibility coefficient*, and is defined as

$$R_{\mathrm{J}}(\boldsymbol{A}, \boldsymbol{B}) = \frac{|\boldsymbol{A} \cap \boldsymbol{B}|}{|\boldsymbol{A} \cup \boldsymbol{B}|}. \tag{5.22}$$

Its main properties are given below:

- R_J is a $t_Ł$-similarity,
- $R_J(\boldsymbol{A}, \boldsymbol{B}) = 1 \Leftrightarrow \boldsymbol{A} = \boldsymbol{B}$,
  $R_J(\boldsymbol{A}, \boldsymbol{B}) = 0 \Leftrightarrow \boldsymbol{A} \cap \boldsymbol{B} = \varnothing \ \& \ \boldsymbol{A} \cup \boldsymbol{B} \neq \varnothing$,
- $1 - R_J(\boldsymbol{A}, \boldsymbol{B})$ is a normed metric, called the *Jaccard metric*,
- $R_J(\boldsymbol{A}, \boldsymbol{B}) \neq R_J(\boldsymbol{A}', \boldsymbol{B}')$.

(B) Matching coefficient. It is given by

$$R_m(\boldsymbol{A}, \boldsymbol{B}) = \frac{|((\boldsymbol{A} \setminus \boldsymbol{B}) \cup (\boldsymbol{B} \setminus \boldsymbol{A}))'|}{|\boldsymbol{U}|} = 1 - \frac{|\boldsymbol{A}| + |\boldsymbol{B}| - 2|\boldsymbol{A} \cap \boldsymbol{B}|}{|\boldsymbol{U}|}. \tag{5.23}$$

Let us look at its basic properties:

- R_m is a $t_Ł$-similarity,
- $R_m(\boldsymbol{A}, \boldsymbol{B}) = 1 \Leftrightarrow \boldsymbol{A} = \boldsymbol{B}$, $R_m(\boldsymbol{A}, \boldsymbol{B}) = 0 \Leftrightarrow \boldsymbol{A} = \boldsymbol{B}'$,
- $1 - R_m(\boldsymbol{A}, \boldsymbol{B})$ forms a normed metric,
- $R_m(\boldsymbol{A}, \boldsymbol{B}) = R_m(\boldsymbol{A}', \boldsymbol{B}')$.

(C) Overlap coefficient. It is defined as

$$R_o(\boldsymbol{A}, \boldsymbol{B}) = \frac{|\boldsymbol{A} \cap \boldsymbol{B}|}{\min(|\boldsymbol{A}|, |\boldsymbol{B}|)}. \tag{5.24}$$

Its basic properties are the following:

- R_o is only a resemblance relation in 2^U,
- $R_o(\boldsymbol{A}, \boldsymbol{B}) = 1 \Leftrightarrow \boldsymbol{A} \subset \boldsymbol{B} \ \# \ \boldsymbol{B} \subset \boldsymbol{A}$, $R_o(\boldsymbol{A}, \boldsymbol{B}) = 0 \Leftrightarrow \boldsymbol{A} \cap \boldsymbol{B} = \varnothing \ \& \ \boldsymbol{A}, \boldsymbol{B} \neq \varnothing$,
- $R_o(\boldsymbol{A}, \boldsymbol{B}) \neq R_o(\boldsymbol{A}', \boldsymbol{B}')$.

Example 5.8. Consider two finite fuzzy sets $\boldsymbol{A}$ and $\boldsymbol{B}$ shown below.

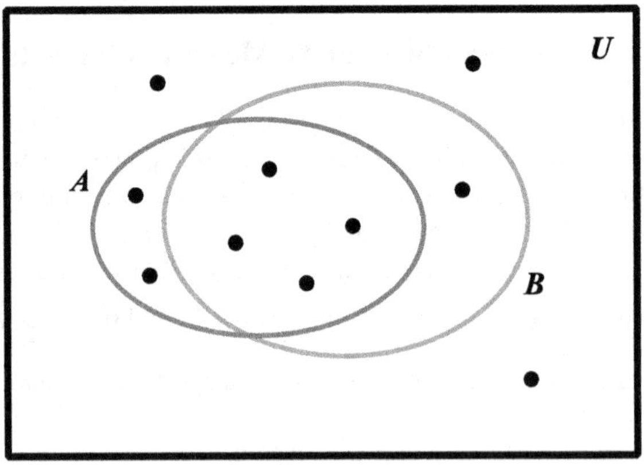

Fig. 5.1 Graphical presentation of $\boldsymbol{A}$ and $\boldsymbol{B}$

Using (5.22)-(5.24), we get

$$R_J(A, B) = \tfrac{4}{7}, \quad R_J(A', B') = \tfrac{3}{6},$$

$$R_m(A, B) = R_m(A', B') = \tfrac{7}{10},$$

$$R_o(A, B) = \tfrac{4}{5}, \quad R_o(A', B') = \tfrac{3}{4}. \qquad \square$$

We refer the reader to DE BAETS/DE MEYER (2005a) for further examples of cardinality-based similarity measures of sets (see also DE BAETS/DE MEYER (2005b)). In Section 10.1, extensions of (5.22)-(5.24) forming cardinality-based similarity measures of fuzzy sets will be presented and discussed.

Finishing this subsection, we like to present one more cardinality-based coefficient. It refers to the inclusion relation between two data sets and is widely used in practice. Its extension to fuzzy sets is relevant to our discussion in this book (see Sections 8.3 and 10.3). Let

$$R_{\subset}(A, B) = \frac{|A \cap B|}{|A|}, \tag{5.25}$$

which is called the *inclusion coefficient*. It is nothing else than the conditional probability prob($B \mid A$) that an element drawn at random from A is simultaneously an element of B. $R_{\subset}$ is reflexive, $R_{\subset}(A, A) = 1$, but is not symmetrical as $R_{\subset}(A, B) \neq R_{\subset}(B, A)$ (cf. (5.24)). Clearly, $R_o(A, B) = R_{\subset}(A, B) \vee R_{\subset}(B, A)$. In contrast to the inclusion relation for sets, which is transitive, one can easily give counterexamples showing that $R_{\subset}$ is not t-transitive for any t, even if $t = t_d$. $R_{\subset}$ is thus not a similarity relation in the sense of Definition 5.3. The following properties hold true:

- $R_{\subset}(A, B) = 1 \Leftrightarrow A \subset B$,
- $R_{\subset}(A, B) = 0 \Leftrightarrow A \cap B = \varnothing \ \& \ A \neq \varnothing$,
- $R_{\subset}(A, B) \neq R_{\subset}(B', A')$.

For instance, $R_{\subset}(A, B) = \tfrac{4}{6}$ and $R_{\subset}(B', A') = \tfrac{3}{5}$ for fuzzy sets from Example 5.8. The reader interested in other cardinality-based inclusion measures for sets is referred to DE BAETS *et al.* (2002) (cf. also BOSTEELS/KERRE (2009)).

5.1.6 Inclusion and Equality Measures for Fuzzy Sets

Again, we assume that fuzzy sets A, B: $U \to [0, 1]$ and our universe U are quite arbitrary. We like to construct some measures of inclusion and equality between A and B inspired by Łukasiewicz logic $Ł_\infty$ (see Subsections 2.1.3 and 2.2.3). They are not cardinality-based, and make it possible to quantify informal relationships such as "A is *almost* contained in B" and "A is *almost* equal to B". Cardinality-based measures for fuzzy sets will be presented in Section 10.1.

Let us introduce two many-valued relationships between fuzzy sets, namely $A \subset_m B$, *many-valued inclusion of A in B*, and $A =_m B$, *many-valued equality of A and B*, defined via the conditions

$$\forall_m x \in \boldsymbol{U}\colon x \in_m A \Rightarrow_m x \in_m B$$

and

$$A \subset_m B \mathbin{\&_m} B \subset_m A,$$

respectively. So, one has

$$[\![A \subset_m B]\!] = [\![\forall_m x \in \boldsymbol{U}\colon x \in_m A \Rightarrow_m x \in_m B]\!] \tag{5.26}$$

and

$$[\![A =_m B]\!] = [\![A \subset_m B \mathbin{\&_m} B \subset_m A]\!] \tag{5.27}$$

or, equivalently,

$$[\![A =_m B]\!] = [\![\forall_m x \in \boldsymbol{U}\colon x \in_m A \Leftrightarrow_m x \in_m B]\!]. \tag{5.28}$$

We thus get (see Subsection 2.2.3 and (2.34))

$$A \subset B \Leftrightarrow [\![A \subset_m B]\!] = 1 \quad \text{and} \quad A = B \Leftrightarrow [\![A =_m B]\!] = 1. \tag{5.29}$$

In a quite natural way, this suggests to define the following graded, fuzzy relations of inclusion, R_{in}, and equality, R_{eq}, for fuzzy sets:

$$R_{\text{in}}(A, B) = [\![A \subset_m B]\!] \quad \text{and} \quad R_{\text{eq}}(A, B) = [\![A =_m B]\!]. \tag{5.30}$$

We will say that $R_{\text{in}}(A, B)$ is the *degree to which A is included in B*. $R_{\text{eq}}(A, B)$ is said to be the *degree to which A equals B*.

First, let us focus on R_{in}. Applying (2.9), (2.7) and (2.3) to (5.26), one obtains

$$\begin{aligned} R_{\text{in}}(A, B) &= \bigwedge_{x \in \boldsymbol{U}} A(x) \rightarrow_{Ł} B(x) \\ &= \bigwedge_{x \in \boldsymbol{U}} 1 \wedge (1 - A(x) + B(x)) \\ &= 1 \wedge (1 - \bigvee_{x \in \boldsymbol{U}} A(x) - B(x)). \end{aligned}$$

Hence

$$\begin{aligned} R_{\text{in}}(A, B) &= 1 - (0 \vee \bigvee_{x \in \boldsymbol{U}} A(x) - B(x)) \\ &= \begin{cases} 1, & \text{if } A \subset B, \\ 1 - \bigvee_{x \in \boldsymbol{U}} A(x) - B(x) < 1 & \text{otherwise.} \end{cases} \end{aligned} \tag{5.31}$$

We easily see that $R_{\text{in}}(A, B) \in \{0, 1\}$ whenever A and B collapse to ordinary sets. $1 - R_{\text{in}}(A, B)$ expresses the maximum size of violation of the condition $A(x) \le B(x)$ by the membership degrees in A.

Example 5.9. Consider two simple fuzzy sets $A = 0.8/x_2 + 0.1/x_3 + 0.4/x_4 + 0.1/x_5$ and $B = 0.1/x_1 + 0.6/x_2 + 0.5/x_3 + 1/x_4$. By (5.31), we have $R_{in}(A, B) = 1 - 0.2 \vee 0.1 = 0.8$ and $R_{in}(B, A) = 1 - 0.1 \vee 0.4 \vee 0.6 = 0.4$. □

Worth emphasizing are the following properties of fuzzy relation R_{in} in $[0, 1]^U$:

- $R_{in}(A, A) = 1$,
- $R_{in}(A, B) \neq R_{in}(B, A)$,
- R_{in} is $t_Ł$-transitive.

This $t_Ł$-transitivity is a consequence of the inequality

$$a \to_Ł c + c \to_Ł b \leq a \to_Ł b + 1, \tag{5.32}$$

which leads to

$$R_{in}(A, C) + R_{in}(C, B) - 1 \leq R_{in}(A, B). \tag{5.33}$$

Graded inclusions of fuzzy sets can be defined in a bit different way as

$$A \subset^t B \Leftrightarrow R_{in}(A, B) \geq t \tag{5.34}$$

with $t \in [0, 1]$. Again, if $A \subset^t B$, we will say that *A is included in B to degree t*. This time, however, one has

$$A \subset^t B \Leftrightarrow \forall x \in \boldsymbol{U}: A(x) \leq B(x) + (1 - t). \tag{5.35}$$

The following properties are thus satisfied:

- $A \subset^1 A$,
- if $A \subset^t B$, then $A \subset^r B$ for each $r \leq t$,
- if $A \subset^r C$ and $C \subset^s B$, then $A \subset^{r t_Ł s} B$, (5.36)

which follows from (5.33). Referring to A and B from Example 5.9, we easily notice that $A \subset^t B$ with an arbitrary $t \leq 0.8$, and $B \subset^t A$ with any $t \leq 0.4$.

Let us move on to the fuzzy relation R_{eq}. By (5.27) and (5.30),

$$\begin{aligned} R_{eq}(A, B) &= R_{in}(A, B) \wedge R_{in}(B, A) \\ &= 1 - \bigvee_{x \in \boldsymbol{U}} |A(x) - B(x)|. \end{aligned} \tag{5.37}$$

An immediate consequence is that

- $R_{eq}(A, B) = 1 \Leftrightarrow A = B$,
- R_{eq} is a $t_Ł$-similarity

as $d(A, B) = \bigvee_{x \in U} |A(x) - B(x)|$ forms a normed metric (see Method 1 in Subsection 5.1.4). $R_{eq}(A, B) \in \{0, 1\}$ whenever A and B collapse to ordinary sets.

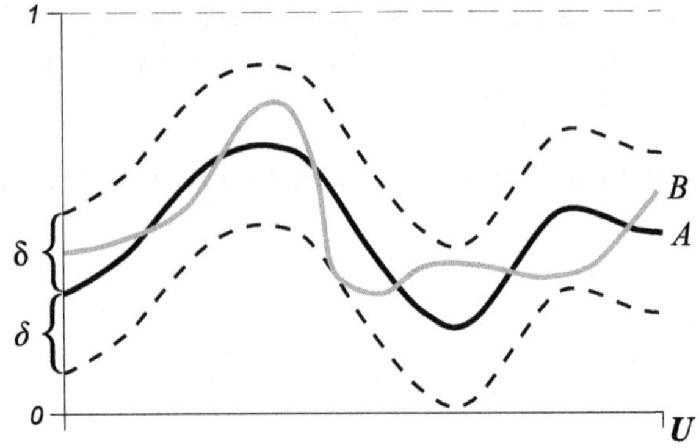

Fig. 5.2 Graphical interpretation of $R_{eq}(A, B)$

The number $\delta = 1 - R_{eq}(A, B)$ is just the maximum size of violation of the condition $A(x) = B(x)$ (see Figure 5.2). Let us refer once more to Example 5.9. We obtain $R_{eq}(A, B) = 0.8 \wedge 0.4 = 0.4$.

Similarly to (5.34), graded equalities of fuzzy sets can be formulated in a slightly different and more flexible manner as

$$A =^t B \Leftrightarrow R_{eq}(A, B) \geq t \tag{5.38}$$

for $t \in [0, 1]$. We then say that *A is equal to B to degree t*. By (5.37) and (5.35),

$$A =^t B \Leftrightarrow \forall x \in \boldsymbol{U}: B(x) - (1 - t) \leq A(x) \leq B(x) + (1 - t). \tag{5.39}$$

For A and B from Example 5.9, we thus have $A =^t B$ with an arbitrary $t \leq 0.4$. The following general properties are easy to check:

- $A =^1 A$,
- if $A =^t B$, then $A =^r B$ for each $r \leq t$,
- if $A =^r C$ and $C =^s B$, then $A =^{r t_L s} B$. (5.40)

A detailed study of graded inclusions and equalities involving more properties and references can be found in Burillo *et al.* (2000) and Gottwald (2001). The reader is also referred to Wygralak (1983b) in which graded equalities

$$A =^{t,s} B \Leftrightarrow A \subset^r B \ \& \ B \subset^s A \tag{5.41}$$

with two parameters are investigated.

Using R_{in}, R_{eq}, $\subset^t$ or $=^t$, maximum violations of suitable conditions on membership degrees are the focal issue. Worth considering, however, seem also to be mean violations. For simplicity, let us restrict ourselves for a moment to finite $\boldsymbol{U}$. Let

$$[\![a \leq_{\mathrm{m}} b]\!] = a \to_{Ł} b, \tag{5.42}$$

where $a, b \in [0, 1]$ and $\leq_{\mathrm{m}}$ is understood as *many-valued inequality*. Put

$$R_{\mathrm{in}*}(A, B) = \frac{1}{|\boldsymbol{U}|} \sum_{x \in \boldsymbol{U}} [\![A(x) \leq_{\mathrm{m}} B(x)]\!]$$

and

$$R_{\mathrm{eq}*}(A, B) = \frac{1}{|\boldsymbol{U}|} \sum_{x \in \boldsymbol{U}} [\![A(x) \leq_{\mathrm{m}} B(x)]\!] \wedge [\![B(x) \leq_{\mathrm{m}} A(x)]\!], \tag{5.43}$$

i.e.

$$R_{\mathrm{eq}*}(A, B) \leq R_{\mathrm{in}*}(A, B) \wedge R_{\mathrm{in}*}(B, A).$$

Elementary transformations lead to (see (5.20))

$$R_{\mathrm{in}*}(A, B) = 1 - \frac{1}{|\boldsymbol{U}|} \sum_{x \in \boldsymbol{U}} 0 \vee (A(x) - B(x))$$

and

$$R_{\mathrm{eq}*}(A, B) = 1 - \frac{1}{|\boldsymbol{U}|} \sum_{x \in \boldsymbol{U}} |A(x) - B(x)| = R_1(A, B). \tag{5.44}$$

Coming back to A and B from Example 5.9 with $\boldsymbol{U} = \{x_1, \dots, x_5\}$, we get

$$R_{\mathrm{in}*}(A, B) = 1 - \tfrac{1}{5}(0.2 + 0.1) = 0.94, \quad R_{\mathrm{in}*}(B, A) = 0.78,$$

$$R_{\mathrm{eq}*}(A, B) = 1 - \tfrac{1}{5}(0.1 + 0.2 + 0.4 + 0.6 + 0.1) = 0.72.$$

Generally, $R_{\mathrm{in}*}$ is a reflexive and $\boldsymbol{t}_{Ł}$-transitive fuzzy relation (see (5.32)) and, clearly, $R_{\mathrm{eq}*}$ forms a $\boldsymbol{t}_{Ł}$-similarity.

5.2 Approximate Reasoning – Basic Issues

Classical logic offers us two basic inference rules known as *modus ponens* and *modus tollens* whose forms are reminded below:

	modus ponens	*modus tollens*
premise	p	$\neg q$
implication	$p \Rightarrow q$	$p \Rightarrow q$
conclusion	q	$\neg p$

(5.45)

One can rewrite them as

$$[p \,\&\, (p \Rightarrow q)] \Rightarrow q \quad \text{and} \quad [\neg q \,\&\, (p \Rightarrow q)] \Rightarrow \neg p, \tag{5.46}$$

respectively. For instance, assume that the Giants Club is a club for people who are at least 200 cm tall ("If someone is a member of the club, he/she is 200 cm tall or more"). John is a member of the club and, thus, he must be at least 200 cm tall. Peter is 178 cm tall. So, he cannot be a member of the club.

The processes of reasoning carried out by humans are usually more sophisticated, and are not of formal character as they refer to the meanings of terms involved. Their general feature is an intrinsic imprecision (fuzziness) of premises and conclusions. Human reasoning is thus *approximate reasoning*. It has many forms, e.g.

(a) *premise*	Tom is *tall*
fuzzy relation	Mark is *a few* centimeters shorter than Tom
conclusion	Mark is *rather tall*

(b) *premise*	The rain is *light*
implication	If rain is *heavy*, puddles are *everywhere*
conclusion	Puddles are *only here and there*

(c) *premise*	The tomato is *very red*
implication	If tomato is *ripe*, it is *red*
conclusion	The tomato is *very ripe*

More types of approximate reasoning can be found in ZADEH (1975d, 1979, 1999b). Referring to the above three examples, we see that approximate reasoning is not a primitive thinking of someone who is ignorant of the principles of logic. On the contrary, it is a subtle, advanced and intelligent reasoning in the presence of information imprecision or if a greater precision is not necessary from the viewpoint of practice. Human beings successfully use (a)-(c) and other types of approximate reasoning in their everyday and professional life to solve problems and make decisions. A formalization of that reasoning is thus important in the context of intelligent systems. However, it cannot be done within the framework of classical logic and its inference rules although (b) and (c) resemble *modus ponens* and *modus tollens*, respectively. Even many-valued logic alone is here insufficient. A suitable tool for computational modeling of approximate reasoning is fuzzy logic, more precisely: fuzzy logic in the narrow sense (see Section 1.1). Focusing on (b) and (c), we like to present basic elements of that modeling.

The implications in (b) and (c) will be presented more formally by means of *fuzzy conditional statements* of the form

$$\text{IF } \alpha \text{ is } A \text{ THEN } \beta \text{ is } B$$

or

$$\text{IF } \alpha = A \text{ THEN } \beta = B \qquad (5.47)$$

with α and β as linguistic variables, whereas A and B denote their values treated as some fuzzy sets (fuzzy numbers, in most cases). So, (5.47) describes a relationship between values of linguistic variables α and β. The "IF" part contains an imprecise, fuzzy premise or condition, while the "THEN" part describes an imprecisely specified conclusion, decision, or action.

Each reasoning of type (b) can then be written in the following form known as *generalized* or *soft modus ponens*:

$$\begin{array}{l} \alpha = \underline{A} \\ \text{IF } \alpha = A \text{ THEN } \beta = B \\ \hline \beta = \underline{B} \end{array} \tag{5.48}$$

A and $\underline{A}$ are fuzzy sets in $\boldsymbol{U}$, whereas B and $\underline{B}$ denote two fuzzy sets in a universe $\boldsymbol{V}$. If $\underline{A} = A$ and $\underline{B} = B$, (5.48) collapses to the usual *modus ponens*. (b) now assumes the form

$$\begin{array}{l} \text{rain} = \textit{light} \\ \text{IF rain} = \textit{heavy} \text{ THEN puddles} = \textit{everywhere} \\ \hline \text{puddles} = \textit{only here and there} \end{array}$$

Fuzzy sets A, B and $\underline{A}$ from (5.48) are given, whereas $\underline{B}$ is unknown and has to be determined. To that end, the fuzzy conditional statement in (5.48) is interpreted by means of an implication operator $\rightarrow$, e.g. one of those discussed in Subsection 2.4.5. More precisely speaking, we define

$$(A \Rightarrow B)(z, y) = A(z) \rightarrow B(y), \tag{5.49}$$

which is equivalent to defining a binary fuzzy relation R: $\boldsymbol{U} \times \boldsymbol{V} \rightarrow [0, 1]$ with

$$R(z, y) = A(z) \rightarrow B(y). \tag{5.50}$$

Looking at (5.46) and (5.48), it is quite reasonable to create $\underline{B}$ as the composition of the fuzzy set $\underline{A}$ and fuzzy relation R with respect to a t-norm $\boldsymbol{t}$, namely

$$\underline{B} = \underline{A} \circ_t (A \Rightarrow B). \tag{5.51}$$

This formula is called the *compositional rule of inference*. More precisely, by (5.7), we have

$$\underline{B}(y) = \bigvee_{z \in \boldsymbol{U}} \underline{A}(z)\, \boldsymbol{t}(A(z) \rightarrow B(y)) \quad \text{for each } y \in \boldsymbol{V}. \tag{5.52}$$

Let us move on to approximate reasoning of type (c). It can be presented in the following form called *generalized* or *soft modus tollens*:

$$\begin{array}{l} \beta = \underline{B} \\ \text{IF } \alpha = A \text{ THEN } \beta = B \\ \hline \alpha = \underline{A} \end{array} \tag{5.53}$$

One has to determine $\underline{A}$ on the basis of given fuzzy sets A, B and $\underline{B}$. Again, we will use the *compositional rule of inference* which is now of the form

$$\underline{A} = (A \Rightarrow B) \circ_t \underline{B}. \tag{5.54}$$

By (5.8),

$$\underline{A}(z) = \bigvee_{y \in V} (A(z) \to B(y))\, t\, \underline{B}(y) \quad \text{for each } z \in \boldsymbol{U}. \tag{5.55}$$

The resulting $\underline{A}$ is thus the composition of a fuzzy relation and a fuzzy set.

5.3 Fuzzy Control and Fuzzy Rule-Based Systems

The idea of *fuzzy control* or *fuzzy logic control*, in other words, was formulated and developed in the 1970s by L. A. Zadeh and E. H. Mamdani (see e.g. ZADEH (1972) and MAMDANI (1976); see also SEISING (2007b) for a detailed historical study). It became probably the most successful application of fuzzy logic and, generally, of fuzzy sets and their methodology.

The conventional approach to control requires a sufficiently adequate (mathematical) model of the process or system to be controlled. In many cases, however, such a model is not available as it is completely unknown or, say, its construction would be too costly or too time consuming. The essence of fuzzy control is to base control on rules rather than on a model. The *control rules* are obtained in a verbal and – thus – imprecise, fuzzy form from experienced operators or experts who know how to control the process or system. Those fuzzy rules are then translated into the language of linguistic variables and fuzzy conditional statements. Since fuzzy rules guarantee a successful manual control, we can expect the same from an automatic control based on them.

This simple and natural idea has turned out to be very fruitful and has led to a successful automation of many control processes which resisted the conventional approach. Speaking more generally, that idea has opened the door to *fuzzy rule-based modeling* in a lot of areas of applications, and to *fuzzy rule-based systems*. We like to present the fundamentals of fuzzy control and fuzzy rule-based systems limiting ourselves to those elements which are relevant to our discussion in this book. The reader is referred to DRIANKOV *et al.* (1993), PEDRYCZ/GOMIDE (2007), PIEGAT (2001), SILER/BUCKLEY (2005), TANAKA (1997) and YAGER/FILEV (1994) for a comprehensive presentation (see also KACPRZYK (1997)).

5.3.1 Computational Approach to Fuzzy Rules

A suitable and convenient language in which fuzzy control rules and, generally, any fuzzy rules can be formulated is that of fuzzy conditional statements from (5.47):

$$\text{IF } \alpha = A \text{ THEN } \beta = B \tag{5.56}$$

The "IF" part as well as the "THEN" one may be compound. For instance, they may be conjunctions or disjunctions of some subconditions as in

IF α = *high* AND β = *medium* THEN γ = *very low*

If necessary, one can use fuzzy conditional statements in an extended form, namely

IF $\alpha = A$ THEN $\beta = B$ ELSE $\gamma = C$

which will not be considered in this book.

How to carry out (5.56)? In other words, we ask which value should be assigned to the output linguistic variable β if $\alpha = \underline{A}$. Two cases are here of special interest for applications.

- Case 1: $\underline{A}$ = *about x*,
  where x is a measured, crisp value of the input variable α. We thus take into consideration that x is affected by measuring error. By (5.48) and (5.52), we get

$$\beta = \underline{B} \text{ with } \underline{B}(y) = \bigvee_{z\in \boldsymbol{U}} \underline{A}(z)\, \boldsymbol{t}(A(z) \rightarrow B(y)) \text{ for each } y \in \boldsymbol{V}. \tag{5.57}$$

- Case 2: $\underline{A} = 1_{\{x\}}$,
  i.e. x is assumed to be precisely determined. Using again (5.52), one obtains

$$\beta = \underline{B} \text{ with } \underline{B}(y) = \bigvee_{z\in \boldsymbol{U}} 1_{\{x\}}(z)\, \boldsymbol{t}(A(z) \rightarrow B(y)) = A(x) \rightarrow B(y). \tag{5.58}$$

  $A(x) \rightarrow B(y)$ is now the degree of fulfilment of the rule (5.56) if y were the value of the output variable β.

In practice, (5.58) is used much more frequently than (5.57) as the former is computationally simpler and gives satisfactory results. Our further presentation, too, will be based on (5.58).

Although, speaking theoretically, an arbitrary implication operator discussed in Subsection 2.4.5 can play the role of $\rightarrow$ in (5.58), a choice dominating in practice is then an operator

$$a \rightarrow b = a\, \boldsymbol{t}\, b, \tag{5.59}$$

called an *engineering implication operator*, where $\boldsymbol{t}$ denotes a t-norm and $a, b \in [0, 1]$. Its two cases are especially important in applications:

- *Mamdani operator*: $$a \rightarrow b = a \wedge b, \tag{5.60}$$
- *Larsen operator*: $$a \rightarrow b = ab. \tag{5.61}$$

Using (5.59), (5.58) collapses to

$$\underline{B}(y) = A(x) \rightarrow B(y) = (A \times_t B)(x, y). \tag{5.62}$$

Let us emphasize that engineering implication operators are not implication operators in the sense of Subsection 2.4.5. Moreover, (5.59) with $a, b \in \{0, 1\}$ gives numerical results differing from those offered by classical logic as $a\,t\,b = 1$ iff $a = b = 1$.

5.3.2 Fuzzy Controller

We like to present in a simple way the original approach to fuzzy control formulated by Mamdani. The general structure of a fuzzy controller is then the following:

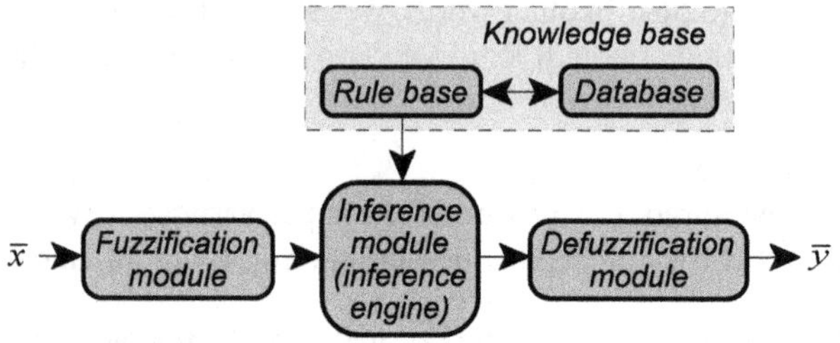

Fig. 5.3 General structure of a fuzzy controller

Actually, Figure 5.3 also presents a more general thing: a fuzzy rule-based system. Besides control issues, applications of such systems encompass expert systems, decision support, modeling and simulation, image processing, etc. Another type of application will be shown in Subsection 9.2.2. Although our further presentation nominally refers to fuzzy controllers, it thus remains valid for arbitrary fuzzy rule-based systems. Let us briefly describe the components shown in Figure 5.3.

- $\bar{x} = (x_1, \ldots, x_n)$ is a vector of measured or observed crisp values of n input variables.
- The *rule base* contains a collection of rules in the form of fuzzy conditional statements. Since their "IF" parts contain some fuzzy sets rather than precisely specified values, the *fuzzification module* transforms $x_1, \ldots, x_n$ into membership values in those fuzzy sets.
- The very inference process is performed by the *inference module* using the rule base and the apparatus of approximate reasoning.
- Since the results of inference are again some fuzzy sets, the *defuzzification module* transforms them into crisp values forming the output vector $\bar{y} = (y_1, \ldots, y_k)$. Clearly, a feedback (skipped in Figure 5.3) may be involved and, then, that vector is used to generate a new input vector.
- The *database* stores all data which are necessary for a proper functioning of the fuzzification, inference, and defuzzification modules. In particular, it stores the fuzzy sets involved in the rules contained in the rule base.
- Finally, the *knowledge base* is composed of the rule base and database.

Presenting the workings of a fuzzy controller, it is convenient to focus on the basic case of two input variables and one output variable ($n = 2$, $k = 1$), which can be easily extended to more variables. Moreover, for clarity, we will refer to a concrete and familiar example of house heating, depending on air temperature and fuel price. A strictly formal and general presentation would not be readable enough.

The input variables *air_temperature* (*AT*) and *fuel_price* (*FP*), and the output variable *heating_level* (*HL*) will be treated as linguistic variables. Assume that their values are

AT: *cool*, *cold*, *frosty*,
FP: *low*, *medium*, *high*,
HL: *low*, *medium*, *high*.

Interpretations of those values by means of fuzzy sets are presented below.

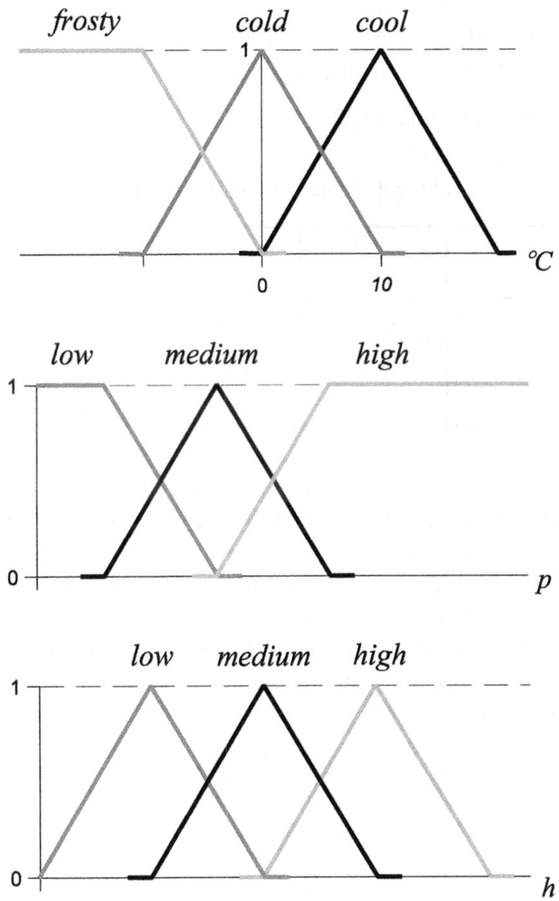

Fig. 5.4 Interpretation of values of *AT* (top), *FP* (middle), and *HL* (bottom)

The real variable *p* denotes price per fuel unit, whereas *h* represents heating understood, say, as heating valve opening level. Interpretations of values of each linguistic variable must always be overlapping as in Figure 5.4.

What one has to do next is to define the rule base. The case of house heating is familiar and we do not need to ask experts. It suffices to use a commonsense approach. The following list of rules seems to be reasonable:

IF *AT* = *cool* AND *FP* = *low* THEN *HL* = *medium*
IF *AT* = *cold* AND *FP* = *low* THEN *HL* = *high*
IF *AT* = *frosty* AND *FP* = *low* THEN *HL* = *high*
IF *AT* = *cool* AND *FP* = *medium* THEN *HL* = *low*
IF *AT* = *cold* AND *FP* = *medium* THEN *HL* = *medium*
IF *AT* = *frosty* AND *FP* = *medium* THEN *HL* = *high*
IF *AT* = *cool* AND *FP* = *high* THEN *HL* = *low*
IF *AT* = *cold* AND *FP* = *high* THEN *HL* = *low*
IF *AT* = *frosty* AND *FP* = *high* THEN *HL* = *medium*

Let us rewrite it in a more concise table form.

Table 5.1 Table form of heating rules

FP↓ *AT*→	*cool*	*cold*	*frosty*
low	*medium*	*high*	*high*
medium	*low*	*medium*	*high*
high	*low*	*low*	*medium*

Speaking generally, the rules in a rule base cannot be contradictory, i.e. the base cannot contain rules with identical "IF" parts and different "THEN" parts. The number of rules should be "suitable": not too small and not too large. Some potential rules are usually skipped whenever they refer to impossible or unimportant combinations of values of input linguistic variables. The rule base in Table 5.1 involving all combinations of linguistic values of *AT* and *FP* is thus redundant in this context. For instance, fuel prices usually rise with the coming of freezing weather and, therefore, at least the rules

IF *AT* = *frosty* AND *FP* = *low* THEN *HL* = *high*

and

IF *AT* = *cool* AND *FP* = *high* THEN *HL* = *low*

could be removed from the base. The corresponding two fields in Table 5.1 would then be blank. Since our discussion aims at giving just the idea of fuzzy control, we will use the redundant rule base.

5.3.3 How Does It Work?

We are ready to present the workings of the fuzzy controller from the previous subsection equipped with the rule base defined in Table 5.1. Generally, the rules from a rule base are grouped with respect to their "THEN" parts and the rules within each group are then performed one after the other.

Assume that the current air temperature and fuel price are 4°C and 35€ per fuel unit, respectively. This combination of temperature and price, as any other, fulfils to a degree the conditions in the "IF" part of each rule (see below).

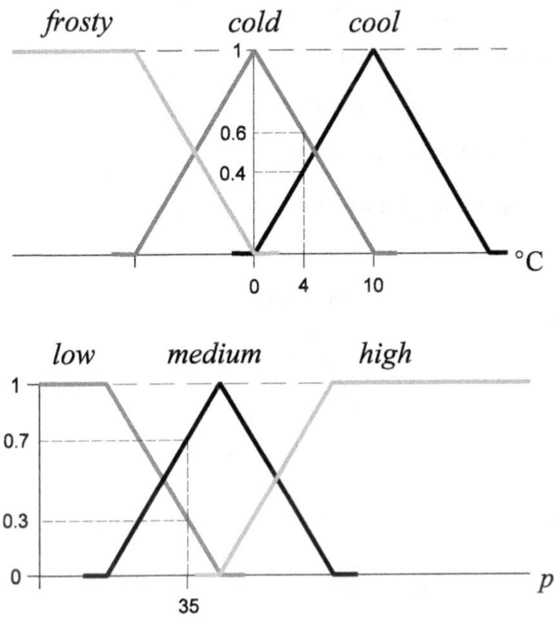

Fig. 5.5 Measured values and their membership degrees in different fuzzy sets of temperatures (top) and prices (bottom)

The fuzzy controller moves on to the execution of the first rule in the base, namely

IF $AT = cool$ AND $FP = low$ THEN $HL = medium$

Step 1: Fuzzification - a transition from measured values to membership degrees. Looking at the above figure, we get

$$cool(4) = 0.4 \quad \text{and} \quad low(35) = 0.3,$$

i.e. it is cool to degree 0.4 and the current fuel price is low to degree 0.3.

Step 2: Inference. One uses the inference mechanism whose computational side is described in (5.58). The key is the degree of fulfilment of the "IF" part, namely

$$cool(4)\, \boldsymbol{t}\, low(35) = 0.4\, \boldsymbol{t}\, 0.3$$

with a t-norm $\boldsymbol{t}$. This degree thus equals

$$0.3 \text{ for } \boldsymbol{t} = \wedge, \quad \text{and } 0.12 \text{ for } \boldsymbol{t} = \boldsymbol{t}_{\mathrm{a}}.$$

According to (5.58), the final result of executing the rule is a fuzzy set $\underline{B}^{(1)}$ in the universe of heating levels, where

$$\underline{B}^{(1)}(h) = (0.4\, \boldsymbol{t}\, 0.3) \rightarrow medium(h).$$

For simplicity, assume that $\boldsymbol{t} = \wedge$ has been chosen, i.e.

$$\underline{B}^{(1)}(h) = 0.3 \rightarrow medium(h).$$

Let us look at $\underline{B}^{(1)}$ with various operators $\rightarrow$.

- $\rightarrow$ = Mamdani operator: $\underline{B}^{(1)}(h) = 0.3 \wedge medium(h)$.

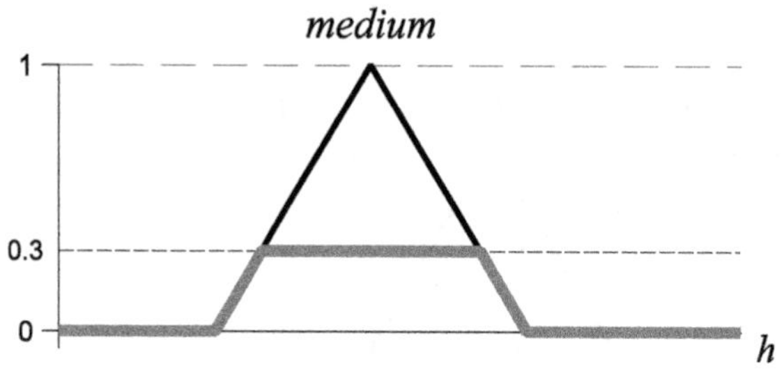

Fig. 5.6 $\underline{B}^{(1)}$ for Mamdani operator

- $\rightarrow$ = Larsen operator: $\underline{B}^{(1)}(h) = 0.3 \cdot medium(h)$.

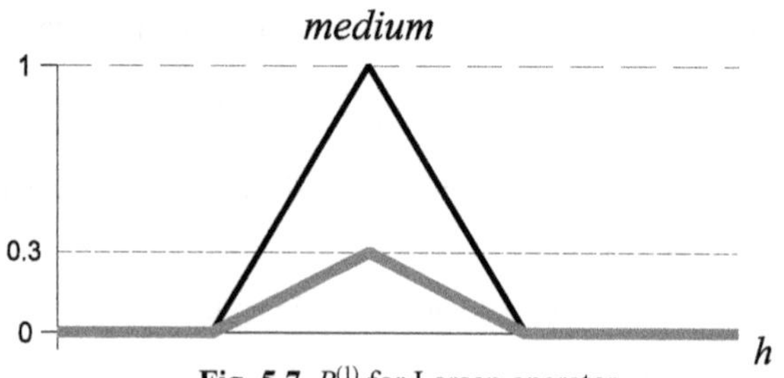

Fig. 5.7 $\underline{B}^{(1)}$ for Larsen operator

- $\rightarrow$ = Gődel implication operator: $\underline{B}^{(1)}(h) = (1, \text{if } medium(h) \geq 0.3, \text{else } medium(h))$.

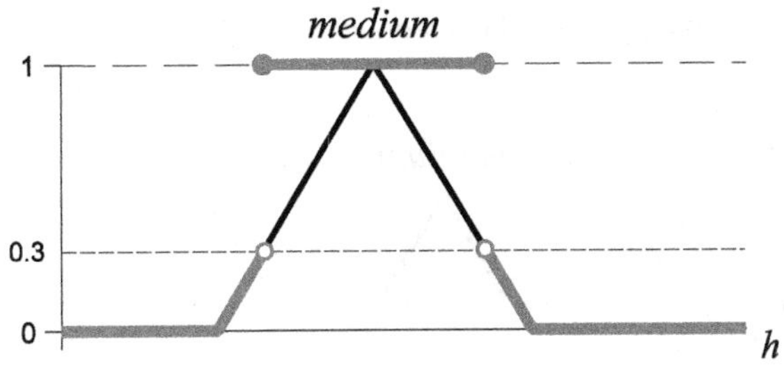

Fig. 5.8 $\underline{B}^{(1)}$ for Gődel implication operator from Subsection 2.4.5

- $\rightarrow$ = Goguen implication operator: $\underline{B}^{(1)}(h) = 1 \wedge medium(h)/0.3$.

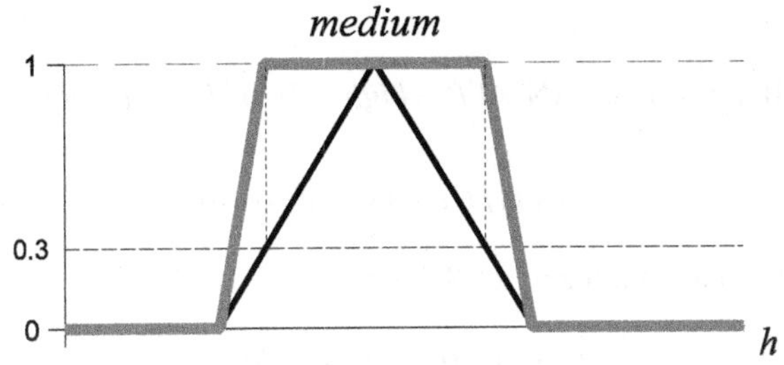

Fig. 5.9 $\underline{B}^{(1)}$ for Goguen implication operator

The fuzzy controller moves on to the next rule with "THEN *HL* = *medium*":

IF *AT* = *cold* AND *FP* = *medium* THEN *HL* = *medium*

Step 1': Fuzzification. Now (see Figure 5.5)

$$cool(4) = 0.6 \quad \text{and} \quad medium(35) = 0.7.$$

Step 2': Inference. Again, choose $t = \wedge$. The degree of fulfilment of the "IF" part is

$$cold(4) \wedge medium(35) = 0.6 \wedge 0.7 = 0.6.$$

The result of executing the second rule is thus a fuzzy set $\underline{B}^{(2)}$ with

$$\underline{B}^{(2)}(h) = 0.6 \rightarrow medium(h).$$

For instance, we thus have (see Figure 5.10)

- $\rightarrow$ = Mamdani operator: $\underline{B}^{(2)}(h) = 0.6 \wedge medium(h)$,
- $\rightarrow$ = Larsen operator: $\underline{B}^{(2)}(h) = 0.6 \cdot medium(h)$.

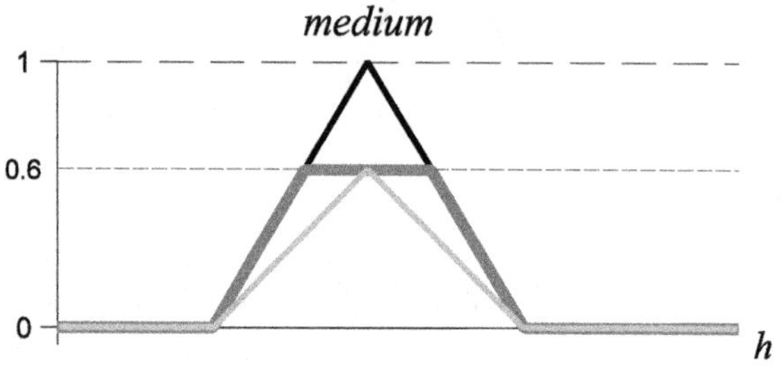

Fig. 5.10 $\underline{B}^{(2)}$ for Mamdani (dark grey line) and Larsen (light grey) operator

Moving on to the rule

$$\text{IF } AT = frosty \text{ AND } FP = high \text{ THEN } HL = medium$$

one gets

$$frosty(4)\, \boldsymbol{t}\, high(35) = 0\, \boldsymbol{t}\, 0 = 0$$

for each $\boldsymbol{t}$. So, it produces a fuzzy set $\underline{B}^{(3)}$ with

$$\underline{B}^{(3)}(h) = 0 \rightarrow medium(h),$$

i.e. $\underline{B}^{(3)} = 1_{\varnothing}$ for each engineering implication operator $\rightarrow$.

The results of the three rules with "THEN $HL = medium$" are then aggregated into one fuzzy set. One assumes that the rules in a rule base are connected by OR. So, the aggregation of $\underline{B}^{(1)}$, $\underline{B}^{(2)}$ and $\underline{B}^{(3)}$ is the sum

$$\underline{B}^{(1)} \cup_{s} \underline{B}^{(2)} \cup_{s} \underline{B}^{(3)} \tag{5.63}$$

with a t-conorm $\boldsymbol{s}$. The most frequently used t-conorm is here $\boldsymbol{s} = \vee$. Let us focus on the case of using $\boldsymbol{t} = \wedge$ and Mamdani operator, which is fundamental for practice. The sum (5.63) then collapses to (see Figures 5.6 and 5.10)

$$\underline{B}^{(1)} \cup \underline{B}^{(2)} \cup \underline{B}^{(3)} = \underline{B}^{(2)}.$$

Exactly the same way of doing will be applied to the rules containing "THEN $HL = low$" and those with "THEN $HL = high$". The reader can easily check that aggregating all the resulting fuzzy sets one obtains a final fuzzy set H illustrated below ($\boldsymbol{s} = \vee$).

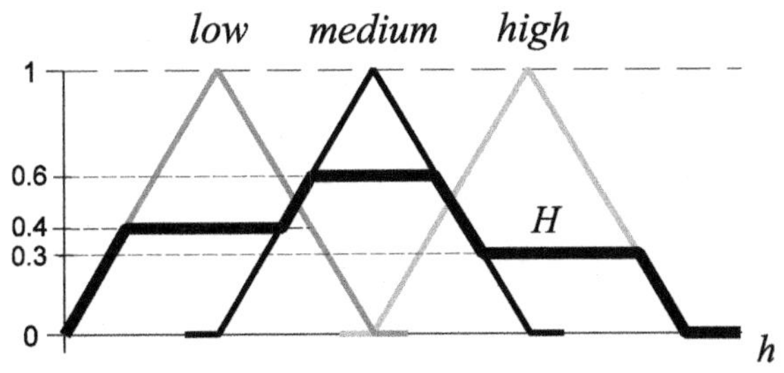

Fig. 5.11 Aggregated result of execution of all the rules

Step 3: Defuzzification – returning from membership degrees to crisp values. According to Figure 5.11, each value of the output parameter h (each setting of the heating valve, in other words) fulfils to a degree all the rules treated *en bloc*. So, different values of h are satisfactory to different degrees $H(h)$. Our fuzzy controller has to choose one of them, h^*, which is the best or most suitable in a sense. This defuzzification can be performed in various ways. For instance, one can take as h^* the smallest or the greatest h for which H attains its maximum. Alternatively, h^* can be defined as the arithmetic mean of those two values. An especially important and frequently used variant of defuzzification is however the COG method from (2.121) and (2.122). Then

$$h^* = \mathrm{COG}_H, \tag{5.64}$$

i.e. h^* becomes the first coordinate of the center of gravity of the figure determined by H and the h-axis.

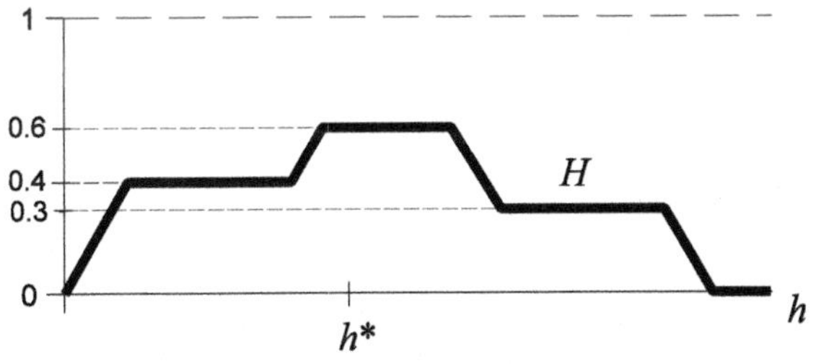

Fig. 5.12 Result of defuzzification by the COG method

h^* is now a compromise, a point of balance taking into account various possible values of h which are satisfactory to different degrees.

Chapter 6

Modeling Incompletely Known Fuzzy Sets

We like to discuss in this chapter a more sophisticated situation when imprecision of information is combined with ignorance. Speaking more precisely, we will deal with incompletely known fuzzy sets, i.e. fuzzy sets with incompletely known membership degrees. Two convenient and effective tools for their modeling will be presented: interval-valued fuzzy sets and I-fuzzy sets, Atanassov's intuitionistic fuzzy sets. Although these constructions are formally equivalent, they form in practice two essentially different approaches to the issue.

6.1 Incompletely Known Sets and Their Modeling

Trying to deal with incompletely known fuzzy sets it is instructive and useful to look first at a simpler case, namely that of incompletely known sets.

Imagine that our information, knowledge about an ordinary set $\boldsymbol{A} \subset \boldsymbol{U}$ is incomplete. The status of some elements $x \in \boldsymbol{U}$ (membership/nonmembership in $\boldsymbol{A}$) is then practically unknown or uncertain. This leads to a natural division of $\boldsymbol{U}$ into three disjoint classes (see Figure 6.1):

A^+ - composed of those x's whose belonging to $\boldsymbol{A}$ is known, sure;

A^- - containing the x's which surely do not belong to $\boldsymbol{A}$;

$A^?$ - the *uncertainty area* of $\boldsymbol{A}$, composed of all x's with unknown or uncertain status.

That an x from $\boldsymbol{A}^?$ belongs to $\boldsymbol{A}$ is thus possible, but not sure. Let us look at a familiar example. Agatha cleans up the fridge trying to leave exactly those food products which are not past the sell-by date (A). This task is not so easy in practice. Besides products which are surely past their sell-by dates (A^-) or are surely not (A^+), Agatha finds products whose sell-by date is illegible ($A^?$) and, then, she does not know what to do: to leave the product or to throw it away.

M. Wygralak: *Intelligent Counting Under Information Imprecision*, STUDFUZZ 292, pp. 139–160.
DOI: 10.1007/978-3-642-34685-9_6

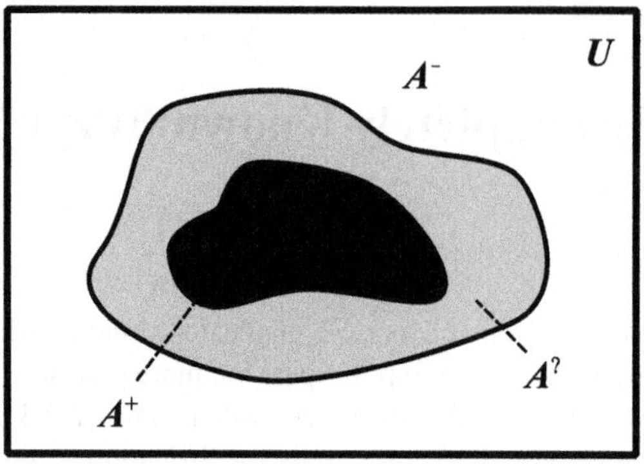

Fig. 6.1 Division of $\boldsymbol{U}$ implied by incompleteness of knowledge about $\boldsymbol{A}$

The above figure suggests that generally

$$A^{+} \subset A \subset (A^{-})' \quad \text{and, hence,} \quad A^{-} \subset A' \subset (A^{+})'. \tag{6.1}$$

A thus lies somewhere between A^{+} and $(A^{-})'$. Moreover, A^{+} is a lower bound on A, A^{-} forms a lower bound on A', and we have

$$A^{+} \cup A^{?} = (A^{-})' \tag{6.2}$$

and

$$A^{?} = (A^{+} \cup A^{-})'. \tag{6.3}$$

Each of the three sets A^{+}, A^{-} and $A^{?}$ is uniquely determined by the remaining two. Our incompleteness of knowledge about A can thus be modeled by specifying any of the pairs

$$(A^{+}, A^{?}),\ (A^{-}, A^{?}),\ (A^{+}, A^{-}).$$

Let us look a bit closer at these three options of modeling.

• **Option 1:** specify $(A^{+}, A^{?})$. It is convenient to rewrite this representation as $(A^{+}, A^{+} \cup A^{?})$. Speaking more formally and generally, the incompleteness of knowledge about A is now expressed by means of a pair

$$(B, C) \text{ with } B \subset C, \tag{6.4}$$

where B is a set containing sure elements of A, and C represents sure elements together with those having unknown or uncertain status. The uncertainty area becomes the set-theoretic difference $C \setminus B$, i.e.

$$A^{?} = B' \cap C. \tag{6.5}$$

It is worth noticing that each pair $(\boldsymbol{B}, \boldsymbol{C})$ from (6.4) can be equivalently presented as a fuzzy set A: $\boldsymbol{U} \to \{0, 0.5, 1\}$ with

$$A(x) = \begin{cases} 1, & \text{if } x \in \boldsymbol{B}, \\ 0.5, & \text{if } x \in \boldsymbol{C} \setminus \boldsymbol{B}, \\ 0, & \text{otherwise.} \end{cases} \tag{6.6}$$

Pairs (6.4) as a tool for representing incompletely known sets were introduced by GENTILHOMME (1968) using the name *flou sets* and, independently, by KLAUA (1968, 1969) as *partial sets* (see also NEGOITA/RALESCU (1975), GOTTWALD (1984), and WYGRALAK (1996a)). Equality and inclusion of two flou vel partial sets, their sum, intersection, cartesian product and complement are defined as follows:

$$(\boldsymbol{B}, \boldsymbol{C}) = (\boldsymbol{D}, \boldsymbol{E}) \Leftrightarrow \boldsymbol{B} = \boldsymbol{D} \;\&\; \boldsymbol{C} = \boldsymbol{E}, \tag{6.7}$$

$$(\boldsymbol{B}, \boldsymbol{C}) \subset (\boldsymbol{D}, \boldsymbol{E}) \Leftrightarrow \boldsymbol{B} \subset \boldsymbol{D} \;\&\; \boldsymbol{C} \subset \boldsymbol{E}, \tag{6.8}$$

$$(\boldsymbol{B}, \boldsymbol{C}) * (\boldsymbol{D}, \boldsymbol{E}) = (\boldsymbol{B} * \boldsymbol{D}, \boldsymbol{C} * \boldsymbol{E}) \text{ for } * \in \{\cup, \cap, \times\}, \tag{6.9}$$

$$(\boldsymbol{B}, \boldsymbol{C})' = (\boldsymbol{C}', \boldsymbol{B}'). \tag{6.10}$$

Let us mention and emphasize that rough sets from PAWLAK (1982) are also constructions collapsing in essence to (6.4). Recollect that a *rough set* or, in other words, a *Pawlak set* $\Re_A$ is a pair

$$\Re_A = (\underline{\boldsymbol{A}}, \bar{\boldsymbol{A}}) \tag{6.11}$$

of approximating sets $\underline{\boldsymbol{A}}$ and $\bar{\boldsymbol{A}}$ in which

$$\underline{\boldsymbol{A}} = \{x \in \boldsymbol{U}\colon [x]_\rho \subset \boldsymbol{A}\} \tag{6.12}$$

is a *lower approximation* of $\boldsymbol{A}$, whereas

$$\bar{\boldsymbol{A}} = \{x \in \boldsymbol{U}\colon [x]_\rho \cap \boldsymbol{A} \neq \varnothing\} \tag{6.13}$$

forms its *upper approximation.* As usual, $[x]_\rho$ denotes the equivalence class of x with respect to a fixed equivalence relation ρ in $\boldsymbol{U}$. That relation is treated as an *indiscernibility relation* of elements described in terms of given attributes and their values (e.g. the equivalence relation of individuals being of the same age, height and weight). The uncertainty area (6.5),

$$\underline{\boldsymbol{A}}' \cap \bar{\boldsymbol{A}}, \tag{6.14}$$

is then called the *boundary* of $\boldsymbol{A}$ with respect to ρ. Contrary to flou sets with (6.9), sums and intersections of rough sets cannot be constructed by performing suitable set-theoretic operations on the components (see however WYGRALAK (1989)). For details on rough sets and their applications the reader is referred to PAWLAK (1991), SŁOWIŃSKI (1992) and SŁOWIŃSKI *et al.* (2005).

A noteworthy generalization of the representation from (6.4) is the concept of a *twofold fuzzy set*

$$T_A = (S_A, P_A) \text{ with } S_A \subset (P_A)_1 \tag{6.15}$$

introduced in DUBOIS/PRADE (1987a). S_A is a fuzzy set of more or less sure elements of A, whereas P_A forms a fuzzy set of its more or less possible elements. The inclusion in (6.15), implying $S_A \subset P_A$, reflects a simple intuition saying that an element being sure to a positive degree must be fully possible. The corresponding uncertainty area

$$S_A' \cap P_A \tag{6.16}$$

is interpreted as a fuzzy set of *dubious elements* of $\boldsymbol{A}$.

• **Option 2:** specify $(\boldsymbol{A}^-, \boldsymbol{A}^?)$. It seems that this form of representation, dual in a way to $(\boldsymbol{A}^+, \boldsymbol{A}^?)$ and focusing on negative information, is not used in the subject literature. Formally, it can be rewritten as (6.4), too, but the semantics of the components is then different: $\boldsymbol{B}$ becomes a set of impossible elements of $\boldsymbol{A}$, and $\boldsymbol{C}$ contains impossible and uncertain elements.

• **Option 3:** specify $(\boldsymbol{A}^+, \boldsymbol{A}^-)$. This variant of modeling seems to be much more interesting. It is *bipolar* by putting emphasis on *positive* $(\boldsymbol{A}^+)$ and *negative* $(\boldsymbol{A}^-)$ information, on positive and negative examples of elements of $\boldsymbol{A}$, in other words. This way of representing incomplete knowledge has been materialized by the idea of a sub-definite set introduced in NARIN'YANI (1980). In a more formalized notation, incomplete knowledge about $\boldsymbol{A}$ is then represented by a pair (cf. (6.1))

$$(\boldsymbol{B}, \boldsymbol{D}) \text{ with } \boldsymbol{B} \subset \boldsymbol{D}'. \tag{6.17}$$

Similarly to (6.4), $\boldsymbol{B}$ is a set of sure elements of $\boldsymbol{A}$, but $\boldsymbol{D}$ contains sure elements of the complement $\boldsymbol{A}'$. $\boldsymbol{B}$ and $\boldsymbol{D}$ are thus lower bounds on $\boldsymbol{A}$ and $\boldsymbol{A}'$, respectively, and

$$\boldsymbol{B} \subset \boldsymbol{A} \subset \boldsymbol{D}', \text{ i.e. } \boldsymbol{D} \subset \boldsymbol{A}' \subset \boldsymbol{B}'. \tag{6.18}$$

The set

$$\boldsymbol{A}^? = \boldsymbol{B}' \cap \boldsymbol{D}', \tag{6.19}$$

the difference $\boldsymbol{D}' \setminus \boldsymbol{B}$, becomes the uncertainty area of $\boldsymbol{A}$.

6.2 Interval-Valued Fuzzy Sets

If one likes to model fuzzy sets with incompletely known membership degrees, a reasonable and natural way of doing is to pattern oneself upon the representation of incompletely known sets from Section 6.1 with special reference to (6.4) and (6.17). And this idea finds its materialization in the concepts of interval-valued fuzzy sets and I-fuzzy sets. Beginning with interval-valued fuzzy sets, we will present those two concepts drawing special attention to uncertainty areas.

6.2.1 The Concept of an Interval-Valued Fuzzy Set

An *interval-valued fuzzy set* (*IVFS*, in short) is a pair $\mathcal{E} = (A_l, A_u)$ of fuzzy sets A_l, A_u: $\boldsymbol{U} \to [0, 1]$ with

$$A_l \subset A_u. \tag{6.20}$$

One says that $\mathcal{E}$ is *finite* whenever A_u is finite. What we deal with is thus an extension of (6.4) to fuzzy sets. A_l and A_u, respectively, form a lower bound and an upper bound of an incompletely known fuzzy set A, i.e. $A(x) \in [A_l(x), A_u(x)]$. And this interpretation will be primary for our discussion. Worth mentioning is however a slightly different interpretation in the language of tolerance in which $[A_l(x), A_u(x)]$ is an interval of equally acceptable membership degrees of x in A. Anyway, $\mathcal{E}$ can be identified with a vague collection of elements whose membership degrees are closed subintervals of $[0, 1]$, $\mathcal{E}(x) = [A_l(x), A_u(x)]$ (see also Section 11.5 and the notion of a vaguely defined object). It is clear that each fuzzy set B can be presented in the language of IVFSs as (B, B). So, $\mathcal{E}_\varnothing = (1_\varnothing, 1_\varnothing)$ and $\mathcal{E}_U = (1_U, 1_U)$, respectively, represent in that language the empty set and the whole universe $\boldsymbol{U}$.

For a t-norm $\boldsymbol{t}$ and a strong negation ν, let us define (WYGRALAK (2007))

$$U_{\mathcal{E}} = A_l^{\nu} \cap_t A_u, \tag{6.21}$$

the *area of uncertainty* of $\mathcal{E}$ (cf. (6.5)). We thus have

$$U_{\mathcal{E}}(x) = \nu(A_l(x))\, \boldsymbol{t}\, A_u(x) \tag{6.22}$$

which is the *degree of uncertainty* concerning x. By (2.68), $U_{\mathcal{E}}(x) = 1$ iff $A_l(x) = 0$ and $A_u(x) = 1$. Let us look at the following three simplest and probably most important particular cases of $U_{\mathcal{E}}(x)$.

- $$U_{\mathcal{E}}(x) = A_u(x) - A_l(x) \tag{6.23}$$

  for $\boldsymbol{t} = \boldsymbol{t}_Ł$ and $\nu = \nu_Ł$. So, $U_{\mathcal{E}}(x)$ becomes the length of the interval $[A_l(x), A_u(x)]$, the size of ignorance as to $A(x)$.

- $$U_{\mathcal{E}}(x) = (1 - A_l(x)) \wedge A_u(x) = A_u(x) - A_l(x) + A_l(x) \wedge (1 - A_u(x)) \tag{6.24}$$

  if $\boldsymbol{t} = \wedge$ and $\nu = \nu_Ł$. The uncertainty degree $U_{\mathcal{E}}(x)$ is now the sum of two values: the size of ignorance, $A_u(x) - A_l(x)$, and the minimum possible fuzziness index of $A(x)$, $A_l(x) \wedge (1 - A_u(x))$, i.e. the minimum possible entropy measure of fuzziness of the singleton $A(x)/x$ (see also Subsection 6.3.4). Indeed,

$$\varphi(a) = a \wedge (1 - a) \tag{6.25}$$

is a fuzziness index of $a \in [0, 1]$ (cf. (2.50)). It is also an entropy measure of fuzziness of a/x in the sense of Definition 2.3. The minimum possible fuzziness index of $A(x) \in [A_l(x), A_u(x)]$ is thus

$$\varphi(A_l(x)) \wedge \varphi(A_u(x)) = A_l(x) \wedge (1 - A_u(x)). \tag{6.26}$$

- $$U_{\mathcal{E}}(x) = (1 - A_l(x)) \cdot A_u(x) = A_u(x) - A_l(x) + A_l(x) \cdot (1 - A_u(x)) \tag{6.27}$$

  for $\boldsymbol{t} = \boldsymbol{t}_a$ and $\nu = \nu_Ł$. Again, what we get as $U_{\mathcal{E}}(x)$ is the sum of two values. The first one is the size of ignorance as to $A(x)$. Let us look at the product. We see that

  $$\varphi(a) = a \cdot (1 - a) \quad \text{with } a \in [0, 1] \tag{6.28}$$

  is a fuzziness index of a, and

  $$A_l(x) \cdot (1 - A_u(x)) \leq A_l(x) \cdot (1 - A_l(x)) \wedge A_u(x) \cdot (1 - A_u(x)). \tag{6.29}$$

  Consequently, $A_l(x) \cdot (1 - A_u(x))$ forms a lower evaluation of the minimum possible fuzziness index (6.28) of $A(x) \in [A_l(x), A_u(x)]$.

The approach to uncertainty areas of IVFSs offered by (6.21) is based on the idea of replacing the difference $\boldsymbol{C} \setminus \boldsymbol{B}$ in (6.5) by the difference $A_u \setminus A_l = A_u \cap A_l'$ and, then, $A_u \cap A_l'$ is generalized to $A_u \cap_t A_l^{\nu}$. A potential alternative seems to be to use the bounded difference, which leads to

$$U_{\mathcal{E}} = A_u \ominus A_l, \tag{6.30}$$

i.e.

$$U_{\mathcal{E}}(x) = A_u(x) - A_l(x).$$

So, this variant of doing is less flexible and collapses to (6.22) with $\boldsymbol{t} = \boldsymbol{t}_Ł$ and $\nu = \nu_Ł$.

6.2.2 General Properties of Uncertainty Degrees

The examples of uncertainty degrees presented in the previous subsection suggest that one should look at those degrees and their interpretation in a more general way taking into account at least two important classes of t-norms. We mean the class of nilpotent t-norms and, on the other hand, that of t-norms having no zero divisors (see Subsection 3.2.1).

First, we will assume that $\boldsymbol{t}$ in (6.21) is nilpotent and the negation $\nu = \nu_t$ induced by $\boldsymbol{t}$ is involved. For this case, the following theorem presents a general form of the uncertainty area of an IVFS.

Theorem 6.1. *Let $\mathcal{E} = (A_l, A_u)$ be an IVFS. Assume $\boldsymbol{t}$ is nilpotent, $\nu = \nu_t$, and h is the normed generator of the complementary t-conorm $\boldsymbol{t}°$. For each $x \in \boldsymbol{U}$, we then have:*

(a) $U_{\mathcal{E}}(x) = h^{-1}(h(A_u(x)) - h(A_l(x)))$,

(b) $A_l(x)\, \boldsymbol{t}°\, U_{\mathcal{E}}(x) = A_u(x)$.

Indeed, as to (a), if $\boldsymbol{t}$ is nilpotent and g is its normed generator, then $\nu(a) = \nu_t(a) = g^{-1}(1 - g(a))$, whereas $h(a) = 1 - g(a)$ becomes the normed generator of $\boldsymbol{t}^\circ$ (see Theorem 3.12(c) and (3.17)). So, $h^{-1}(y) = g^{-1}(1 - y)$. By (6.22) and Theorem 3.8(a), we thus get

$$\begin{aligned} U_{\mathcal{E}}(x) &= \nu(A_l(x))\, \boldsymbol{t}\, A_u(x) \\ &= g^{-1}(1 \wedge (g(\nu(A_l(x))) + g(A_u(x)))) \\ &= g^{-1}(1 - g(A_l(x)) + g(A_u(x))) \\ &= h^{-1}(h(A_u(x)) - h(A_l(x))). \end{aligned}$$

Using (a) and Theorem 3.8(b), it is a routine matter to show (b).

What Theorem 6.1(a) says is that the uncertainty degree $U_{\mathcal{E}}(x)$ with a nilpotent $\boldsymbol{t}$ is always isomorphic to $A_u(x) - A_l(x)$, the length of the interval $[A_l(x), A_u(x)]$. For instance, if $\boldsymbol{t} = \boldsymbol{t}_{\mathrm{S},p}$ with $p > 0$ (see Example 2.6), then the complementary t-conorm is $\boldsymbol{t}^\circ = \boldsymbol{s}_{\mathrm{Y},p}$, $h(a) = a^p$, and

$$U_{\mathcal{E}}(x) = [(A_u(x))^p - (A_l(x))^p]^{1/p}. \tag{6.31}$$

By (b), aggregating the left endpoint $A_l(x)$ and $U_{\mathcal{E}}(x)$ via $\boldsymbol{t}^\circ$, one gets $A_u(x)$. $U_{\mathcal{E}}(x)$ is thus (a sort of) a "pure" size of ignorance as to $A(x)$. By (a), $h(A_l(x)) + h(U_{\mathcal{E}}(x))$ is equal to $h(A_u(x))$. Using $\boldsymbol{t} = \boldsymbol{t}_{\text{Ł}}$, (a) and (b) collapse to the standard relationship $A_l(x) + U_{\mathcal{E}}(x) = A_u(x)$. Another consequence of Theorem 6.1(a) is that

$$\begin{aligned} &U_{\mathcal{E}}(x) = 0 \Leftrightarrow A_l(x) = A_u(x), \\ \text{i.e.} \quad &U_{\mathcal{E}} = 1_{\varnothing} \Leftrightarrow \mathcal{E} \text{ collapses to a fuzzy set.} \end{aligned} \tag{6.32}$$

The situation becomes more sophisticated whenever we use in (6.21) a strong negation ν together with a t-norm $\boldsymbol{t}$ having no zero divisors, say, a strict $\boldsymbol{t}$ or $\boldsymbol{t} = \wedge$. This time $U_{\mathcal{E}}(x)$ seems to be the size of ignorance as to $A(x)$ combined with a fuzziness index of $A(x)$. Indeed, we then get

$$\begin{aligned} &U_{\mathcal{E}}(x) = 0 \Leftrightarrow (A_l(x), A_u(x)) \in \{(0,0), (1,1)\}, \\ \text{i.e.} \quad &U_{\mathcal{E}} = 1_{\varnothing} \Leftrightarrow \mathcal{E} \text{ collapses to a set.} \end{aligned} \tag{6.33}$$

If our knowledge about A is complete, $\mathcal{E}$ becomes an ordinary fuzzy set, $\mathcal{E} = (A, A)$, and (6.22) gives

$$U_{\mathcal{E}}(x) = A(x)\, \boldsymbol{t}\, \nu(A(x)), \tag{6.34}$$

which is a $\boldsymbol{t}$-based fuzziness index of $A(x)$ (cf. (6.25) and (6.28)).

6.2.3 Operations on IVFSs

Let $\mathcal{E} = (A_l, A_u)$ with $A_l \subset A_u$ and $\mathcal{F} = (B_l, B_u)$ with $B_l \subset B_u$ be two IVFSs. As previously, let ν denote a strong negation, whereas $\boldsymbol{T}$ and $\boldsymbol{S}$ will denote a t-norm and a t-conorm, respectively. Basic relationships between and operations on IVFSs are defined as follows:

(*equality*)	$\mathcal{E} = \mathcal{F} \Leftrightarrow A_l = B_l \;\&\; A_u = B_u,$	(6.35)
(*inclusion*)	$\mathcal{E} \subset \mathcal{F} \Leftrightarrow A_l \subset B_l \;\&\; A_u \subset B_u,$	(6.36)
(*intersection*)	$\mathcal{E} \cap_{\boldsymbol{T}} \mathcal{F} = (A_l \cap_{\boldsymbol{T}} B_l, A_u \cap_{\boldsymbol{T}} B_u),$	(6.37)
(*sum*)	$\mathcal{E} \cup_{\boldsymbol{S}} \mathcal{F} = (A_l \cup_{\boldsymbol{S}} B_l, A_u \cup_{\boldsymbol{S}} B_u),$	(6.38)
(*cartesian product*)	$\mathcal{E} \times_{\boldsymbol{T}} \mathcal{F} = (A_l \times_{\boldsymbol{T}} B_l, A_u \times_{\boldsymbol{T}} B_u),$	(6.39)
(*complement*)	$\mathcal{E}^{\nu} = (A_u^{\nu}, A_l^{\nu}).$	(6.40)

As the choice of symbols suggests, ν is the negation used in (6.21), while the t-norm $\boldsymbol{T}$ possibly differs from that in (6.21). We will say that (6.37)-(6.40) are operations *induced* by $\boldsymbol{T}$, $\boldsymbol{S}$ and ν, respectively. Similarly to usual fuzzy sets, if $\boldsymbol{T} = \wedge$, $\boldsymbol{S} = \vee$ and $\nu = \nu_Ł$, a simplified notation will be used, namely $\mathcal{E} \cap \mathcal{F}$, $\mathcal{E} \cup \mathcal{F}$, $\mathcal{E} \times \mathcal{F}$, and $\mathcal{E}'$. This choice of t-operations and negation together with $\boldsymbol{t} = \boldsymbol{t}_Ł$ in (6.21) gives the standard formulation of operations on IVFSs. It seems to be basic for applications. $\mathcal{E}$ and $\mathcal{F}$ are said to be *disjoint* whenever their intersection is empty, i.e. $\mathcal{E} \cap \mathcal{F} = \mathcal{E}_{\varnothing}$.

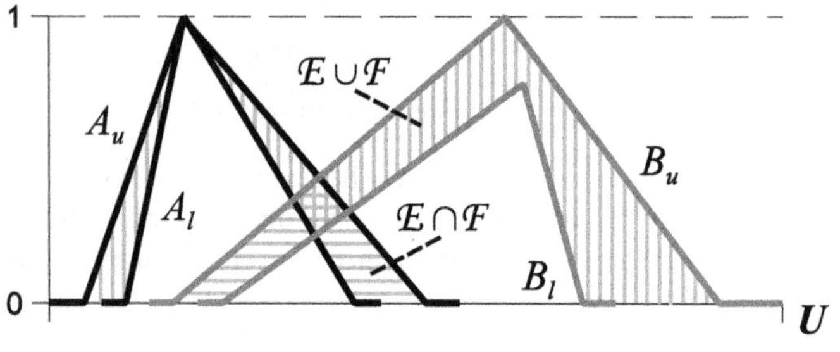

Fig. 6.2 Example of $\mathcal{E} \cap \mathcal{F}$ and $\mathcal{E} \cup \mathcal{F}$

Properties of operations on IVFSs are direct consequences of properties of operations on fuzzy sets and, therefore, we skip their explicit formulation. Let us mention only De Morgan laws:

$$(\mathcal{E} \cap_{\boldsymbol{T}} \mathcal{F})^{\nu} = \mathcal{E}^{\nu} \cup_{\boldsymbol{S}} \mathcal{F}^{\nu} \quad \text{and} \quad (\mathcal{E} \cup_{\boldsymbol{S}} \mathcal{F})^{\nu} = \mathcal{E}^{\nu} \cap_{\boldsymbol{T}} \mathcal{F}^{\nu}, \tag{6.41}$$

which are satisfied whenever $\boldsymbol{T}$ and $\boldsymbol{S}$ are ν-dual, $\boldsymbol{S} = \boldsymbol{T}^{\nu}$.

Finally, we like to outline a more general and sophisticated approach to operations on IVFSs. Then

$$(\mathcal{E} \cap_{\boldsymbol{T}} \mathcal{F})(x) = \mathcal{E}(x)\, \boldsymbol{T}\, \mathcal{F}(x)$$

and

$$(\mathcal{E} \cup_{\boldsymbol{S}} \mathcal{F})(x) = \mathcal{E}(x)\, \boldsymbol{S}\, \mathcal{F}(x) \tag{6.42}$$

for each element $x \in \boldsymbol{U}$. $\boldsymbol{T}$ and $\boldsymbol{S}$ are here understood as an *interval-valued t-norm and t-conorm*, respectively. One defines them by means of axioms (T1)-(T5) from Subsection 2.4.3 in which, however, numbers $a, b, c, d \in [0, 1]$ are replaced by arbitrary closed subintervals of $[0, 1]$. Moreover, the one-element intervals $[1, 1]$ and $[0, 0]$ are used as neutral elements in (T4) and (T5) instead of 1 and 0, respectively. Inequalities in (T3) are treated as a partial order relation defined by (2.118). An analogous way of doing applied to the definition of negation leads to *interval-valued negations*.

It is a routine task to show that if $\boldsymbol{t}_1 \le \boldsymbol{t}_2$ and $\boldsymbol{s}_1 \le \boldsymbol{s}_2$ are two (ordinary) t-norms and t-conorms, respectively, then $\boldsymbol{T}$ such that

$$[a, b]\, \boldsymbol{T}\, [c, d] = [a\boldsymbol{t}_1 c, b\boldsymbol{t}_2 d] \tag{6.43}$$

forms an interval-valued t-norm, whereas $\boldsymbol{S}$ given by

$$[a, b]\, \boldsymbol{S}\, [c, d] = [a\boldsymbol{s}_1 c, b\boldsymbol{s}_2 d] \tag{6.44}$$

is an interval-valued t-conorm. $\boldsymbol{T}$ and $\boldsymbol{S}$ are then said to be *representable*. Speaking more precisely, they are representable via pairs $(\boldsymbol{t}_1, \boldsymbol{t}_2)$ and $(\boldsymbol{s}_1, \boldsymbol{s}_2)$ of usual t-norms and t-conorms, respectively. What one actually uses in (6.37)-(6.38) is thus a particular case with $\boldsymbol{t}_1 = \boldsymbol{t}_2$ and $\boldsymbol{s}_1 = \boldsymbol{s}_2$.

It is worth emphasizing that there exist *non-representable* interval-valued t-norms and t-conorms. Simple examples are

$$\begin{aligned}
&[a, b]\, \boldsymbol{T}\, [c, d] = [a \wedge c, (a \wedge d) \vee (b \wedge c)],\\
&[a, b]\, \boldsymbol{S}\, [c, d] = [(a \vee d) \wedge (b \vee c), b \vee d],\\
&[a, b]\, \boldsymbol{T}\, [c, d] = [0 \vee (a + c - (1 - b)(1 - d) - 1), 0 \vee (b + d - 1)],\\
&[a, b]\, \boldsymbol{S}\, [c, d] = [1 \wedge (a + c), 1 \wedge (b + d + ac)].
\end{aligned} \tag{6.45}$$

The reader interested in further aspects of IVFSs and their operations and, especially, in further details about interval-valued triangular norms and conorms is referred e.g. to BUSTINCE *et al.* (2008), CORNELIS *et al.* (2004), DESCHRIJVER (2006a, b), DESCHRIJVER *et al.* (2004), DESCHRIJVER/KRÁL (2007), VLACHOS/SERGIADIS (2007a); see also Sections 8.4 and 9.5, and Subsection 9.3.2.

6.3 I-Fuzzy Sets

We move on to I-fuzzy sets, another tool for the modeling of incompletely known fuzzy sets. In comparison with interval-valued fuzzy sets, it offers a different, bipolar optics of looking at information. The reader is also referred to Section 1.2 for terminological remarks on I-fuzzy sets.

6.3.1 Basic Notions

By an *I-fuzzy set* (*IFS*, in short) one means a pair $\mathscr{E} = (A^+, A^-)$ of fuzzy sets A^+, A^-: $U \to [0, 1]$ such that

$$A^+ \subset (A^-)^\nu, \tag{6.46}$$

where ν denotes a strong negation. This construction thus forms a direct extension of (6.17) to fuzzy sets. $\mathscr{E}$ is said to be *finite* whenever so is $(A^-)^\nu$. (B, B^ν) represents an ordinary fuzzy set B in the language of IFSs.

By (6.46), IFSs and IVFSs from the previous section are formally equivalent concepts: (A^+, A^-) is an IFS iff $(A^+, (A^-)^\nu)$ is an IVFS. The results from Section 6.2 can thus be transferred to IFSs, and we will take advantage of this fact when dealing with IFSs. Again, $\mathscr{E}$ models an incompletely known fuzzy set A. The concepts of IVFSs and IFSs, however, are practically different because, unlike IVFSs, $\mathscr{E}$ forms a bipolar representation by using and putting emphasis on positive (A^+) and negative (A^-) information, on imprecisely specified positive and negative examples of elements from A. In essence, $\mathscr{E}$ models incomplete knowledge about A and, on the other hand, about its complement A^ν which are such that (cf. (6.18))

$$A^+ \subset A \subset (A^-)^\nu \quad \text{and, hence,} \quad A^- \subset A^\nu \subset (A^+)^\nu. \tag{6.47}$$

A^+ thus forms a lower bound on A, whereas A^- is a lower bound on A^ν:

$$A(x) \in [A^+(x), \nu(A^-(x))] \quad \text{and} \quad A^\nu(x) \in [A^-(x), \nu(A^+(x))]. \tag{6.48}$$

In the original and commonly used terminology from ATANASSOV (1986, 1999), A^+ is called a *membership function*, while A^- is interpreted as a *nonmembership function*. Consequently, $A^+(x)$ and $A^-(x)$, respectively, form the *membership degree* and the *nonmembership degree* of x in $\mathscr{E}$. $\mathscr{E}$ is viewed as a vague collection of elements whose membership is characterized by means of pairs $(A^+(x), A^-(x))$. So, formally, one can write $\mathscr{E}(x) = (A^+(x), A^-(x))$. Condition (6.46) guarantees that

$$A^+(x) \le e(\nu) \quad \text{or/and} \quad A^-(x) \le e(\nu) \tag{6.49}$$

for each $x \in U$ as $A^+(x), A^-(x) > e(\nu)$ would imply $\nu(A^-(x)) < e(\nu) < A^+(x)$ (see Subsection 2.4.1). In other words, the membership and nonmembership degrees of x must not be "too large" at the same time.

In contrast to IVFSs, IFSs make it possible and even force us to think about and look at given decision alternatives in the language of advantages and disadvantages, positive and negative features, satisfaction and dissatisfaction, trust and distrust, etc. Psychological investigations suggest that such a bipolar optics is very suitable as decision makers have a tendency to focus on positive sides of decisions they consider and, thereby, to forget about negative sides (see also DUBOIS/PRADE (2009)).

Example 6.2. Assume U is a set of $n \geq 2$ alternatives $x_1, \ldots, x_n$. Each of them is evaluated with respect to $k \geq 1$ conditions $cond_1, \ldots, cond_k$ by marking "yes" (fulfils) or "no" (does not fulfil). A familiar example is here the evaluation of holiday offers taking into account conditions such as "affordable price", "nice beach", "good sports facilities", etc. with the aim of finding satisfactory offers. Let us look at two evaluations, say, of alternatives x_4 and x_5:

	x_4: **yes**	**no**	x_5: **yes**	**no**
$cond_1$	✓		✓	
$cond_2$		✓		
$cond_3$	✓			✓
$cond_4$	✓			
$cond_5$		✓		✓
$cond_6$	✓			✓

Our information about x_4 is complete: we are able to evaluate that alternative with respect to all conditions. As to x_5, we deal with incompleteness of information and, therefore, we are not able to evaluate x_5 with respect to $cond_2$ and $cond_4$. Satisfactory alternatives can thus be viewed as an incompletely known fuzzy set B modeled as an I-fuzzy set (B^+, B^-) with $\nu = \nu_Ł$ and

$$B^+(x_4) = \frac{4}{6}, \quad B^-(x_4) = \frac{2}{6}, \quad B^+(x_5) = \frac{1}{6}, \quad B^-(x_5) = \frac{3}{6}.$$

$B^+(x_i)$ forms a degree of satisfaction with x_i, whereas $B^-(x_i)$ is a degree of dissatisfaction.

The above binary evaluation "yes/no" of x_i with respect to $cond_j$ can be replaced by a more subtle evaluation, namely

	x_i: **yes**	**no**
⋮		
$cond_j$	a	b
⋮		

with $a, b \in [0, 1]$ and $a + b \leq 1$. (a, b) forms a numerical answer to the question whether x_i fulfils the jth condition. Basic examples of linguistic interpretations of pairs (a, b) are:

$$(1, 0) \text{ - definitely yes}, \quad (\tfrac{3}{4}, \tfrac{1}{4}) \text{ - rather yes}, \quad (\tfrac{1}{2}, \tfrac{1}{2}) \text{ - fifty-fifty},$$

$$(\tfrac{1}{4}, \tfrac{3}{4}) \text{ - rather no}, \quad (0, 1) \text{ - definitely no}.$$

Let us look at (1/4, 3/4) and (1/4, 0) in the context of the condition "good sports facilities" in the task of finding satisfactory holiday offers. (1/4, 3/4) means that choosing holiday at x_i, say, only a tennis court will be available. (1/4, 0) represents incomplete knowledge. All we know is that a tennis court is available. We have no idea if x_i offers any other sports facilities we are interested in, e.g. a basketball field, a swimming pool, a volleyball field. □

The following figure illustrates the difference between the modeling of imprecision via fuzzy sets, IVFSs and IFSs. It refers to the case of trying to determine the membership degree of a decision alternative x in a fuzzy set B of satisfactory alternatives.

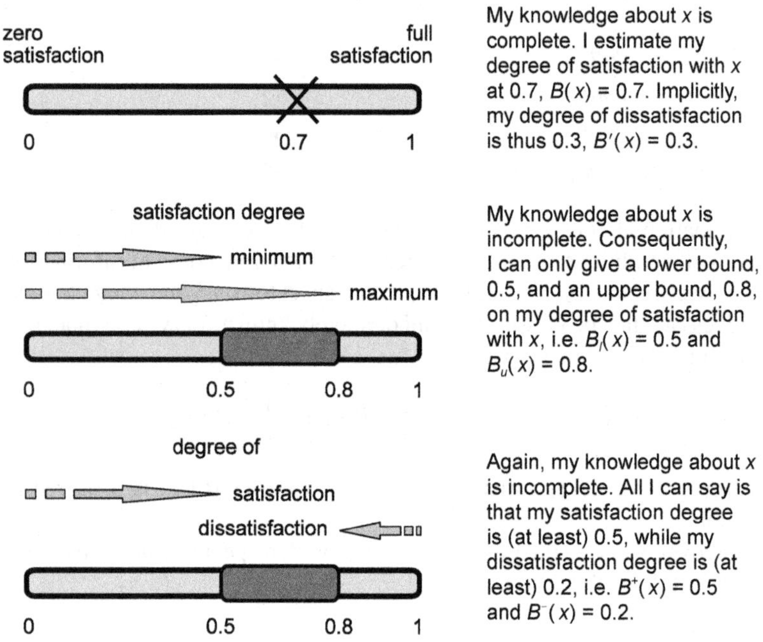

Fig. 6.3 Three variants of membership degree modeling

The bold interval [0.5, 0.8] in Figure 6.3(middle, bottom) represents our uncertainty. Generally, the uncertainty area of an I-fuzzy set $\mathscr{E} = (A^+, A^-)$ with $A^+ \subset (A^-)^\nu$ and a strong negation ν is called its *hesitation area*. We define it as a fuzzy set

$$\chi_{\mathscr{E}} = (A^+)^\nu \cap_t (A^-)^\nu = (A^+ \cup_{t^\nu} A^-)^\nu \tag{6.50}$$

with a t-norm t (see PANKOWSKA/WYGRALAK (2003, 2004b, 2006) and WYGRALAK (2007, 2010); cf. (6.19)). Each membership degree

$$\chi_{\mathscr{E}}(x) = \nu(A^+(x))\, t\, \nu(A^-(x)) = \nu(A^+(x)\, t^\nu A^-(x)) \tag{6.51}$$

is viewed as the *degree of hesitation* (or *hesitation margin, hesitation index*) concerning x. (6.51) reflects an intuitive understanding of hesitation as "not *yes* and not *no*", "not *pro* and not *contra*". By (6.46), we have

$$\chi_{\mathscr{E}} \supset A^{+} \cap_{t} A^{-}. \tag{6.52}$$

Let us notice in this context that

$$A^{+} \subset (A^{-})^{\nu} \Leftrightarrow A^{+} \cap_{t} A^{-} = 1_{\varnothing} \tag{6.53}$$

whenever t is nilpotent and $\nu = \nu_t$ (see (3.19)). If $\mathscr{E}$ collapses to an ordinary fuzzy set, $\mathscr{E} = (A, A^{\nu})$, we always get

$$\chi_{\mathscr{E}} = A \cap_{t} A^{\nu}. \tag{6.54}$$

We like to look closer at three important particular cases of hesitation degrees. A general discussion will be presented in the next subsection.

- $t = t_{Ł}$ and $\nu = \nu_t$, i.e. $\nu = \nu_{Ł}$. Then

$$\chi_{\mathscr{E}}(x) = 1 - A^{+}(x) - A^{-}(x), \tag{6.55}$$

whereas (6.46) comes down to

$$A^{+}(x) + A^{-}(x) \leq 1 \tag{6.56}$$

and, trivially,

$$A^{+}(x) + A^{-}(x) + \chi_{\mathscr{E}}(x) = 1 \tag{6.57}$$

for each $x \in U$. What one gets is thus the standard Atanassov's formulation of hesitation degrees (see e.g. ATANASSOV (1999, 2003)). $\chi_{\mathscr{E}}(x)$ forms the length of the interval $[A^{+}(x), 1 - A^{-}(x)]$, the size of our ignorance as to $A(x)$ (see (6.48)).

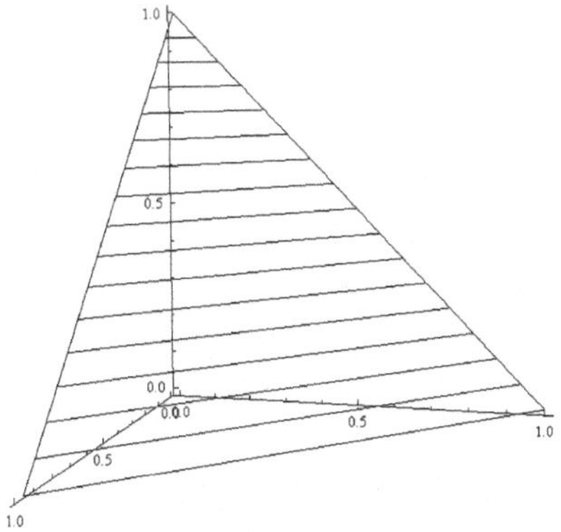

Fig. 6.4 3D plot of $\chi_{\mathscr{E}}$ for $t = t_{Ł}$ and $\nu = \nu_{Ł}$. The horizontal axes represent $A^{+}(x)$ (left) and $A^{-}(x)$ (right), while the vertical one shows $\chi_{\mathscr{E}}(x)$

- $\boldsymbol{t} = \wedge$ and $\nu = \nu_{Ł}$. Now

$$\chi_{\mathcal{E}}(x) = 1 - A^{+}(x) \vee A^{-}(x) = 1 - A^{+}(x) - A^{-}(x) + A^{+}(x) \wedge A^{-}(x). \qquad (6.58)$$

Using the fuzziness index $\varphi(a) = a \wedge (1 - a)$ from (6.25), we obtain

$$\varphi(A^{+}(x)) \wedge \varphi(1 - A^{-}(x)) = A^{+}(x) \wedge A^{-}(x). \qquad (6.59)$$

$A^{+}(x) \wedge A^{-}(x)$ is thus the minimum possible fuzziness index of $A(x)$, the minimum possible entropy measure of fuzziness of $A(x)/x$. $\chi_{\mathcal{E}}(x)$ becomes the sum of the size of ignorance as to $A(x)$ and that minimum possible entropy measure of fuzziness (see also Subsection 6.3.4).

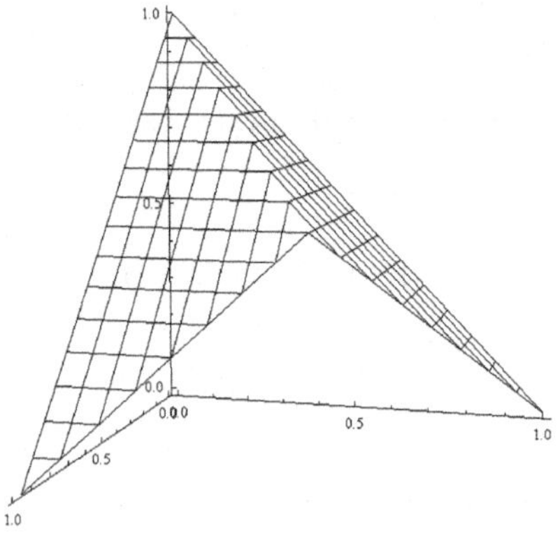

Fig. 6.5 3D plot of $\chi_{\mathcal{E}}$ for $\boldsymbol{t} = \wedge$ and $\nu = \nu_{Ł}$. The description of the axes is as in Figure 6.4

- $\boldsymbol{t} = \boldsymbol{t}_{a}$ and $\nu = \nu_{Ł}$.

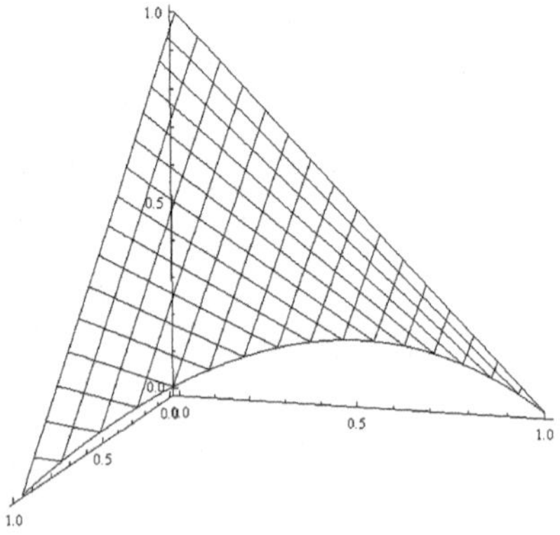

Fig. 6.6 3D plot of $\chi_{\mathcal{E}}$ for $\boldsymbol{t} = \boldsymbol{t}_{a}$ and $\nu = \nu_{Ł}$. The description of the axes is as in Figure 6.4

This time

$$\chi_{\mathscr{E}}(x) = 1 - A^{+}(x) - A^{-}(x) + A^{+}(x)\cdot A^{-}(x). \tag{6.60}$$

One has

$$A^{+}(x)\cdot A^{-}(x) \le \varphi(A^{+}(x)) \wedge \varphi(1 - A^{-}(x)) \tag{6.61}$$

with the fuzziness index $\varphi(a) = a\cdot(1 - a)$ from (6.28). Consequently, $\chi_{\mathscr{E}}(x)$ is the sum of the size of ignorance as to $A(x)$ and a lower evaluation of the minimum possible fuzziness index (6.28) of $A(x)$.

Finally, we see that (6.50) is a generalization of the difference $\boldsymbol{D}'\setminus \boldsymbol{B} = \boldsymbol{B}'\cap \boldsymbol{D}'$ from (6.19). An alternative variant of doing is to define $\chi_{\mathscr{E}}$ via the bounded difference of $(A^{-})^{\nu}$ and A^{+}, namely

$$\chi_{\mathscr{E}} = (A^{-})^{\nu} \ominus A^{+}$$

and, hence, (6.62)

$$\chi_{\mathscr{E}}(x) = \nu(A^{-}(x)) - A^{+}(x)$$

for each $x \in \boldsymbol{U}$. This gives $\chi_{\mathscr{E}}(x) = 1 - A^{+}(x) - A^{-}(x)$ if $\nu = \nu_{Ł}$. In our further discussion we will use (6.50) as it is more flexible. The reader is also referred to BARRENECHEA *et al.* (2009) for another general concept of hesitation degrees which can be characterized in terms of implication operators.

6.3.2 General Properties of Hesitation Degrees

Let us look more generally at I-fuzzy sets with (6.46) and their hesitation areas from (6.50) constructed via a t-norm $\boldsymbol{t}$ and a strong negation ν. By (6.51) and (2.68), we always have

$$\chi_{\mathscr{E}}(x) = 1 \Leftrightarrow A^{+}(x) = A^{-}(x) = 0. \tag{6.63}$$

Our first subject is the case of nilpotent t-norms (cf. Theorem 6.1).

Theorem 6.3. *Let* $\mathscr{E} = (A^{+}, A^{-})$ *with* $A^{+} \subset (A^{-})^{\nu}$. *For each* $x \in \boldsymbol{U}$, *if* $\boldsymbol{t}$ *is a nilpotent t-norm,* $\nu = \nu_{t}$, *and* h *is the normed generator of* $\boldsymbol{t}^{\circ}$, *then*

(a) $\chi_{\mathscr{E}}(x) = h^{-1}(1 - h(A^{+}(x)) - h(A^{-}(x)))$,

(b) $A^{+}(x)\, \boldsymbol{t}^{\circ}\, \chi_{\mathscr{E}}(x) = \nu(A^{-}(x))$.

Hesitation degrees are now isomorphic to $1 - A^{+}(x) - A^{-}(x)$, the length of the interval $[A^{+}(x), 1 - A^{-}(x)]$. So, again, $\chi_{\mathscr{E}}(x)$ is (a sort of) a "pure" size of ignorance as to $A(x)$. An immediate consequence of (a) is that (cf. (6.57))

$$h(A^{+}(x)) + h(A^{-}(x)) + h(\chi_{\mathscr{E}}(x)) = 1 \tag{6.64}$$

for each $x \in \boldsymbol{U}$. By (b) and (3.19), we also obtain

$$A^{+}(x)\, \boldsymbol{t}^{\circ} A^{-}(x)\, \boldsymbol{t}^{\circ}\, \chi_{\mathscr{E}}(x) = 1. \tag{6.65}$$

If $\boldsymbol{t} = \boldsymbol{t}_{\mathrm{S},p}$ with $p > 0$, one gets

$$\chi_{\mathscr{E}}(x) = [1 - (A^{+}(x))^{p} - (A^{-}(x))^{p}]^{1/p}, \tag{6.66}$$

whereas (6.46) is equivalent to

$$(A^{+}(x))^{p} + (A^{-}(x))^{p} \leq 1 \quad \text{for each } x \in \boldsymbol{U}. \tag{6.67}$$

Obviously, this condition can be satisfied even if $A^{+}(x) + A^{-}(x) > 1$ for some x's. For $\boldsymbol{t} = \boldsymbol{t}_{\mathrm{Y},p}$ with $p > 0$, we obtain

$$\chi_{\mathscr{E}}(x) = 1 - [2 - (1 - A^{+}(x))^{p} - (1 - A^{-}(x))^{p}]^{1/p} \tag{6.68}$$

and (6.46) collapses to

$$(1 - A^{+}(x))^{p} + (1 - A^{-}(x))^{p} \geq 1 \quad \text{for each } x \in \boldsymbol{U}. \tag{6.69}$$

By Theorem 6.3(a), one has for nilpotent $\boldsymbol{t}$ and $\nu = \nu_{\boldsymbol{t}}$

$$\chi_{\mathscr{E}}(x) = 0 \Leftrightarrow A^{+}(x) = \nu(A^{-}(x)),$$

i.e.

$$\chi_{\mathscr{E}} = 1_{\varnothing} \Leftrightarrow \mathscr{E} \text{ collapses to a fuzzy set.} \tag{6.70}$$

Similarly to IVFSs, the task of interpreting $\chi_{\mathscr{E}}(x)$ becomes more advanced when $\boldsymbol{t}$ is strict or $\boldsymbol{t} = \wedge$, which means that $\boldsymbol{t}$ has no zero divisors. $\chi_{\mathscr{E}}(x)$ is then the size of ignorance as to $A(x)$ combined with a fuzziness index of $A(x)$. Really, by (6.51), we have

$$\chi_{\mathscr{E}}(x) = 0 \Leftrightarrow (A^{+}(x), A^{-}(x)) \in \{(1,0), (0,1)\}$$

and, hence,

$$\chi_{\mathscr{E}} = 1_{\varnothing} \Leftrightarrow \mathscr{E} \text{ collapses to a set.} \tag{6.71}$$

If $\mathscr{E}$ is an ordinary fuzzy set, $\mathscr{E} = (A, A^{\nu})$, one obtains

$$\chi_{\mathscr{E}}(x) = A(x)\, \boldsymbol{t}\, \nu(A(x)), \tag{6.72}$$

i.e. $\chi_{\mathscr{E}}(x)$ forms a $\boldsymbol{t}$-based fuzziness index of $A(x)$. Finally, for $\nu = \nu_{\mathrm{L}}$ and $\boldsymbol{t}$ being strict or equal to $\wedge$, (6.51) implies

$$A^{+}(x)\, \boldsymbol{t}^{*} A^{-}(x) + \chi_{\mathscr{E}}(x) = 1 \quad \text{for each } x \in \boldsymbol{U}. \tag{6.73}$$

6.3.3 Operations on IFSs

Assume $\mathscr{E} = (A^{+}, A^{-})$ and $\mathscr{F} = (B^{+}, B^{-})$ are two IFSs, i.e. $A^{+} \subset (A^{-})^{\nu}$ and $B^{+} \subset (B^{-})^{\nu}$ with a strong negation ν. Let (*equality*) $$\mathscr{E} = \mathscr{F} \Leftrightarrow A^+ = B^+ \ \&\ A^- = B^-, \tag{6.74}$$

(*inclusion*) $$\mathscr{E} \subset \mathscr{F} \Leftrightarrow A^+ \subset B^+ \ \&\ B^- \subset A^-. \tag{6.75}$$

Drawing inspiration from Pawlak's rough set theory (PAWLAK (1982, 1991)), one can define two types of *approximate equalities and inclusions* in which we focus on positive or negative information. Namely,

($\mathscr{E}$ *positively equals* $\mathscr{F}$) $$\mathscr{E} =_{pos} \mathscr{F} \Leftrightarrow A^+ = B^+, \tag{6.76}$$

($\mathscr{E}$ *negatively equals* $\mathscr{F}$) $$\mathscr{E} =_{neg} \mathscr{F} \Leftrightarrow A^- = B^-, \tag{6.77}$$

($\mathscr{E}$ *positively contained in* $\mathscr{F}$) $$\mathscr{E} \subset_{pos} \mathscr{F} \Leftrightarrow A^+ \subset B^+, \tag{6.78}$$

($\mathscr{E}$ *negatively contained in* $\mathscr{F}$) $$\mathscr{E} \subset_{neg} \mathscr{F} \Leftrightarrow B^- \subset A^-. \tag{6.79}$$

It is obvious that $=$, $=_{pos}$, and $=_{neg}$ are all equivalence relations, and one has

$$\mathscr{E} = \mathscr{F} \Leftrightarrow \mathscr{E} =_{pos} \mathscr{F} \ \&\ \mathscr{E} =_{neg} \mathscr{F}. \tag{6.80}$$

Further, let us define

(*intersection*) $$\mathscr{E} \cap_{\boldsymbol{T},\boldsymbol{S}} \mathscr{F} = (A^+ \cap_{\boldsymbol{T}} B^+, A^- \cup_{\boldsymbol{S}} B^-), \tag{6.81}$$

(*sum*) $$\mathscr{E} \cup_{\boldsymbol{T},\boldsymbol{S}} \mathscr{F} = (A^+ \cup_{\boldsymbol{S}} B^+, A^- \cap_{\boldsymbol{T}} B^-), \tag{6.82}$$

(*cartesian product*) $$\mathscr{E} \times_{\boldsymbol{T}} \mathscr{F} = (A^+ \times_{\boldsymbol{T}} B^+, ((A^-)^\nu \times_{\boldsymbol{T}} (B^-)^\nu)^\nu), \tag{6.83}$$

(*complement*) $$\mathscr{E}' = (A^-, A^+). \tag{6.84}$$

$\boldsymbol{T}$ is here a t-norm possibly differing from that in (6.50). ν is the negation from (6.46). $\boldsymbol{S}$ denotes a t-conorm. A unique suitable choice seems to be $\boldsymbol{S} = \boldsymbol{T}^\nu$ which guarantees that $\mathscr{E} \cap_{\boldsymbol{T},\boldsymbol{S}} \mathscr{F}$ as well as $\mathscr{E} \cup_{\boldsymbol{T},\boldsymbol{S}} \mathscr{F}$ are still IFSs and, moreover, that the intersection and sum of two IFSs collapsing to usual fuzzy sets are also fuzzy sets expressed in terms of IFSs (see PANKOWSKA/WYGRALAK (2006)). We will refer to (6.81)-(6.83) as *operations induced by* $\boldsymbol{T}$ *and* $\boldsymbol{S}$. As usual, if $\boldsymbol{T} = \wedge$, $\boldsymbol{S} = \vee$ and $\nu = \nu_Ł$, one simply writes $\mathscr{E} \cap \mathscr{F}$, $\mathscr{E} \cup \mathscr{F}$ and $\mathscr{E} \times \mathscr{F}$, i.e. we have

$$\mathscr{E} \cap \mathscr{F} = (A^+ \cap B^+, A^- \cup B^-), \tag{6.85}$$

$$\mathscr{E} \cup \mathscr{F} = (A^+ \cup B^+, A^- \cap B^-), \tag{6.86}$$

$$\mathscr{E} \times \mathscr{F} = (A^+ \times B^+, ((A^-)' \times (B^-)')'). \tag{6.87}$$

This combination of $\boldsymbol{T}$, $\boldsymbol{S}$ and ν together with $\boldsymbol{t} = \boldsymbol{t}_Ł$ leads to the standard formulation of IFSs and their operations.

Properties of (6.81)-(6.84) are simple consequences of basic properties of operations on fuzzy sets. For instance, $\mathscr{E}_U = (1_U, 1_\varnothing)$ and $\mathscr{E}_\varnothing = (1_\varnothing, 1_U)$, respectively, the representations of the whole universe and the empty set in the language of IFSs, are neutral elements of $\cap_{\boldsymbol{T},\boldsymbol{S}}$ and $\cup_{\boldsymbol{T},\boldsymbol{S}}$, respectively. It is easy to check that De Morgan laws hold true:

$$(\mathscr{E} \cap_{T,S} \mathscr{F})' = \mathscr{E}' \cup_{T,S} \mathscr{F}' \quad \text{and} \quad (\mathscr{E} \cup_{T,S} \mathscr{F})' = \mathscr{E}' \cap_{T,S} \mathscr{F}'. \tag{6.88}$$

$\mathscr{E}$ and $\mathscr{F}$ are said to be *disjoint* whenever $\mathscr{E} \cap \mathscr{F} = \mathscr{E}_{\emptyset}$. It is a routine task to verify the equivalence

$$\mathscr{E}, \mathscr{F} \text{ disjoint} \Leftrightarrow A^{-} \cup B^{-} = 1_{U}, \tag{6.89}$$

which forms a counterpart of the elementary property of sets saying that $\boldsymbol{A}$ and $\boldsymbol{B}$ are disjoint iff $\boldsymbol{A}' \cup \boldsymbol{B}' = \boldsymbol{U}$.

6.3.4 Model Examples

Our point of departure is the case of usual fuzzy sets. Imagine a collection of eight bottles of water: two of them are full, one is empty, and the others are 80%, 50%, 10%, 90%, and 30% full. In terms of fuzzy sets, we thus deal with a universe of eight bottles, $\boldsymbol{U} = \{b_1, b_2, \dots, b_8\}$, and a fuzzy set A of full bottles.

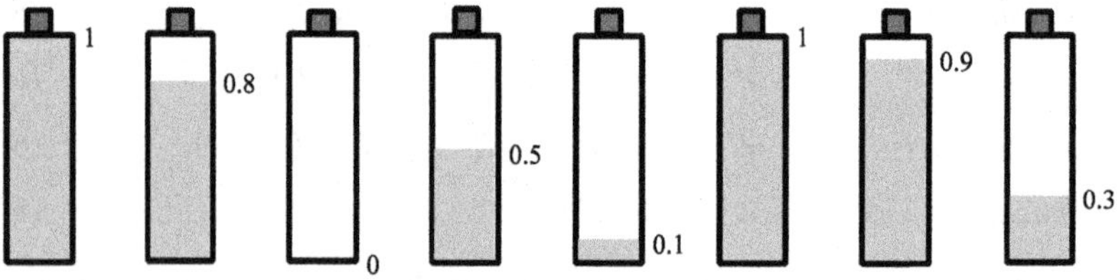

Fig. 6.7 Fuzzy set of full bottles

As to the membership degrees in A, one has

$$\begin{gathered} A(b_1) = 1, \quad A(b_2) = 0.8, \quad A(b_3) = 0, \quad A(b_4) = 0.5, \\ A(b_5) = 0.1, \quad A(b_6) = 1, \quad A(b_7) = 0.9, \quad A(b_8) = 0.3. \end{gathered} \tag{6.90}$$

These membership degrees are degrees of saturation of a property. Looking dually, we deal with a fuzzy set A' of empty bottles with $A'(b_i) = 1 - A(b_i)$, $i = 1, \dots, 8$. The reader sees that the eight-bottle example is a model example which can be easily transferred to a collection of some resources, to modules of a system with different extent of faults, etc.

Our information about the membership degrees in A is complete. Nevertheless, A can be expressed in the language of IVFSs as $\mathcal{E} = (A, A)$ or in the language of IFSs as $\mathscr{E} = (A, A')$. For convenience, we will focus on the latter representation. By (6.51), the hesitation degree for bottle b_i is equal to $\chi_{\mathscr{E}}(b_i) = A(b_i)\, \boldsymbol{t}\, (1 - A(b_i))$, where $\boldsymbol{t}$ denotes a t-norm. Let us look at three cases involving the three basic t-norms.

- Use $t_Ł$. Then

$$\chi_{\mathscr{E}}(b_i) = A(b_i)\, t_Ł\, (1 - A(b_i)) = 0 \quad \text{for } i = 1, \ldots, 8$$

as $\chi_{\mathscr{E}}(b_i)$ is now the size of ignorance as to $A(b_i)$, while our information about $A(b_i)$ is full.

- Take $t = \wedge$. $\chi_{\mathscr{E}}(b_i)$ becomes the fuzziness index (6.25) of $A(b_i)$. Its values are

$$\begin{aligned} &\chi_{\mathscr{E}}(b_1) = 0, \;\; \chi_{\mathscr{E}}(b_2) = 0.2, \;\; \chi_{\mathscr{E}}(b_3) = 0, \;\; \chi_{\mathscr{E}}(b_4) = 0.5, \\ &\chi_{\mathscr{E}}(b_5) = 0.1, \;\; \chi_{\mathscr{E}}(b_6) = 0, \;\; \chi_{\mathscr{E}}(b_7) = 0.1, \;\; \chi_{\mathscr{E}}(b_8) = 0.3. \end{aligned} \tag{6.91}$$

Hence

$$\sum_{i=1}^{8} \chi_{\mathscr{E}}(b_i) = \sum_{i=1}^{8} (A \cap A')(b_i) = 1.2,$$

which is the entropy measure of fuzziness Fuzz(A) of A created by means of (6.25). This sum is also a kind of cardinality of $\chi_{\mathscr{E}}$ (see Subsections 2.3.4 and 8.1.2). Let us look a bit closer at the interpretation via entropy measures of fuzziness. We will say that bottle b_i is *classifiable* - as full or empty - whenever it is totally full or totally empty, $A(b_i) \in \{0, 1\}$ (see also Subsection 9.1.4). Only b_1, b_3 and b_6 are then classifiable; $\chi_{\mathscr{E}}(b_i) = 0$ for $i = 1, 3, 6$. The remaining bottles are not. However, one can make them classifiable by pouring more water to the full (if $A(b_i) > 0.5$) or pouring it out (if $A(b_i) \le 0.5$). Speaking very practically, Fuzz(A) = 1.2 means that the total amount of water one has to pour out from or pour into the bottles in order to make all of them classifiable is just 1.2 (bottles of water).

- Put $t = t_a$. $\chi_{\mathscr{E}}(b_i)$ is now the fuzziness index (6.28) of $A(b_i)$. We get

$$\begin{aligned} &\chi_{\mathscr{E}}(b_1) = 0, \;\; \chi_{\mathscr{E}}(b_2) = 0.16, \;\; \chi_{\mathscr{E}}(b_3) = 0, \;\; \chi_{\mathscr{E}}(b_4) = 0.25, \\ &\chi_{\mathscr{E}}(b_5) = 0.09, \;\; \chi_{\mathscr{E}}(b_6) = 0, \;\; \chi_{\mathscr{E}}(b_7) = 0.09, \;\; \chi_{\mathscr{E}}(b_8) = 0.21. \end{aligned} \tag{6.92}$$

The sum Fuzz(A) = $\Sigma_{i=1,8}\, \chi_{\mathscr{E}}(b_i) = 0.8$ is again an entropy measure of fuzziness of A. However, its interpretation in a style similar to $t = \wedge$ is not clear.

Let us complicate the situation in Figure 6.7 and assume that the water level in some bottles is incompletely known as they are partially covered up and, thus, invisible.

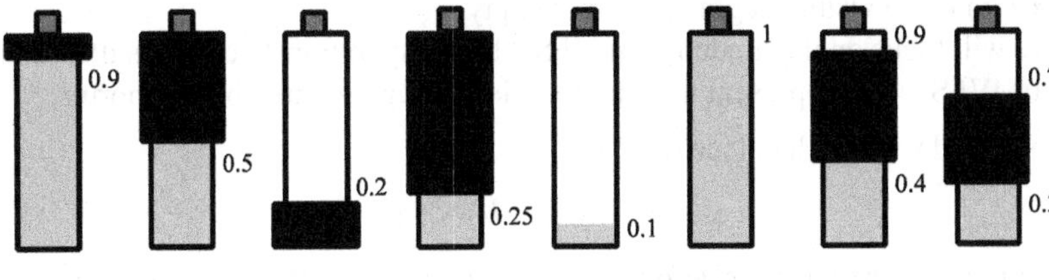

Fig. 6.8 Incompletely known fuzzy set of full bottles

That water level, $A(b_i)$, is somewhere between the horizontal sides of the rectangles (curtains) covering up the bottles. Looking more formally at this modified eight-bottle example, the fuzzy set A of full bottles is now incompletely known: our knowledge about its membership degrees is only partial. Again, Figure 6.8 presents a model situation which can be transferred, say, to

- a collection of resources (financial, natural, ...) whose levels are partially unknown,
- preferences with incompletely known degrees of intensity,
- modules of a system with incompletely known extent of faults,
- decisions with incompletely known consequences,
- partially unknown levels of trust, satisfaction, ...

Using the language of IVFSs we look at the bottles from one perspective: in terms of advantages, of being full, and we get:

$$\begin{gathered} A_l(b_1)=0.9,\ \ A_l(b_2)=0.5,\ \ A_l(b_3)=0,\ \ A_l(b_4)=0.25, \\ A_l(b_5)=0.1,\ \ A_l(b_6)=1,\ \ A_l(b_7)=0.4,\ \ A_l(b_8)=0.3, \\ A_u(b_1)=1,\ \ A_u(b_2)=1,\ \ A_u(b_3)=0.2,\ \ A_u(b_4)=1, \\ A_u(b_5)=0.1,\ \ A_u(b_6)=1,\ \ A_u(b_7)=0.9,\ \ A_u(b_8)=0.7. \end{gathered} \tag{6.93}$$

For bottle b_i with $i = 1, \dots, 8$, $A_l(b_i)$ and $A_u(b_i)$, respectively, are the minimum and maximum possible degree of being full (pessimistic and optimistic evaluation of $A(b_i)$, respectively).

In the approach through IFSs, we look at each bottle from two opposite perspectives: in terms of advantages and disadvantages, of being full and being empty. We thus obtain:

$$\begin{gathered} A^+(b_1)=0.9,\ \ A^+(b_2)=0.5,\ \ A^+(b_3)=0,\ \ A^+(b_4)=0.25, \\ A^+(b_5)=0.1,\ \ A^+(b_6)=1,\ \ A^+(b_7)=0.4,\ \ A^+(b_8)=0.3, \\ A^-(b_1)=0,\ \ A^-(b_2)=0,\ \ A^-(b_3)=0.8,\ \ A^-(b_4)=0, \\ A^-(b_5)=0.9,\ \ A^-(b_6)=0,\ \ A^-(b_7)=0.1,\ \ A^-(b_8)=0.3. \end{gathered} \tag{6.94}$$

$A^+(b_i)$ is a lower bound on $A(b_i)$, whereas $A^-(b_i)$ forms a lower bound on $A'(b_i)$ for $i = 1, \dots, 8$. Speaking more practically, $A^+(b_i)$ is the visible degree to which b_i is full, and $A^-(b_i)$ is the visible degree of being empty.

We will focus on the modeling via IFSs. It is easy to translate the results into the case of IVFSs. Let us present a short study involving the three basic t-norms.

- $t = t_Ł$ and $\nu = \nu_Ł$. Then (see (6.55))

  $$\chi_{\mathscr{E}}(b_i) = 1 - A^+(b_i) - A^-(b_i)$$

  with $\chi_{\mathscr{E}}(b_i)$ being the size of ignorance as to $A(b_i)$, the height of the curtain covering up bottle b_i. Using (6.94), we get the following numerical results:

$$\chi_{\mathscr{E}}(b_1)=0.1,\ \ \chi_{\mathscr{E}}(b_2)=0.5,\ \ \chi_{\mathscr{E}}(b_3)=0.2,\ \ \chi_{\mathscr{E}}(b_4)=0.75,$$
$$\chi_{\mathscr{E}}(b_5)=0,\ \ \chi_{\mathscr{E}}(b_6)=0,\ \ \chi_{\mathscr{E}}(b_7)=0.5,\ \ \chi_{\mathscr{E}}(b_8)=0.4. \qquad (6.95)$$

Hence

$$\sum_{i=1}^{8}\chi_{\mathscr{E}}(b_i) = 2.45.$$

This is a total size of ignorance as to the content of the bottles. That lack of knowledge is thus equivalent to a complete lack of knowledge about the content of 2.45 bottles.

- $t=\wedge$ and $\mathrm{v}=\mathrm{v}_{Ł}$. Now (see (6.58))

$$\chi_{\mathscr{E}}(b_i)=1-A^+(b_i)-A^-(b_i)+A^+(b_i)\wedge A^-(b_i).$$

This hesitation degree becomes the sum of the size of ignorance as to $A(b_i)$ and the minimum possible fuzziness index of $A(b_i)$. Using (6.94), we obtain:

$$\chi_{\mathscr{E}}(b_1)=0.1,\ \ \chi_{\mathscr{E}}(b_2)=0.5,\ \ \chi_{\mathscr{E}}(b_3)=0.2,\ \ \chi_{\mathscr{E}}(b_4)=0.75,$$
$$\chi_{\mathscr{E}}(b_5)=0.1,\ \ \chi_{\mathscr{E}}(b_6)=0,\ \ \chi_{\mathscr{E}}(b_7)=0.6,\ \ \chi_{\mathscr{E}}(b_8)=0.7. \qquad (6.96)$$

So, $\chi_{\mathscr{E}}(b_i)$ gives us more information (2 in 1) in comparison with the case of $t = t_{Ł}$. To illustrate this, let us look at bottle b_1 and two other bottles with the same size of ignorance.

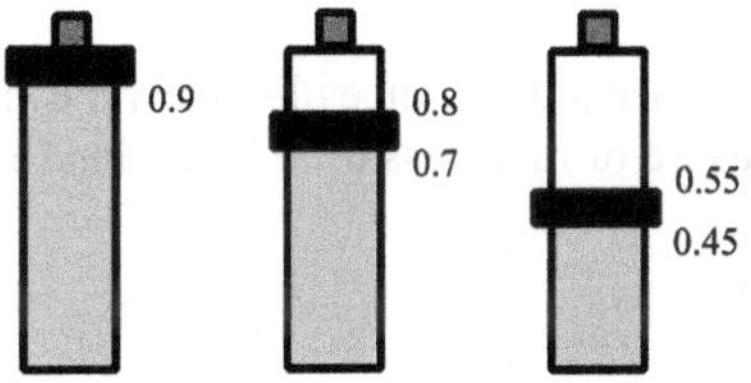

Fig. 6.9 Hesitation – ignorance – fuzziness index

We thus get the following results:

	left	**middle**	**right bottle**
hesitation	0.1	0.3	0.55
ignorance	0.1	0.1	0.1
minimum possible fuzziness index	0	0.2	0.45

As we see, the same ignorance concerns three quite different situations. They are distinguishable thanks to the use of $\chi_{\mathcal{E}}(b_i)$ with $t = \wedge$. On account of (6.96),

$$\sum_{i=1}^{8} \chi_{\mathcal{E}}(b_i) = \sum_{i=1}^{8} 1 - A^{+}(b_i) - A^{-}(b_i) + \sum_{i=1}^{8} A^{+}(b_i) \wedge A^{-}(b_i) = 2.45 + 0.5 = 2.95.$$

This forms a total size of ignorance plus the minimum possible entropy measure of fuzziness, the minimum possible amount of water which has to be poured out from or poured into the bottles if one likes to make all of them classifiable. The second component, 0.5, can thus be viewed as the minimum possible distance of the system from a totally classifiable one.

- $t = t_a$ and $v = v_Ł$. Then (see (6.60))

$$\chi_{\mathcal{E}}(x) = 1 - A^{+}(x) - A^{-}(x) + A^{+}(x) \cdot A^{-}(x).$$

$\chi_{\mathcal{E}}(b_i)$ is now the sum of the size of ignorance as to $A(b_i)$ and a lower evaluation of the minimum possible fuzziness index $\varphi(a) = a \cdot (1 - a)$ of $A(b_i)$. By (6.94),

$$\begin{aligned} &\chi_{\mathcal{E}}(b_1) = 0.1, \quad \chi_{\mathcal{E}}(b_2) = 0.5, \quad \chi_{\mathcal{E}}(b_3) = 0.2, \quad \chi_{\mathcal{E}}(b_4) = 0.75, \\ &\chi_{\mathcal{E}}(b_5) = 0.09, \quad \chi_{\mathcal{E}}(b_6) = 0, \quad \chi_{\mathcal{E}}(b_7) = 0.54, \quad \chi_{\mathcal{E}}(b_8) = 0.49. \end{aligned} \tag{6.97}$$

Consequently,

$$\sum_{i=1}^{8} \chi_{\mathcal{E}}(b_i) = \sum_{i=1}^{8} 1 - A^{+}(b_i) - A^{-}(b_i) + \sum_{i=1}^{8} A^{+}(b_i) \cdot A^{-}(b_i) = 2.45 + 0.22 = 2.67.$$

What we have is a total size of ignorance plus a lower evaluation of the minimum possible entropy measure of fuzziness of A involving (6.28).

Part II

Methods of Intelligent Counting under Information Imprecision

Chapter 7

General Remarks and Motivations

Counting belongs to the most basic and frequent mental activities of humans. This is hardly surprising as cardinality seems to be one of the most fundamental characteristics of a given collection of elements. It forms an important type of information about that collection, a basis for coming to a decision in a lot of situations and in many dimensions of our life. However, speaking about counting one should distinguish between two essentially different cases described below.

- The objects of counting are *precisely specified*, e.g.

"How many apples are there in the basket?".

The counting process then collapses to the usual counting in a set by means of the natural numbers. There is no problem what to count ("do not think, just count the apples!") and, thus, this task is trivial.

- On the other hand, in many practical situations, the objects of counting are *imprecisely, fuzzily specified*, e.g.

"How many *big* apples are there in the basket?".

Each apple in the basket is big to a degree and, consequently, we have to face a qualitatively new problem: what and how to count? also only to a degree, partially? This counting thus requires intelligence, and what we deal with can be termed *intelligent counting*. The counting process is now a process going on in a fuzzy set. It can also be called *counting under imprecision of information about the objects of counting* or, in short, *counting under information imprecision*, or simply *counting under imprecision*. Let us mention further familiar examples of queries/tasks requiring to count in a fuzzy set or related to such counting (see also Subsection 8.1.3):

"How many records in the data file are *similar* to a given record?",

"How many modules of the system are *faulty*?",

"How many *affordable* hotels are there in the *vicinity* of the center of Paris?",

"Are there more objects which are *small* than objects which are *medium*?".

The subject of Part II of this book is just intelligent counting. We will present counting methods in and the resulting cardinalities of fuzzy sets, including theoretical aspects and applications (Chapters 8-11). Needless to say, the main issue will be finite

M. Wygralak: *Intelligent Counting Under Information Imprecision*, STUDFUZZ 292, pp. 163–165.
DOI: 10.1007/978-3-642-34685-9_7

fuzzy sets as they play a key role in practice. Nevertheless, for the sake of completeness, cardinalities of infinite fuzzy sets will be discussed in Chapter 11.

Presenting and studying various counting methods, we will pay attention to their true technical sense and motivations in everyday practice. The reader will see that those methods reflect and formalize real, human counting procedures under information imprecision. Two of these procedures seem to be absolutely fundamental as they are used by human beings most frequently and inspire two types of approach to counting in fuzzy sets (see below).

- **Counting by thresholding.** Assume we ask

"How many *warm* days were there last (calendar) summer?".

The standard human procedure of answering is then, first, to establish a threshold temperature for a day to be counted as warm, say, 22°C. Second, one counts up all the summer days with temperature ≥ 22°C or > 22°C. The resulting sum is treated as the cardinality $|A|$ of the fuzzy set A of warm days. Speaking more formally, we thus define

$$|A| = |A_t| \quad \text{or} \quad |A| = |A^t|$$

with a predefined membership degree threshold t. This *counting by thresholding*, in other words: this *cut-and-count method* (CAC), is an elementary instance of the *scalar approach* to intelligent counting. The result of counting in a fuzzy set is then a single usual number called its *scalar cardinality*. Chapter 8 will present a detailed study of that approach. Some related remarks are also placed in Subsections 9.1.2, 9.2.2 and 9.2.3.

- **Counting by multiple thresholding.** Look at the following typical news item:

"The whirlwind came over two villages causing damages to 48 buildings of which 15 are *seriously* damaged and 8 are *completely* destroyed".

We guess that the damages to the remaining 25 buildings are rather slight. Speaking more formally, the collection of damaged buildings is treated in the news item as a fuzzy set, A, since the damages are more or less severe. In total, 48 buildings are thus damaged at least slightly. 23 of them are damaged at least seriously. Finally, 8 of these 23 buildings are completely destroyed. So, in essence, the number of damaged buildings is expressed as a fuzzy set of natural numbers with linguistically specified membership degrees:

$$|A| = \mathit{completely}/8 + \mathit{seriously}/23 + \mathit{slightly}/48.$$

This can be transformed into numerical membership degrees, say, as

$$|A| = 1/8 + 0.6/23 + 0.1/48,$$

i.e.

$$|A| = 1/|A_1| + 0.6/|A_{0.6}| + 0.1/|A_{0.1}|.$$

One sees that this combination of the counting results for three (generally: many) thresholds gives us more detailed as well as more interesting and advanced information than CAC with any single threshold value.

The above counting by multiple thresholding (MCAC, *multiple cut-and-count method*) is a fundamental example of and forms a natural motivation for the *fuzzy approach* to intelligent counting. The result of counting in a fuzzy set is now itself a fuzzy set of nonnegative integers, termed its *fuzzy cardinality*, rather than a single number. This approach will be investigated in Chapter 9.

Also an even more sophisticated case of intelligent counting is discussed in Chapters 8 and 9. We mean counting methods in incompletely known fuzzy sets modeled by IVFSs and IFSs presented in Chapter 6. Our knowledge about the membership degrees in a fuzzy set is then only partial and, thus, what we deal with is *counting under imprecision and incompleteness of information* (about the objects of counting). Both the scalar and fuzzy approaches will be extended to IVFSs and IFSs. The resulting extension of the scalar approach leads to cardinalities which are intervals of nonnegative real numbers. As to the fuzzy approach, cardinalities of IVFSs and IFSs become IVFSs and IFSs in the universe of nonnegative integers. We will show in each case that those extensions reflect and formalize human counting methods performed under imprecision combined with incompleteness of information.

Chapter 8

Scalar Approach

This chapter presents the scalar approach to intelligent counting. Our main subject will be the concept of the sigma *f*-count of a fuzzy set. Further, we will extend it to interval-valued fuzzy sets and I-fuzzy sets. Sigma *f*-counts and their extensions form a formalization of human counting processes under imprecision possibly combined with incompleteness of information. In other words, they offer human-consistent counting procedures. This will be illustrated by a set of examples which refer to the eight-bottle example from Subsection 6.3.4 and show that counting by means of sigma *f*-counts makes a true practical sense. Those examples can be easily adapted to various problems from the areas of computer science and decision support.

8.1 Sigma *f*-Counts and Counting in Fuzzy Sets

In this discussion and Chapters 9-10 as well, we focus our attention on finite fuzzy sets. If not emphasized otherwise, the phrase "fuzzy set" will mean "finite fuzzy set". FFS and FCS will denote the family of all (finite) fuzzy sets and all finite crisp sets in $\boldsymbol{U}$, respectively. So,

$$\mathrm{FFS} = \{B \in [0, 1]^{\boldsymbol{U}}\colon \mathrm{supp}(B) \text{ is finite}\}$$

and

$$\mathrm{FCS} = \{B \in \{0, 1\}^{\boldsymbol{U}}\colon \mathrm{supp}(B) \text{ is finite}\}.$$

8.1.1 Sigma *f*-Counts

Speaking generally and formally, a *scalar cardinality* can be understood as some function FFS → $[0, \infty)$. How to define it in a reasonable way? We like to present a simple axiomatic framework for a large class of possible scalar cardinalities (however, see remarks closing Subsection 8.1.2). It was proposed in WYGRALAK (2000b) and developed in WYGRALAK (2003a). The reader is also referred to WYGRALAK (1998b, 1999b, c, 2000a). This section offers a new exposition of the topic.

A function σ: FFS → $[0,\infty)$ will be viewed as a scalar cardinality if the following postulates are satisfied for each $a, b \in [0, 1]$, $x, y \in \boldsymbol{U}$ and $A, B \in \mathrm{FFS}$:

M. Wygralak: *Intelligent Counting Under Information Imprecision*, STUDFUZZ 292, pp. 167–186.
DOI: 10.1007/978-3-642-34685-9_8

(P1) $\sigma(1/x) = 1$, (*consistency*)

(P2) $\sigma(a/x) \leq \sigma(b/y)$ whenever $a \leq b$, (*monotonicity*)

(P3) $\sigma(A \cup B) = \sigma(A) + \sigma(B)$ whenever $A \cap B = 1_{\emptyset}$. (*additivity*)

The number $\sigma(A)$ is then called the *scalar cardinality of* $A \in$ FFS. So, in the scalar approach, the cardinality of A will be denoted by $\sigma(A)$ besides and usually instead of the standard notation $|A|$. Obviously, comparisons of scalar cardinalities collapse to ordinary comparisons of reals.

(P1) will guarantee correct results of counting in fuzzy sets being usual sets. According to (P2), the larger the membership degree a, the larger the cardinality $\sigma(a/x)$ of a/x. Applying twice that postulate, one gets

(P2)′ $\sigma(a/x) = \sigma(b/y)$ whenever $a = b$,

i.e. $\sigma(a/x)$ depends on a only. By (P3), the cardinality of the sum of $k \geq 2$ pairwise disjoint fuzzy sets equals the sum of cardinalities of those fuzzy sets:

(P3)′ $\sigma(A_1 \cup \ldots \cup A_k) = \sum_{i=1}^{k} \sigma(A_i)$ if $A_i \cap A_j = 1_{\emptyset}$ for each $i \neq j$.

This *finite additivity* of σ means that, speaking practically, the counting process in a fuzzy set can be performed "part by part". Finally, the use of any σ may be combined with the use of operations $\cap_t$, $\times_t$ and $\cup_s$ induced by a t-norm $\boldsymbol{t}$ and t-conorm $\boldsymbol{s}$ (see Section 8.2). This is because (P1) and (P2) do not involve t-operations at all, whereas what (P3) says in essence is only that additivity holds if the standard operations $\cap$ and $\cup$ are used.

The following theorem gives a practical characterization of scalar cardinalities defined via (P1)-(P3), and suggests how to create a concrete scalar cardinality.

Theorem 8.1. *A mapping* σ: FFS $\to [0, \infty)$ *is a scalar cardinality defined by* (P1)-(P3) *iff there exists a function* f: $[0,1] \to [0,1]$ *fulfilling the conditions*

$$f(0) = 0, \ f(1) = 1$$

and

$$f(a) \leq f(b) \ \textit{whenever} \ a \leq b,$$

and such that

$$\sigma(A) = \sum_{x \in \mathrm{supp}(A)} f(A(x)) \ \textit{for each} \ A \in \mathrm{FFS}. \tag{8.1}$$

Indeed, it is a routine matter to check that (8.1) satisfies (P1)-(P3). Conversely, assume σ: FFS $\to [0, \infty)$ is a scalar cardinality fulfilling (P1)-(P3). By (P3)′,

$$\sigma(A) = \sum_{x \in \mathrm{supp}(A)} \sigma(A(x)/x)$$

for each $A \in \mathrm{FFS}$. Let f: $[0, 1] \rightarrow [0, 1]$ be a function such that $f(a) = \sigma(a/x)$ for $a \in [0, 1]$ and $x \in \boldsymbol{U}$ (see (P2)′). $\sigma(A)$ is thus of the form (8.1). By (P1), $f(1) = 1$, whereas (P2) guarantees that f must be nondecreasing. Finally, (P3) implies $\sigma(A) = \sigma(1_\emptyset) + \sigma(A)$, i.e. $\sigma(1_\emptyset) = \sigma(0/x) = f(0) = 0$.

Since $f(0) = 0$, the summation in (8.1) can be extended to all $x \in \boldsymbol{U}$. Each function f from Theorem 8.1 will be interpreted as a *weighting function*. $f(t)$ is a weight, a degree of participation in the counting process in A assigned to each element x whose membership degree in A is $t \in [0, 1]$. We will use the enhanced notation $\sigma_f(A)$ instead of $\sigma(A)$ in order to emphasize which weighting function is involved. So,

$$\sigma_f(A) = \sum_{x \in \mathrm{supp}(A)} f(A(x)) \tag{8.2}$$

for $A \in \mathrm{FFS}$. The simplest and most frequently used weighting function is the identity function $f = id$ and, then,

$$\sigma_{id}(A) = \sum_{x \in \mathrm{supp}(A)} A(x), \tag{8.3}$$

which is known as the *sigma count of A*. Consequently, $\sigma_f(A)$ will be called the *sigma f-count of A* or, less specifically, a *generalized sigma count of A*. It is worth noticing that (8.3) and (8.2) form a natural generalization of the relationship

$$|\boldsymbol{D}| = \sum_{x \in \boldsymbol{D}} 1_D(x) \tag{8.4}$$

for sets, which says that the cardinality of a set $\boldsymbol{D}$ is just the sum of values of the characteristic function of $\boldsymbol{D}$. The sigma f-count of $A \in \mathrm{FFS}$ is the sum of weighted membership degrees in A. By (8.2), for each f, we have:

- *consistency*: $\sigma_f(1_D) = |\boldsymbol{D}|$, (8.5)
- *monotonicity*: $\sigma_f(A) \le \sigma_f(B)$ if $A \subset B$, (8.6)
- *boundedness*: $|\mathrm{core}(A)| \le \sigma_f(A) \le |\mathrm{supp}(A)|$, (8.7)
- *shiftability*: $\sigma_f(A) = \sigma_f(B)$ whenever there exists a bijection $\boldsymbol{b}$: $\mathrm{supp}(A) \rightarrow \mathrm{supp}(B)$ such that $A(x) = B(\boldsymbol{b}(x))$ for each $x \in \mathrm{supp}(A)$.

Formulating the last property in a practical way, if A and B are functions attaining the same values, possibly at different points of $\boldsymbol{U}$ and including possible repetitions of those values, then A and B are of the same (scalar) cardinality, are *equipotent*. The inverse implication is not generally true (see also Section 9.4).

Notice that (8.2) has a natural integral generalization to arbitrary fuzzy sets in $\boldsymbol{U}$, namely

$$\sigma_f(A) = \int_{\boldsymbol{U}} (f \circ A)(x)\, dx \tag{8.8}$$

provided that the above integral does make sense. Worth mentioning is also a possible generalization of postulate (P3) to

(P3)′′ $\sigma(A \cup_s B) = \sigma(A) + \sigma(B)$ whenever $A \cap_t B = 1_\emptyset$

with a t-norm $\boldsymbol{t}$ and t-conorm $\boldsymbol{s}$. Clearly, for $\boldsymbol{t} = \wedge$ with any $\boldsymbol{s}$, (P3)′′ collapses to (P3). Scalar cardinalities defined via the system (P1)-(P2)-(P3)′′ are investigated in CASASNOVAS/TORRENS (2003b). Their characterization is then identical to that given in (8.1). However, the weighting function has to satisfy one additional assumption:

$$f(a\boldsymbol{s}b) = f(a) + f(b) \quad \text{whenever } a\boldsymbol{t}b = 0.$$

We will limit ourselves to the original approach based on (P1)-(P3) as it seems to be quite sufficient from the viewpoint of applications.

8.1.2 Main Weighting Functions and Cases of Sigma *f*-Counts

Let us present main instances of weighting functions together with the resulting sigma *f*-counts of fuzzy sets. Inequalities between those functions will be understood in the usual pointwise way: $f \le f^*$ iff $f(a) \le f^*(a)$ for each $a \in [0, 1]$.

- **Counting by thresholding:** $f = f_{1,t}$ with $t \in (0, 1]$, where (see Figure 8.1)

$$f_{1,t}(a) = \begin{cases} \mathbf{1}, & \textbf{if } a \ge t, \\ \mathbf{0}, & \textbf{otherwise.} \end{cases} \tag{8.9}$$

By (8.2), we then get

$$\sigma_f(A) = \sum_{x \in \mathrm{supp}(A)} f_{1,t}(A(x)) = |A_t| \tag{8.10}$$

for $A \in \mathrm{FFS}$, i.e. the sigma $f_{1,t}$-count of A is simply the cardinality of its t-cut A_t. Notice that $f = f_{1,1}$ forms the smallest possible weighting function. It gives

$$\sigma_f(A) = |\,\mathrm{core}(A)|. \tag{8.11}$$

- **Counting by sharp thresholding:** $f = f_{2,t}$ with $t \in [0, 1)$ and

$$f_{2,t}(a) = \begin{cases} \mathbf{1}, & \textbf{if } a > t, \\ \mathbf{0}, & \textbf{otherwise.} \end{cases} \tag{8.12}$$

This time

$$\sigma_f(A) = |A^t|. \tag{8.13}$$

$f_{2,0}$ is the largest possible weighting function: $f_{1,1} \le f \le f_{2,0}$ for each weighting f. $f = f_{2,0}$ generates the largest sigma f-count of A (see (8.7)), namely

$$\sigma_f(A) = |\,\mathrm{supp}(A)|. \tag{8.14}$$

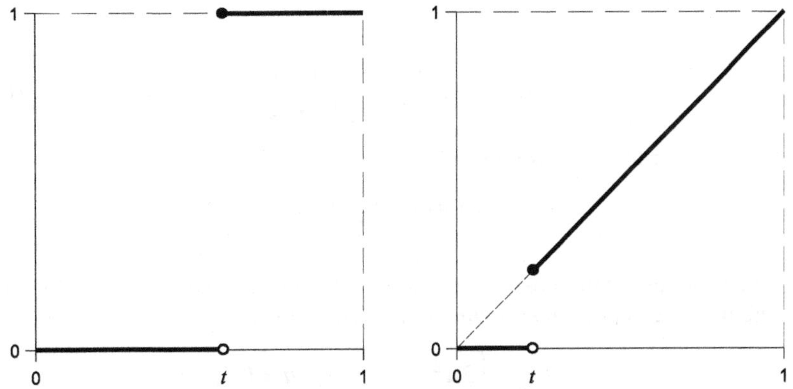

Fig. 8.1 Weighting functions (left) $f_{1,t}$ and (right) $f_{4,t,1}$

As emphasized in Chapter 7, counting by thresholding (sharp or not) forming the CAC method seems to be the most basic counting procedure used by humans when having to count under information imprecision. It collapses to counting up those elements whose quality, $A(x)$, is sufficiently "high", is not lower than or exceeds a predefined quality threshold t. $f_{1,t}$ and $f_{2,t}$ will therefore be called the *basic weighting functions*. They generate integer scalar cardinalities. The following examples of weighting functions lead to generally noninteger values of cardinality.

- **Counting by joining:** $f = f_{3,p}$, where $p > 0$ and

$$f_{3,p}(a) = a^p. \tag{8.15}$$

  Then

$$\sigma_f(A) = \sum_{x \in \mathrm{supp}(A)} (A(x))^p \tag{8.16}$$

  with $A \in \mathrm{FFS}$. Obviously, $p = 1$ gives the usual sigma count of A from (8.3).

- **Counting by thresholding and joining:** $f = f_{4,t,p}$ with $t \in (0, 1]$, $p > 0$ and

$$f_{4,t,p}(a) = \begin{cases} a^p, & \text{if } a \ge t, \\ 0, & \text{otherwise.} \end{cases} \tag{8.17}$$

The particular case with $p = 1$ is presented in Figure 8.1. It leads to

$$\sigma_f(A) = \sum_{x \in A_t} A(x). \tag{8.18}$$

We then ignore the elements having "small" membership degrees ($< t$) and, consequently, $\sigma_f(A)$ becomes free from the effect of possible cumulation of such degrees.

- **Counting by sharp thresholding and joining:** $f = f_{5,t,p}$, where $t \in [0, 1)$, $p > 0$ and

$$f_{5,t,p}(a) = \begin{cases} a^p, & \text{if } a > t, \\ 0, & \text{otherwise.} \end{cases} \tag{8.19}$$

What we get for $p = 1$ is now

$$\sigma_f(A) = \sum_{x \in A^t} A(x). \tag{8.20}$$

- Finally, let us mention two more sophisticated instances of weighting functions. The first one is the *contrast enhancement function* f_6 with

$$f_6(a) = \begin{cases} 2a^2, & \text{if } a \le 0.5, \\ 1 - 2(1-a)^2, & \text{otherwise.} \end{cases} \tag{8.21}$$

Second, by definition, the normed generator of a nilpotent t-conorm is always a weighting function (see Example 3.10). The reader is also referred to WYGRALAK (2003a) for further examples of weighting functions, called therein *cardinality patterns*, and for a theoretical discussion devoted to them.

We realize that different weighting functions can generate very different values of the cardinality $\sigma_f(A)$ of a fuzzy set $A \in \text{FFS}$, especially if the support of A is large in comparison with its core (see (8.7)). The choice of a suitable f is thus essential. It determines our way of doing when performing the counting process in A. If two persons count in a set and obtain different results, then at least one of them made a mistake. Different results of counting in a fuzzy set do not necessarily signify a mistake. Instead, they are possible reflections of using different counting procedures, different counting optics based on different weighting functions (see also the next section).

The concept of the sigma f-count (8.2) of a fuzzy set brings together as particular examples all ideas of scalar cardinality proposed in the existing subject literature (however, see further remarks). Looking historically, the first attempt at defining the cardinality of a fuzzy set in a scalar way was the sigma count (8.3) introduced in DE LUCA/TERMINI (1972) and developed as well as applied in ZADEH (1983a) (see also ZADEH (1999a, b)). A bit forgotten but worth emphasizing fact is that ZADEH (1983a) also proposes two other concepts of scalar cardinality, namely (8.18) and a weighted approach to sigma counts, similar to (8.2), in which

$$\sigma(A) = \sum_{i=1}^{n} w_i \cdot A(x_i) \tag{8.22}$$

with real weights $w_i \geq 0$ and $\boldsymbol{U} = \{x_1, \ldots, x_n\}$. Furthermore, we like to mention that (8.14) was proposed in GOTTWALD (1980), whereas the idea of (8.16) was introduced in KAUFMANN (1977) as the *p-power of A*.

Finally, although the axiomatic approach from Subsection 8.1.1 leading to sigma *f*-counts seems to be quite general and reasonable, it does not encompass all counting methods and forms of scalar cardinalities which are possible to imagine and could be useful in practice. Especially, we mean non-additive scalar cardinalities. They preserve the consistency and monotonicity conditions of type (8.5)-(8.6), which seem to be indispensable with special emphasis on consistency, and violate the additivity condition (P3). At least in some situations, this lack of additivity is acceptable when counting under information imprecision. Examples of non-additive scalar cardinalities are (9.49) and (10.39) in Subsection 9.2.2 and Section 10.3, respectively, forming scalar cardinalities derived from fuzzy cardinalities.

8.1.3 The Eight-Bottle Example Continued

We like to return to the eight-bottle example from Subsection 6.3.4 involving a fuzzy set, *A*, of full bottles (see Figure 6.7). For convenience, let us recollect that

$$\begin{aligned} &A(b_1) = 1, \quad A(b_2) = 0.8, \quad A(b_3) = 0, \quad A(b_4) = 0.5, \\ &A(b_5) = 0.1, \quad A(b_6) = 1, \quad A(b_7) = 0.9, \quad A(b_8) = 0.3. \end{aligned} \tag{8.23}$$

One can say that the entropy measures of fuzziness Fuzz(A) = 1.2 and Fuzz(A) = 0.8 obtained in Subsection 6.3.4 for $\boldsymbol{t} = \wedge$ and $\boldsymbol{t} = \boldsymbol{t}_a$, respectively, are simply particular cases of the sigma count $\sigma_{id}(\chi_{\mathscr{E}})$.

Let us ask a simple question: how many bottles of water do we have? Speaking more formally, what is the cardinality of *A*? Various reasonable counting procedures and, thus, various answers are then possible. A few examples are given below.

- Peter decided to pour the water from one bottle into another trying to make full to the brim as many bottles as possible, which gave

  $$1 + 0.8 + 0.5 + 0.1 + 1 + 0.9 + 0.3 = 4.6 \text{ bottles of water.}$$

  This corresponds to computing $\sigma_{id}(A) = 4.6$.

- Chris does as above. However, he ignores the bottles containing just scraps of water, say, those being less than 25% full. He gets

  $$1 + 0.8 + 0.5 + 1 + 0.9 + 0.3 = 4.5 \text{ bottles of water,}$$

  which is equivalent to computing $\sigma_f(A)$ with $f = f_{4,\,0.25,\,1}$.

- Ann decided to take into account only untouched (totally full) bottles. This collapses to determining

$$\sigma_f(A) = 2 \text{ with } f = f_{1,1}.$$

- Mary is less demanding than Ann and counts up all the bottles which are totally or almost full, say, which are at least 75% full. This counting procedure is equivalent to computing

$$\sigma_f(A) = 4 \text{ with } f = f_{1,\,0.75}.$$

- Mat does as Mary, but he pours together the water in totally or almost full bottles. So, he obtains

$$1 + 0.8 + 1 + 0.9 = 3.7 \text{ bottles,}$$

  which is just $\sigma_f(A)$ with $f = f_{4,\,0.75,1}$.

- Tom decided to count up all the bottles being at least half full. This is equivalent to computing

$$\sigma_f(A) = 5 \text{ with } f = f_{1,\,0.5}.$$

  Clearly, the use of $f_{2,\,0.5}$ would collapse to counting up all the bottles which are more than half full (4 bottles).

- John does as Tom, but he pours together the water in the bottles. The resulting value is just

$$\sigma_f(A) = 4.2 \text{ with } f = f_{4,\,0.5,1}.$$

- Finally, Bob is extremely liberal and counts all the bottles containing even a little bit of water. So, formally, he computes

$$\sigma_f(A) = 7 \text{ with } f = f_{2,\,0}.$$

Analogous options of doing refer to many other practical tasks which require counting under information imprecision.

We see that sigma f-counts offer an adequate formalization of human counting procedures performed under information imprecision, at least of a large class of such procedures. In other words, counting by means of sigma f-counts is human-consistent and makes a true practical and technical sense. Another important conclusion from the eight-bottle example is that counting in a fuzzy set, in contrast to counting in a set, is really a more or less subjective process, and generally leads to different numerical results. What is essential, each of those results can be viewed as a correct and reasonable one. This non-uniqueness is a consequence of using different reasonable optics in the counting process. Peter , Chris, Ann, ... in the above examples are surely able to give a convincing justification for his/her choice of the way of counting. This means, on the other hand, that there is no one, universal and "best" counting procedure under information imprecision in exactly the same way as, say, there is no universal and "best" metric. The choice of a suitable way of measuring the distance between two objects is always problem-dependent. By the way, $\sigma_f(A)$ forms the Hamming distance between $f \circ A$ and the empty fuzzy set.

Finally, let us move on to two other examples of counting in a fuzzy set. First, assume A is a fuzzy set of faulty modules of a system (information, technological, ...). $A(x)$ denotes the degree of failure of module x:

- $A(x) = 0$ means that x works fully properly;
- $A(x) = 1$ signifies a critical failure, x does not work at all;
- $A(x) \in (0, 1)$ means a partial failure of x, i.e. it works but some of its functions are, say, inaccessible or slowed down.

How many modules are faulty? The counting procedure can be based on the following weighting functions:

- $f = f_{2,0}$ - we take into account all, even slight faults,
- $f = f_{1,1}$ - only critical failures are counted up,
- $f = f_{1,0.5}$ - we count up all the modules whose degree of failure is at least 0.5.

In the second example, imagine we like to organize a big scientific conference. We ask (cf. Section 3.1): how many *affordable* hotel single rooms are there in the *vicinity* of the conference venue? The answer can be constructed using the weighting functions $f_{1,t}$ and $f_{2,t}$ with a suitable threshold t, e.g. $t = 0.6$. In other words, we count up all the hotel rooms whose aggregated evaluation with respect to price and distance is at least t or is greater than t, respectively. The use of $f_{3,p}$, $f_{4,t,p}$, $f_{5,t,p}$ or f_6 is not recommended in this case as it does not have a practical justification. The same concerns the previous example, too. It is essential in this context to distinguish between two quite different situations: the membership degrees in a specific fuzzy set can be *joinable* or *non-joinable*, i.e. their summation can make a practical sense or not. As to the former, say, two bottles of water being full to degree 0.5 are more or less the same as one totally full bottle. The counting methods used in the eight-bottle example thus encompass not only counting by thresholding, but also counting by joining or joining combined with thresholding. In the two other examples presented in this subsection we deal with non-joinable membership degrees. For instance, two hotel rooms being affordable and in the vicinity of the conference venue to degree 0.5 are definitely not the same as one room satisfying these two conditions to degree 1. Consequently, counting by thresholding (sharp or not) is then the only appropriate counting method.

8.2 Arithmetic Aspects

This section is devoted to the operations of addition and multiplication of sigma f-counts. We will discuss the question of interpretability of the resulting sums and products. Moreover, selected properties such as compensation, valuation and complementarity will be investigated and characterized. Detailed proofs and other details can be found in WYGRALAK (2003a).

8.2.1 Addition of Sigma *f*-Counts

By (P3) from Subsection 8.1.1, the addition of two sigma *f*-counts always leads to meaningful results having a classical-like interpretation. Speaking more specifically, $\sigma_f(A) + \sigma_f(B)$ is the scalar cardinality $\sigma_f(A \cup B)$ of the sum $A \cup B$ whenever A and B are disjoint.

Another important question one has to look at is whether the scalar cardinalities from (8.2) satisfy the following *compensation property*:

$$\sigma_f(A) < \sigma_f(C) \Rightarrow \exists B \in \mathrm{FFS}\colon \sigma_f(A) + \sigma_f(B) = \sigma_f(C). \tag{8.24}$$

The answer is affirmative whenever f is continuous, $f = f_{1,t}$ or $f = f_{2,t}$ for some t. However, one should be careful if another weighting function is involved. For instance, take $f = f_{4,\,0.25,\,1}$ from (8.17),

$$A = 0.9/x + 0.9/y \quad \text{and} \quad C = 1/x + 0.9/y.$$

Then

$$\sigma_f(A) = 1.8 \quad \text{and} \quad \sigma_f(C) = 1.9,$$

whereas a fuzzy set B such that $\sigma_f(B) = 0.1$ does not exist and, thus, (8.24) is not satisfied.

Similarly, $\sigma_f(A) < \sigma_f(C)$ does not generally imply that $\sigma_f(A) = \sigma_f(C^*)$ for a fuzzy set $C^* \subset C$ if a discontinuous weighting function differing from $f_{1,t}$ and $f_{2,t}$ is used. Indeed, let

$$A = 0.4/x \quad \text{and} \quad C = 0.3/x + 0.3/y$$

and, again, $f = f_{4,\,0.25,\,1}$. Now

$$\sigma_f(A) = 0.4 \quad \text{and} \quad \sigma_f(C) = 0.6,$$

but $C^* \subset C$ with $\sigma_f(C^*) = 0.4$ does not exist.

An important feature of sets and their cardinalities is the well-known *valuation property* saying that $|\boldsymbol{A} \cap \boldsymbol{B}| + |\boldsymbol{A} \cup \boldsymbol{B}| = |\boldsymbol{A}| + |\boldsymbol{B}|$ for each $\boldsymbol{A}, \boldsymbol{B} \subset \boldsymbol{U}$. We like to ask whether and when its extension to fuzzy sets and their sigma *f*-counts is still true. Our task will be to find triples $(f, \boldsymbol{t}, \boldsymbol{s})$ composed of a weighting function f, a t-norm $\boldsymbol{t}$ and a t-conorm $\boldsymbol{s}$ such that

$$\sigma_f(A \cap_t B) + \sigma_f(A \cup_s B) = \sigma_f(A) + \sigma_f(B) \quad \text{for each } A, B \in \mathrm{FFS}. \tag{8.25}$$

First of all, notice that (8.25) is not common in the world of sigma *f*-counts. A counterexample is the triple $(f_{3,\,2}, \boldsymbol{t}_a, \boldsymbol{s}_a)$ with $A = 1/x + 0.9/y$ and $B = 0.7/y + 1/z$. Then

$$A \cap_t B = 0.63/y, \quad A \cup_s B = 1/x + 0.97/y + 1/z$$

and

$$\sigma_f(A \cap_t B) + \sigma_f(A \cup_s B) = 3.34, \quad \sigma_f(A) + \sigma_f(B) = 3.3.$$

Theorem 8.2. *The valuation property holds true iff f, $\boldsymbol{t}$ and $\boldsymbol{s}$ are such that*

$$f(a\boldsymbol{t}b)+f(a\boldsymbol{s}b)=f(a)+f(b) \quad \textit{for each } a, b\in[0,1]. \tag{8.26}$$

Indeed, (⇐) is a consequence of (8.2). On the other hand, suppose (8.26) does not hold, i.e.

$$f(a\boldsymbol{t}b)+f(a\boldsymbol{s}b) \neq f(a)+f(b) \text{ for some } a, b\in(0,1).$$

Putting $A = a/x$ and $B = b/x$ with any $x\in \boldsymbol{U}$, we obtain

$$\sigma_f(A\cap_t B)=f(a\boldsymbol{t}b),\ \sigma_f(A\cup_s B)=f(a\boldsymbol{s}b),\ \sigma_f(A)=f(a) \text{ and } \sigma_f(B)=f(b).$$

The valuation property is thus not fulfilled. Consequently, (8.25) and (8.26) are equivalent, which completes the proof.

Example 8.3. Let us present a few instances of triples (f, $\boldsymbol{t}$, $\boldsymbol{s}$) satisfying (8.26) and, thus, fulfilling the valuation property (8.25):

(a) (f, $\wedge$, $\vee$) with any weighting function f;

(b) (id, $\boldsymbol{t}_a$, $\boldsymbol{s}_a$), (id, $\boldsymbol{t}_L$, $\boldsymbol{s}_L$) and, generally, (id, $\boldsymbol{t}_{F,\lambda}$, $\boldsymbol{s}_{F,\lambda}$) with any $\lambda\in[0,\infty]$, which follows from Theorem 3.6;

(c) (h, $\boldsymbol{s}^\circ$, $\boldsymbol{s}$) with a nilpotent t-conorm $\boldsymbol{s}$ and h being its normed generator, which is a consequence of Theorems 3.8(b) and 3.14. □

Assume in this part of the subsection that $\boldsymbol{U}$ is finite. Another fundamental and well-known property of sets and their cardinalities is that $|\boldsymbol{A}|+|\boldsymbol{A}'| = |\boldsymbol{U}|$ for each $\boldsymbol{A}\subset\boldsymbol{U}$. For fuzzy sets and their sigma f-counts, an analogue of this *complementarity rule* is

$$\sigma_f(A)+\sigma_f(A^\nu) = |\boldsymbol{U}| \quad \text{for each } A\in\text{FFS} \tag{8.27}$$

with a negation ν. Our task now is to find all pairs (f, ν) satisfying this rule.

Theorem 8.4. *The complementarity rule holds true for a weighting function f and a negation* ν *iff*

$$f(a)+f(\nu(a)) = 1 \quad \textit{for each } a\in[0.1]. \tag{8.28}$$

Again, (8.28) ⇒ (8.27) follows from (8.2). If (8.28) is not satisfied, then $f(a)+f(\nu(a)) \neq 1$ for some $a\in(0,1)$. Taking A with $A(x) = a$ for each $x\in\boldsymbol{U}$, one gets

$$\sigma_f(A) = |\boldsymbol{U}|f(a) \text{ and } \sigma_f(A^\nu) = |\boldsymbol{U}|f(\nu(a)),$$

which means that $\sigma_f(A)+\sigma_f(A^\nu) \neq |\boldsymbol{U}|$. This completes the proof.

If ν is a strict negation with equilibrium point $e(\nu)$, then

$$f(e(\nu)) = 0.5 \tag{8.29}$$

whenever (8.28) is satisfied.

Example 8.5. As previously, we like to give some examples of pairs (f, ν) fulfilling (8.28) and, hence, satisfying the complementarity rule (8.27).

(a) $(f, \nu_Ł)$ satisfies (8.28) iff f is such that (0.5, 0.5) is the symmetry point of its diagram. So, for instance, $(id, \nu_Ł)$ and $(f_6, \nu_Ł)$ satisfy (8.28).

(b) Each pair (h, ν_s), where $\boldsymbol{s}$ denotes a nilpotent t-conorm with normed generator h, fulfils (8.28), which follows from Theorem 3.12(c). For instance, if $\boldsymbol{s} = \boldsymbol{s}_{Y,p}$ with $p > 0$, then $\nu_s(a) = (1 - a^p)^{1/p}$ and $h(a) = a^p$ (see Examples 3.10 and 3.13). □

Looking at (8.27) practically, it makes it possible to replace - if more convenient - direct counting in A with indirect counting performed via counting in A^ν (see also Subsection 9.1.1). The result is then subtracted from the cardinality of $\boldsymbol{U}$. By Example 8.5, this can be safely done if, say, the usual sigma count and $\nu = \nu_Ł$ are used, whereas counting by thresholding cannot be performed via counting in A' as they lead to different results.

8.2.2 Multiplication

In contrast to the operation of addition, (P1)-(P3) from Subsection 8.1.1 do not guarantee that multiplication of sigma f-counts must lead to meaningful products. This is why one has to ask about the fulfilment of the *cartesian product rule*

$$\sigma_f(A \times_t B) = \sigma_f(A) \cdot \sigma_f(B) \quad \text{for each } A, B \in \text{FFS} \tag{8.30}$$

with a weighting function f and a t-norm $\boldsymbol{t}$. Our task will be to find all pairs $(f, \boldsymbol{t})$ satisfying (8.30) and, thus, leading to products having a classical-like interpretation in which $\sigma_f(A) \cdot \sigma_f(B)$ is just the sigma f-count of $A \times_t B$. A simple negative example is here $(f, \boldsymbol{t}) = (id, \wedge)$. For $A = 0.4/x$ and $B = 0.5/y$ with $x, y \in \boldsymbol{U}$, we get $\sigma_{id}(A \times B) = 0.4$ and $\sigma_{id}(A) \cdot \sigma_{id}(B) = 0.2$. As one sees, this product has no clear interpretation in terms of A, B and $\wedge$ (see also Example 8.7(a)).

Theorem 8.6. *The cartesian product rule holds true iff f and $\boldsymbol{t}$ are such that*

$$f(a\boldsymbol{t}b) = f(a) \cdot f(b) \quad \text{for each } a, b \in [0, 1]. \tag{8.31}$$

The proof is analogous to those of Theorems 8.2 and 8.4.

Example 8.7. Let us present instances of pairs $(f, \boldsymbol{t})$ satisfying (8.31) and, consequently, fulfilling the cartesian product rule.

(a) $(f, \wedge)$ satisfies (8.31) iff $f = f_{1,t}$ or $f = f_{2,t}$ with any t.

(b) Each pair $(f_{1,1}, \boldsymbol{t})$ with any t-norm $\boldsymbol{t}$ as well as $(f_{2,0}, \boldsymbol{t})$ with $\boldsymbol{t}$ having no zero divisors do satisfy (8.31).

(c) If $\boldsymbol{t}$ is a strict t-norm with generator g, then e^{-g} with $(e^{-g})(a) = e^{-g(a)}$ forms a weighting function and $(e^{-g}, \boldsymbol{t})$ fulfils (8.31) (see Theorem 3.8(a)). A simple instance of a pair $(f, \boldsymbol{t})$ satisfying the cartesian product rule is thus $(id, \boldsymbol{t}_a)$.

(d) If $\boldsymbol{t} = \boldsymbol{t}_a$, (8.31) collapses to

$$f(ab) = f(a) \cdot f(b) \quad \text{for each } a, b \in [0, 1],$$

which is the Cauchy functional equation. Its unique continuous solutions are the weighting functions $f = f_{3,p}$ with $p > 0$. By (b), the extreme weighting functions $f_{1,1}$ and $f_{2,0}$ are examples of discontinuous solutions of that equation.

(e) Finally, worth mentioning is the following negative example. If $\boldsymbol{t}$ is nilpotent, then it has zero divisors and, hence, no pair $(f, \boldsymbol{t})$ with a weighting function f fulfilling the equivalence

$$f(a) = 0 \Leftrightarrow a = 0$$

can satisfy (8.31). In particular, no $(f, \boldsymbol{t})$ with a strictly increasing f can satisfy (8.31). Thus, the cartesian product rule cannot be fulfilled, say, by $(id, \boldsymbol{t}_Ł)$ and $(f_6, \boldsymbol{t}_Ł)$. □

Moving on to conclusions, let us look at the standard case when the usual sigma count is used in combination with one of the basic t-norms and the Łukasiewicz negation. As pointed out in Example 8.7(a, e), if $\boldsymbol{t} = \wedge$ or $\boldsymbol{t} = \boldsymbol{t}_Ł$, then the cartesian product rule does not work and, thus, products of scalar cardinalities – in contrast to sums – do not have a clear interpretation. That rule is satisfied by $\boldsymbol{t} = \boldsymbol{t}_a$. Speaking more generally, the quadruple $(f, \boldsymbol{t}, \boldsymbol{s}, \nu) = (id, \boldsymbol{t}_a, \boldsymbol{s}_a, \nu_Ł)$ seems to be a unique system in which f is not $\{0, 1\}$-valued and which guarantees that sums and products of sigma f-counts have classical-like interpretations and, moreover, both the valuation property and complementarity rule are satisfied. As to $\{0, 1\}$-valued weighting functions, $(f_{1,t}, \wedge, \vee, \nu_Ł)$ and $(f_{2,t}, \wedge, \vee, \nu_Ł)$ have all these features, too, except for the complementarity rule.

8.3 Relative Cardinality

As previously, let f denote a weighting function and $A, B \in \text{FFS}$. The *relative cardinality of A* or, more precisely, the *cardinality of A in relation to B* is the ratio

$$\sigma_f(A \mid B) = \frac{\sigma_f(A \cap B)}{\sigma_f(B)}. \tag{8.32}$$

This value will also be called the *relative sigma f-count of A* or the *sigma f-count of A in relation to B*. $\sigma_f(A \mid B)$ expresses the proportion of the number of elements in A which are also in B to the cardinality of B (cf. (5.25)). By (8.2),

$$\sigma_f(A\,|\,B) = \frac{\sum_{x\in U} f(A(x)\wedge B(x))}{\sum_{x\in U} f(B(x))}. \tag{8.33}$$

If $\boldsymbol{U}$ is a finite universe, (8.5) leads to

$$\sigma_f(A\,|\,1_U) = \frac{\sigma_f(A)}{|U|} = \frac{1}{|U|}\sum_{x\in \mathrm{supp}(A)} f(A(x)), \tag{8.34}$$

which collapses to the proportion of the sigma f-count of A to the cardinality of $\boldsymbol{U}$.

The construction of $\sigma_f(A\,|\,B)$ is analogous to that of conditional probability. Its properties, however, are generally different as fuzzy sets do not form a boolean algebra. An instance is the inequality (ZADEH (1983a))

$$\sigma_{id}(A\,|\,B) + \sigma_{id}(A'\,|\,B) \geq 1 \tag{8.35}$$

for relative sigma counts.

Finally, we like to mention a triangular norm-based approach to relative sigma f-counts. Then

$$\sigma_{f,t}(A\,|\,B) = \frac{\sigma_f(A\cap_t B)}{\sigma_f(B)} \tag{8.36}$$

with a t-norm $\boldsymbol{t}$. Properties of (8.36) are investigated in PILARSKI (2005).

8.4 Counting in IVFSs and IFSs

The subject of this section is an extension of sigma f-counts to interval-valued fuzzy sets (IVFSs) and I-fuzzy sets (IFSs) from Chapter 6.

For a finite interval-valued fuzzy set $\mathcal{E} = (A_l, A_u)$ with $A_l \subset A_u$, let us define its *sigma f-count* $\sigma_f(\mathcal{E})$ as the interval

$$\sigma_f(\mathcal{E}) = [\sigma_f(A_l), \sigma_f(A_u)] \tag{8.37}$$

of nonnegative real numbers, where f denotes a weighting function (see (8.2), (8.6)). This way of doing can be applied to IFSs, too. If $\mathscr{E} = (A^+, A^-)$ with $A^+ \subset (A^-)^\nu$ and a strong negation ν is a finite IFS, its *sigma f-count* is defined by

$$\sigma_f(\mathscr{E}) = [\sigma_f(A^+), \sigma_f((A^-)^\nu)]. \tag{8.38}$$

This cardinality, again, forms a closed interval of nonnegative reals. As in (8.37), its left endpoint is a minimum possible scalar cardinality of an incompletely known fuzzy set A modelled by $\mathscr{E}$, whereas the right endpoint forms a maximum possible scalar cardinality $\sigma_f(A)$ of A (see (6.47)). If $\sigma_f(\mathscr{E}) = \sigma_f(\mathscr{F})$, we will say that $\mathscr{E}$ and $\mathscr{F}$ are *equipotent* IFSs. Equipotency of two IVFSs is defined in the same way.

The cardinality concepts for IVFSs and IFSs presented in (8.37) and (8.38), respectively, are thus identical. This is not surprising as IVFSs and IFSs are formally equivalent constructions. For convenience, we will focus our discussion on the case of IFSs. The reader can easily translate the results into the language of IVFSs and their cardinalities.

Worth emphasizing is that the concept of sigma f-counts can be extended to IVFSs and, thereby, to IFSs in an alternative manner proposed and investigated in DESCHRIJVER/KRÁL (2007) (see also KRÁL (2004, 2005a, b)). That extension is then done already at the level of axioms (P1)-(P3) from Subsection 8.1.1. σ is understood as a mapping assigning to (A_l, A_u) or (A^+, A^-) a closed interval of nonnegative real numbers. The membership degrees and fuzzy sets appearing in (P1)-(P3) are replaced with closed subintervals of $[0, 1]$ and interval-valued fuzzy sets, respectively (see (2.109), (2.118), (6.38)). The resulting characterization is then identical to that in Theorem 8.1. However, the weighting function f it involves is now an interval function whose domain and range is the family of all closed subintervals of $[0, 1]$. The conditions $f(0) = 0$ and $f(1) = 1$ are replaced by $f([0, 0]) = [0, 0]$ and $f([1, 1]) = [1, 1]$, respectively. An interesting theoretical aspect of this alternative approach is that it leads to conditions and weighting functions which are or are not representable in terms of (pairs of) cardinalities and weighting functions from (P1)-(P3) and Theorem 8.1 (cf. Subsection 6.2.3). However, our further study will be focused on the approach offered in (8.37)-(8.38) as it seems to be quite sufficient in applications.

8.4.1 Main Properties

Let us rewrite (8.38) in a more detailed form, namely (see also Theorem 6.3(b))

$$\sigma_f(\mathscr{E}) = \left[\sum_{x \in U} f(A^+(x)),\ \sum_{x \in U} f(\nu(A^-(x)))\right]. \tag{8.39}$$

For the standard negation $\nu_Ł$, we thus get

$$\sigma_f(\mathscr{E}) = \left[\sum_{x \in U} f(A^+(x)),\ \sum_{x \in U} f(1 - A^-(x))\right], \tag{8.40}$$

and the weighting function $f = id$ gives the way of counting proposed in SZMIDT/KACPRZYK (2001) (see also VLACHOS/SEGIADIS (2007a)).

If our knowledge about the fuzzy set A modeled by $\mathscr{E}$ is complete, then $\mathscr{E} = (A, A^\nu)$ and (8.38) becomes a one-element interval:

$$\sigma_f(\mathscr{E}) = [\sigma_f(A), \sigma_f(A)].$$

What we obtain is practically the sigma f-count of A. $\sigma_f(\mathscr{E})$ from (8.38) is always monotonic with respect to $\mathscr{E}$ and f. Indeed,

$$\sigma_f(\mathscr{E}) \le \sigma_f(\mathscr{F}) \text{ for } \mathscr{E} \subset \mathscr{F} \tag{8.41}$$

with $\le$ defined via (2.118), and

$$\sigma_f(\mathscr{E}) \le \sigma_{f^*}(\mathscr{E}) \text{ for } f \le f^*. \tag{8.42}$$

Since $f_{1,1} \le f \le f_{2,0}$ for each weighting function f, we have

$$[|\operatorname{core}(A^+)|, |\operatorname{core}((A^-)^\nu)|] \le \sigma_f(\mathscr{E}) \le [|\operatorname{supp}(A^+)|, |\operatorname{supp}((A^-)^\nu)|]. \tag{8.43}$$

Addition and multiplication of sigma f-counts of IFSs will be defined as (see (2.109) and (2.113))

$$\sigma_f(\mathscr{E}) + \sigma_f(\mathscr{F}) = [\sigma_f(A^+) + \sigma_f(B^+), \sigma_f((A^-)^\nu) + \sigma_f((B^-)^\nu)] \tag{8.44}$$

and

$$\sigma_f(\mathscr{E}) \cdot \sigma_f(\mathscr{F}) = [\sigma_f(A^+) \cdot \sigma_f(B^+), \sigma_f((A^-)^\nu) \cdot \sigma_f((B^-)^\nu)] \tag{8.45}$$

for $\mathscr{E} = (A^+, A^-)$ and $\mathscr{F} = (B^+, B^-)$. It is a routine task to check that

$$\sigma_f(\mathscr{E}) + \sigma_f(\mathscr{F}) = \sigma_f(\mathscr{E} \cup \mathscr{F}) \text{ whenever } \mathscr{E} \cap \mathscr{F} = \mathscr{E}_\varnothing. \tag{8.46}$$

Sums of sigma f-counts of IFSs can thus be interpreted in a classical-like way. Let us notice in this context that, for each t-norm $\boldsymbol{T}$ and $\boldsymbol{S} = \boldsymbol{T}^\nu$, one has $\mathscr{E} \cup_{\boldsymbol{T},\boldsymbol{S}} \mathscr{F} = \mathscr{E} \cup \mathscr{F}$ if $\mathscr{E}$ and $\mathscr{F}$ are disjoint (cf. (P3)'' in Subsection 8.1.1). A consequence of (8.38), (8.44) and (2.85) is the following fact.

Theorem 8.8. *If* $\boldsymbol{T}$ *and* $\boldsymbol{S}$ *are* ν-*dual and the triple* $(f, \boldsymbol{T}, \boldsymbol{S})$ *satisfies* (8.26), *then*

$$\sigma_f(\mathscr{E} \cap_{\boldsymbol{T},\boldsymbol{S}} \mathscr{F}) + \sigma_f(\mathscr{E} \cup_{\boldsymbol{T},\boldsymbol{S}} \mathscr{F}) = \sigma_f(\mathscr{E}) + \sigma_f(\mathscr{F}) \text{ for each } \mathscr{E} \text{ and } \mathscr{F}. \tag{8.47}$$

So, the valuation property (8.25) has a direct extension to IFSs and their sigma f-counts.

Theorem 8.9. *If* $(f, \boldsymbol{T})$ *is a pair fulfilling* (8.31), *then*

$$\sigma_f(\mathscr{E}) \cdot \sigma_f(\mathscr{F}) = \sigma_f(\mathscr{E} \times_{\boldsymbol{T}} \mathscr{F}) \text{ for each } \mathscr{E} \text{ and } \mathscr{F}. \tag{8.48}$$

The proof is a routine matter (see (8.45) and (6.83)). If the pair $(f, \boldsymbol{T})$ satisfies (8.31), the cartesian product rule thus works and, hence, multiplication of sigma f-counts of IFSs leads to meaningful products.

8.4.2 Relativization

This subsection is devoted to an extension of the concept of relative sigma *f*-counts to IFSs. An analogous way of doing can be applied to IVFSs.

Assume $\mathscr{E} = (A^+, A^-)$ and $\mathscr{F} = (B^+, B^-)$ with $A^+ \subset (A^-)^\nu$, $B^+ \subset (B^-)^\nu$, and a strong negation ν. Modeling oneself on (8.32) and using (8.38), let us define

$$\sigma_f(\mathscr{E} \mid \mathscr{F}) = \frac{\sigma_f(\mathscr{E} \cap \mathscr{F})}{\sigma_f(\mathscr{F})} \tag{8.49}$$

with a weighting function *f*. We will say that $\sigma_f(\mathscr{E} \mid \mathscr{F})$ is the *relative sigma f-count of* $\mathscr{E}$, the *sigma f-count of* $\mathscr{E}$ *in relation to* $\mathscr{F}$. $\sigma_f(\mathscr{E} \mid \mathscr{F})$ forms a quotient of two intervals of nonnegative reals with the operation of division performed via (2.114). By (6.85), (8.38) and (2.85), we get

$$\sigma_f(\mathscr{E} \mid \mathscr{F}) = \left[\frac{\sigma_f(A^+ \cap B^+)}{\sigma_f((B^-)^\nu)}, \frac{\sigma_f((A^-)^\nu \cap (B^-)^\nu)}{\sigma_f(B^+)}\right]. \tag{8.50}$$

If U is finite, one can compute

$$\sigma_f(\mathscr{E} \mid \mathscr{E}_U) = \frac{\sigma_f(\mathscr{E})}{[|U|, |U|]} \subset [0, 1], \tag{8.51}$$

the sigma *f*-count of $\mathscr{E}$ in relation to U. More precisely,

$$\sigma_f(\mathscr{E} \mid \mathscr{E}_U) = \left[\frac{\sigma_f(A^+)}{|U|}, \frac{\sigma_f((A^-)^\nu)}{|U|}\right] = [\sigma_f(A^+ \mid 1_U), \sigma_f((A^-)^\nu \mid 1_U)]. \tag{8.52}$$

$\sigma_f(\mathscr{E} \mid \mathscr{E}_U)$ is thus the interval of all ratios $\sigma_f(A)/|U|$ involving all possible forms of an incompletely known fuzzy set $A^+ \subset A \subset (A^-)^\nu$ modeled by $\mathscr{E}$.

One should be careful with (8.50) if $\mathscr{F} \neq \mathscr{E}_U$. The resulting closed interval is not generally contained in [0, 1]. Although its left endpoint always belongs to [0, 1], the right one can be greater than 1. For instance, this is the case of

$$A^+ = 0.5/x + 0.6/y, \quad (A^-)' = 0.6/x + 0.8/y,$$
$$B^+ = 0.5/x + 0.7/y, \quad (B^-)' = 0.8/x + 1/y$$

in $U = \{x, y\}$. Then

$$\sigma_{id}(\mathscr{E} \mid \mathscr{F}) = \frac{[1.1, 1.4]}{[1.2, 1.8]} = \left[\frac{1.1}{1.8}, \frac{1.4}{1.2}\right].$$

This difficulty disappears if $\sigma_f((A^-)^\nu) \le \sigma_f(B^+)$ as $\sigma_f((A^-)^\nu \cap (B^-)^\nu) \le \sigma_f((A^-)^\nu)$. Anyway, a slight modification in (8.50) is required, namely

$$\sigma_f(\mathscr{E} \mid \mathscr{F}) = \left[\frac{\sigma_f(A^+ \cap B^+)}{\sigma_f((B^-)^\nu)}, 1 \wedge \frac{\sigma_f((A^-)^\nu \cap (B^-)^\nu)}{\sigma_f(B^+)}\right]. \tag{8.53}$$

Let $I(\sigma_f(\mathscr{E}|\mathscr{F}))$ denote the interval stretched over all ratios $\sigma_f(A\cap B)/\sigma_f(B)$ from (8.32) with all possible fuzzy sets $A^+ \subset A \subset (A^-)^\nu$ and $B^+ \subset B \subset (B^-)^\nu$. We see that $\sigma_f(\mathscr{E}|\mathscr{F})$ from (8.53) forms an upper evaluation of $I(\sigma_f(\mathscr{E}|\mathscr{F}))$: $I(\sigma_f(\mathscr{E}|\mathscr{F})) \subset \sigma_f(\mathscr{E}|\mathscr{F})$. Indeed, the endpoints of $\sigma_f(\mathscr{E}|\mathscr{F})$ are not generally limit ratios as they are not relative sigma f-counts of fuzzy sets: the denominator of an endpoint involves a fuzzy set which does not appear in the numerator. A simple counterexample is (again, $\boldsymbol{U} = \{x, y\}$)

$$A^+ = (A^-)' = 0.6/x + 0.6/y,$$

$$B^+ = 0.6/x + 0.4/y, \quad (B^-)' = 0.6/x + 0.8/y.$$

Then

$$\sigma_{id}(\mathscr{E}|\mathscr{F}) = \left[\frac{1}{1.4}, 1\right], \quad \text{whereas } I(\sigma_{id}(\mathscr{E}|\mathscr{F})) = \left[\frac{1.2}{1.4}, 1\right].$$

A lower evaluation $L(\sigma_f(\mathscr{E}|\mathscr{F})) \subset I(\sigma_f(\mathscr{E}|\mathscr{F}))$ of $I(\sigma_f(\mathscr{E}|\mathscr{F}))$ is (see also STACHOWIAK (2010))

$$L(\sigma_f(\mathscr{E}|\mathscr{F})) = \left[\frac{\sigma_f(A^+\cap B^+)}{\sigma_f(B^+)} \wedge \frac{\sigma_f(A^+\cap (B^-)^\nu)}{\sigma_f((B^-)^\nu)}, \frac{\sigma_f((A^-)^\nu\cap B^+)}{\sigma_f(B^+)} \vee \frac{\sigma_f((A^-)^\nu\cap (B^-)^\nu)}{\sigma_f((B^-)^\nu)}\right]. \tag{8.54}$$

This evaluation is basically defined as

$$L(\sigma_f(\mathscr{E}|\mathscr{F})) = [a \wedge b \wedge c \wedge d, a \vee b \vee c \vee d]$$

with

$$a = \frac{\sigma_f(A^+\cap B^+)}{\sigma_f(B^+)}, \; b = \frac{\sigma_f(A^+\cap (B^-)^\nu)}{\sigma_f((B^-)^\nu)}, \; c = \frac{\sigma_f((A^-)^\nu\cap B^+)}{\sigma_f(B^+)}, \; d = \frac{\sigma_f((A^-)^\nu\cap (B^-)^\nu)}{\sigma_f((B^-)^\nu)}.$$

That (8.54) generally differs from $I(\sigma_f(\mathscr{E}|\mathscr{F}))$ is exemplified by $\boldsymbol{U} = \{x, y\}$ and

$$A^+ = (A^-)' = 0.1/x + 0.3/y,$$

$$B^+ = 0.2/x + 0.1/y, \quad (B^-)' = 0.6/x + 0.6/y,$$

$$B = 0.6/x + 0.1/y.$$

In this case,

$$L(\sigma_{id}(\mathscr{E}|\mathscr{F})) = \left[\frac{1}{3}, \frac{2}{3}\right], \quad \text{while } \frac{\sigma_{id}(A\cap B)}{\sigma_{id}(B)} = \frac{2}{7} < \frac{1}{3}.$$

Finally, we like to mention a generalization of (8.49) to IFSs equipped with intersections induced by a t-norm $\boldsymbol{T}$ and the t-conorm $\boldsymbol{S} = \boldsymbol{T}^\nu$. Then

$$\sigma_{f,\boldsymbol{T}}(\mathscr{E}|\mathscr{F}) = \sigma_f(\mathscr{E} \cap_{\boldsymbol{T},\boldsymbol{S}} \mathscr{F})/\sigma_f(\mathscr{F}). \tag{8.55}$$

The reader can easily extend to (8.55) the whole discussion and constructions related to (8.50) and (8.53).

8.4.3 The Eight-Bottle Example Once Again

Let us come back to the eight-bottle example in Subsection 6.3.4 involving an incompletely known fuzzy set A of full bottles presented in Figure 6.8. We model A as an I-fuzzy set $\mathscr{E} = (A^+, A^-)$. Recollect that

$$\begin{gathered} A^+(b_1) = 0.9,\quad A^+(b_2) = 0.5,\quad A^+(b_3) = 0,\quad A^+(b_4) = 0.25,\\ A^+(b_5) = 0.1,\quad A^+(b_6) = 1,\quad A^+(b_7) = 0.4,\quad A^+(b_8) = 0.3,\\ A^-(b_1) = 0,\quad A^-(b_2) = 0,\quad A^-(b_3) = 0.8,\quad A^-(b_4) = 0,\\ A^-(b_5) = 0.9,\quad A^-(b_6) = 0,\quad A^-(b_7) = 0.1,\quad A^-(b_8) = 0.3. \end{gathered} \tag{8.56}$$

The sums of hesitation degrees $\chi_{\mathscr{E}}(b_i)$ from (6.95)-(6.97) are just the sigma counts of $\chi_{\mathscr{E}}$. An interesting type of information is also offered by $\sigma_f(\chi_{\mathscr{E}})$ with $f \neq id$. Let us use the standard hesitation degrees $\chi_{\mathscr{E}}(b_i) = 1 - A^+(b_i) - A^-(b_i)$ from (6.95) and consider a few instances.

- Assume our hesitation as to b_i is viewed as *large* if $\chi_{\mathscr{E}}(b_i) \geq 0.5$. We get

  $$\sigma_f(\chi_{\mathscr{E}}) = 3 \text{ for } f = f_{1,0.5}$$

  and

  $$\sigma_f(\chi_{\mathscr{E}}) = 1.75 \text{ for } f = f_{4,0.5,1}.$$

  This means that the large hesitation involves 3 bottles, which are b_2, b_4 and b_7. Its size is equivalent to a complete lack of knowledge about the content of 1.75 bottles.

- Further, one has

  $$\sigma_f(\chi_{\mathscr{E}}) = 6 \text{ for } f = f_{2,0},$$

  i.e. a positive hesitation encompasses 6 bottles.

Looking at Figure 6.8, let us ask: how many bottles of water do we have? In other words, we ask about the cardinality of incompletely known fuzzy set A of full bottles modeled by $\mathscr{E}$. Different reasonable counting procedures and, consequently, various numerical answers are again possible. This diversity is formally reflected in different possible choices of a weighting function f in (8.39). Use the standard negation $\nu = \nu_Ł$. By (8.40), we obtain

$$\sigma_f(\mathscr{E}) = \left[\sum_i f(A^+(b_i)), \sum_i f(1 - A^-(b_i))\right]. \tag{8.57}$$

Examples of the resulting cardinalities are listed below.

- $\sigma_{id}(\mathscr{E}) = [3.45, 5.9]$.

  The left endpoint of this interval forms a pessimistic, lower evaluation: we have at least 3.45 bottles of water. The right one is an optimistic, upper evaluation according to which we have at most 5.9 bottles of water. Notice that generally

  $$\sigma_{id}(\mathscr{E}) = [\sigma_{id}(A^+), \sigma_{id}(A^+) + \sigma_{id}(\chi_{\mathscr{E}})]. \tag{8.58}$$

- $\sigma_f(\mathscr{E}) = [5, 6]$ for $f = f_{1,\,0.3}$, and $\sigma_f(\mathscr{E}) = [3.1, 5.6]$ for $f = f_{4,\,0.3,\,1}$.

  For some reason, the counting person is now interested in the bottles being full at least to degree 0.3. The above results say that 5 bottles are certainly full to such a degree. The total amount of water in those bottles is equal at least to 3.1 bottles. On the other hand, 6 bottles are surely or possibly full at least to degree 0.3, and they contain at most 5.6 bottles of water in total.

- $\sigma_f(\mathscr{E}) = [7, 8]$ for $f = f_{2,\,0}$.

  This time the interest is in the bottles containing even a little bit of water. $\sigma_f(\mathscr{E})$ says that 7 bottles surely meet that requirement, whereas 8 bottles are surely or possibly not totally empty.

- $\sigma_f(\mathscr{E}) = [1, 4]$ for $f = f_{1,1}$.

  Totally full bottles are now of interest to the counting person. $\sigma_f(\mathscr{E})$ indicates that 1 bottle is surely full to the brim, while 4 bottles are surely or possibly totally full.

Finally, by (8.52), one has

$$\sigma_{id}(\mathscr{E}\,|\,\mathscr{E}_U) = \left[\frac{3.45}{8}, \frac{5.9}{8}\right] = [0.43, 0.74].$$

This relative cardinality informs us that, for the bottles in Figure 6.8, the average degree of filling with water lies somewhere between 0.43 and 0.74, between 43% and 74%.

Concluding, sigma f-counts turn out to be useful when dealing with incompletely known fuzzy sets, too. They make it possible to model different real counting procedures performed by humans under imprecision combined with incompleteness of information.

Chapter 9

Fuzzy Approach

We like to present a study of the fuzzy approach to intelligent counting. The result of counting in a fuzzy set is then itself a fuzzy set of nonnegative integers. We will define and investigate various types of fuzzy cardinalities. In each case, similarly to the scalar approach from Chapter 8, we will emphasize that they reflect and formalize the results of real counting methods used by human beings when counting under information imprecision. An especially interesting type of fuzzy cardinalities are so-called FECounts. Their connections with classification and similarity measures as well as a look at FECounts through rule-based systems and the Bellman-Zadeh model of decision making will be presented.

Finally, we like to deal with counting under imprecision combined with incompleteness of information about the objects of counting. To that end, fuzzy cardinalities will be extended to interval-valued fuzzy sets and I-fuzzy sets.

9.1 Related Methods of Counting in Fuzzy Sets

This section discusses various counting methods in which the resulting cardinalities of fuzzy sets are themselves fuzzy sets in the universe $\mathbb{N}$ of nonnegative integers, i.e. are functions $\mathrm{FFS} \to [0, 1]^{\mathbb{N}}$. One of the goals we set ourselves is that of showing that these constructions are not sophisticated and reflect human procedures of counting under imprecision.

Our discussion in this chapter requires some additional symbols. From now on, for compactness of notation, we put

$$m = |\operatorname{core}(A)| \text{ and } n = |\operatorname{supp}(A)| \quad \text{for } A \in \mathrm{FFS}.$$

Moreover, let

$$[A]_k = \bigvee\{t \in (0, 1]: |A_t| \geq k\} \text{ with } k \in \mathbb{N}. \tag{9.1}$$

Since A is finite, $\bigvee$ collapses to the usual maximum operation. We easily notice that $[A]_k$ with $0 < k \leq n$ is just the kth element of the nonincreasingly ordered sequence of all positive membership values $A(x)$, including their possible repetitions. In other words, $[A]_k$ is the kth greatest membership degree in A, including possible repetitions of those degrees. As we see, one has

$$[A]_k = 1 \text{ for } 0 \leq k \leq m, \quad [A]_k = 0 \text{ for } k > n$$

M. Wygralak: *Intelligent Counting Under Information Imprecision*, STUDFUZZ 292, pp. 187–229.
DOI: 10.1007/978-3-642-34685-9_9

and

$$[A]_k \in (0, 1) \ \text{for} \ m < k \le n.$$

Look at an example with

$$A = 0.3/x_1 + 1/x_2 + 0.3/x_3 + 0.8/x_4 + 0.2/x_5.$$

Then $m = 1$, $n = 5$, and

$$[A]_0 = [A]_1 = 1, \ [A]_2 = 0.8, \ [A]_3 = [A]_4 = 0.3, \ [A]_5 = 0.2,$$

whereas $[A]_k = 0$ for $k > 5$.

Generally, $[A]_k$ is not only nonincreasing with respect to k. We have

$$[A]_k \le [B]_k \ \text{for each} \ k \in \mathbb{N} \ \text{whenever} \ A \subset B, \tag{9.2}$$

i.e. it is also monotonic with respect to A.

9.1.1 MCAC and the Basic Fuzzy Count

Let us return to the MCAC method from Chapter 7. As we pointed out, it seems to be more subtle and offers more advanced counting results than CAC with any single threshold t as a weakness of the latter is that $|A_t| = |B_t|$ and $|A^t| = |B^t|$ may hold even if A and B are very different fuzzy sets. Second, we see that the larger the number of the thresholds we employ the better. The final result of counting is then more and more detailed.

For convenience, let us focus on the case of t-cuts A_t. Going to extreme, one can use all possible thresholds from (0, 1] and combine all the results of counting. This collapses to the following idea of cardinality of $A \in \mathrm{FFS}$:

$$|A| = \sum_{t \in (0,1]} t/|A_t|. \tag{9.3}$$

It was proposed in ZADEH (1979) and, looking historically, is the first concept of counting formulated within the fuzzy approach. $|A|$ from (9.3) will be called the *basic fuzzy count of A* (*BFCount of A*, in short), and will be denoted by BF(A). So, formally, BF: FFS $\to [0, 1]^{\mathbb{N}}$ with

$$\begin{aligned}
\mathrm{BF}(A) &= \sum_{t \in (0,1]} t/|A_t| \\
&= \sum_{k \in \mathbb{N}} \bigvee\{t \in (0,1]: |A_t| = k\}/k \\
&= \sum_{k=0}^{n} \bigvee\{t \in (0,1]: |A_t| = k\}/k
\end{aligned} \tag{9.4}$$

and, hence,

$$\mathrm{BF}(A)(k) = \bigvee\{t \in (0,1]: |A_t| = k\} \quad \text{for } k \in \mathbb{N}. \tag{9.5}$$

One can say that what BF(A) forms is an encoded and compact piece of information about all possible values of $\sigma_f(A)$ with $f = f_{1,t}$ and $t \in (0, 1]$, i.e. about all possible results of counting by thresholding in A (see Example 9.1). Let us emphasize once again that counting by thresholding, a variant of the CAC method from Chapter 7, forms one of the fundamental human counting procedures under information imprecision (see also Subsection 8.1.2). If $\mathrm{BF}(A)(k) = 0$, k is impossible as a result of counting by thresholding in A, whereas $\mathrm{BF}(A)(k) > 0$ means that $\mathrm{BF}(A)(k)$ is a maximum threshold t giving $\sigma_f(A) = k$.

Example 9.1. Let

$$A = 0.6/x_1 + 1/x_2 + 0.9/x_3 + 0.8/x_4 + 1/x_5 + 0.6/x_6 + 0.6/x_7 + 0.2/x_8.$$

By (9.5), we get

$$\mathrm{BF}(A) = 1/2 + 0.9/3 + 0.8/4 + 0.6/7 + 0.2/8.$$

It is convenient to rewrite this result as

$$\mathrm{BF}(A) = (\underset{0}{0}, \underset{1}{0}, \underset{2}{1}, \underset{3}{0.9}, \underset{4}{0.8}, \underset{5}{0}, \underset{6}{0}, \underset{7}{0.6}, \underset{8}{0.2}).$$

In this *vector notation,* the fuzzy cardinality of A is expressed as a vector whose coordinates are numbered from 0. The value of the kth component is just $\mathrm{BF}(A)(k)$, whereas the right round bracket signifies that $\mathrm{BF}(A)(k) = 0$ for the remaining k's. That notation will also be applied to other types of fuzzy cardinalities. Taking into account the form of BF(A), we easily guess all possible results of counting by thresholding in A ($f = f_{1,t}$):

$$\sigma_f(A) = \begin{cases} 2, & \text{if } t \in (0.9, 1], \\ 3, & \text{if } t \in (0.8, 0.9], \\ 4, & \text{if } t \in (0.6, 0.8], \\ 7, & \text{if } t \in (0.2, 0.6], \\ 8, & \text{if } t \in (0, 0.2]. \end{cases}$$

□

Looking at BFCounts very practically, BF(A) in the vector notation forms a modification of the nonincreasing list $[A]_0, [A]_1, [A]_2, \ldots, [A]_n$. If a value on that list repeats itself, we leave its last copy, whereas the remaining ones are replaced with zeros. Indeed, in Example 9.1, we have

$$[A]_0 = [A]_1 = [A]_2 = 1, \ [A]_3 = 0.9, \ [A]_4 = 0.8,$$

$$[A]_5 = [A]_6 = [A]_7 = 0.6,\ \ [A]_8 = 0.2,$$

and

$$\mathrm{BF}(A) = (0, 0, [A]_2, [A]_3, [A]_4, 0, 0, [A]_7, [A]_8).$$

Lists of this type are frequently used by humans as a basis for decision making, which is illustrated by the following

Example 9.2. Let us refer to the conference example from Subsection 8.1.3. Imagine that the organizers try to explore accommodation possibilities and inspect various hotels. They are interested in the number of available, *affordable* single rooms in the *vicinity* of the conference venue. For each available room, an aggregated evaluation with respect to the price and distance conditions is formulated. Assume the following discrete scale is used: 1 (perfect), 0.75 (good), 0.5 (fair, acceptable), 0.25 (rather unsuitable, of last resort), 0.1 (available, but definitely to expensive or/and too distant). The results are collected below.

Table 9.1 Information about the number of available rooms

Number of rooms	Aggregated evaluation	Cumulative number of rooms
80	1	80
145	0.75	225
125	0.5	350
60	0.25	410
140	0.1	550

These data are for the organizers a convenient basis for coming to a decision as to an upper limit on the number of participants. They suggest that a reasonable limit is "no more than about 350 participants". A larger number of participants (especially, a number greater than 410) is risky because some of them will be dissatisfied with their accommodation.

Let us realize that what Table 9.1 presents is nothing else than the basic fuzzy count $\mathrm{BF}(R)$ of a fuzzy set R of affordable hotel rooms in the vicinity of the conference venue, implicitly created by the organizers during the evaluation process. More precisely (see the second and third column of Table 9.1),

$$\mathrm{BF}(R) = 1/80 + 0.75/225 + 0.5/350 + 0.25/410 + 0.1/550.$$

And just this fuzzy cardinality, looking formally, is a basis for solving the upper limit problem. □

We move on to general properties of BF(A) with $A \in$ FFS. The following list presents simple consequences of (9.4):

- BF(A) is always normal as BF(A)(m) = 1;
- BF(A) is not convex, unless A does not involve repetitions of membership degrees from (0, 1);
- BF(A)(k) = 0 for $k < m$ and $k > n$;
- BF(A) is strictly decreasing on its support;
- BF(A) = $1/n$ whenever $A \in$ FCS;
- BF(A) = BF(B) iff $[A]_k = [B]_k$ for each $k \in \mathbb{N}$ (see also Section 9.4).

Again, assume for a moment that $\boldsymbol{U}$ is a finite universe. A well-known and simple trick when having to count in a set $\boldsymbol{D} \subset \boldsymbol{U}$ and $|\boldsymbol{U}|$ is given is to make use of the equality $|\boldsymbol{D}| = |\boldsymbol{U}| - |\boldsymbol{D}'|$. If more convenient, direct counting in $\boldsymbol{D}$ can then be replaced with indirect counting: first, we determine $|\boldsymbol{D}'|$ and, second, the result is subtracted from $|\boldsymbol{U}|$. Let us try to apply the same trick to counting in a fuzzy set A. For sigma f-counts, Theorem 8.4 and Example 8.5 describe conditions on which both the ways of counting, direct and indirect, give identical results. If the basic fuzzy count is involved, we will determine BF(A') and subtract it from $|\boldsymbol{U}|$. However, the result of this *indirect counting* differs from BF(A) and, in essence, forms a new type of fuzzy cardinality of A (cf. also GRZEGORZEWSKI (2006)).

Moving on to details, let us define the *dual basic fuzzy count of A* (DBFCount, in short) by

$$\mathrm{DBF}(A) = \mathrm{BF}(1_U) - \mathrm{BF}(A') \tag{9.6}$$

with the subtraction performed via the extension principle (2.104) with $\boldsymbol{t} = \wedge$. As $\mathrm{BF}(1_U) = 1/|\boldsymbol{U}|$, we get

$$\begin{aligned}\mathrm{DBF}(A)(k) &= \bigvee\{\mathrm{BF}(1_U)(i) \wedge \mathrm{BF}(A')(j)\colon\ i - j = k\}\\ &= \mathrm{BF}(A')(|\boldsymbol{U}| - k)\\ &= \bigvee\{t \in (0,1]\colon\ |(A')_t| = |\boldsymbol{U}| - k\}.\end{aligned}$$

Since

$$|(A')_t| = |\boldsymbol{U}| - |A^{1-t}|$$

with $t \in (0, 1]$, this gives

$$\begin{aligned}\mathrm{DBF}(A)(k) &= \bigvee\{t \in (0,1]\colon\ |\boldsymbol{U}| - |(A')_t| = k\}\\ &= \bigvee\{t \in (0,1]\colon\ |A^{1-t}| = k\}\end{aligned} \tag{9.7}$$

for $k = 0, 1, \ldots, |\boldsymbol{U}|$. Hence

$$\begin{aligned}\mathrm{DBF}(A) &= \sum_{k=0}^{|U|} \bigvee\{t\in(0,1]\colon |U| - |(A')_t| = k\}/k \\ &= \sum_{k\in\mathbf{N}} \bigvee\{t\in(0,1]\colon |U| - |(A')_t| = k\}/k\end{aligned} \tag{9.8}$$

and, on the other hand,

$$\begin{aligned}\mathrm{DBF}(A) &= \sum_{k=0}^{|U|} \bigvee\{t\in(0,1]\colon |A^{1-t}| = k\}/k \\ &= \sum_{k\in\mathbf{N}} \bigvee\{t\in(0,1]\colon |A^{1-t}| = k\}/k \\ &= \sum_{t\in(0,1]} t/|A^{1-t}|.\end{aligned} \tag{9.9}$$

DBF(A) is an encoded, compact piece of information about all possible results $|U| - |(A')_t| = |A^{1-t}|$ of indirect counting in A by thresholding. DBF(A)(k) = 0 means that k is impossible as a result of that counting. If DBF(A)(k) > 0, this positive value forms a maximum threshold t giving $|U| - |(A')_t| = |U| - \sigma_f(A') = k$, where $f = f_{1,t}$. Finally, as (9.9) suggests, DBFCounts can also be applied to (finite) fuzzy sets in infinite universes. Our initial assumption that U is finite is thus no longer necessary and can be removed.

Example 9.3. We refer to A from Example 9.1. By (9.9), one gets

$$\begin{aligned}\mathrm{DBF}(A) &= 0.1/2 + 0.2/3 + 0.4/4 + 0.8/7 + 1/8 \\ &= (0, 0, 0.1, 0.2, 0.4, 0, 0, 0.8, 1)\end{aligned}$$

as

$$|A^{1-t}| = \begin{cases} 2, & \text{if } t\in(0, 0.1], \\ 3, & \text{if } t\in(0.1, 0.2], \\ 4, & \text{if } t\in(0.2, 0.4], \\ 7, & \text{if } t\in(0.4, 0.8], \\ 8, & \text{if } t\in(0.8, 1], \end{cases}$$

i.e. DBF(A) ≠ BF(A). □

We see that DBF(A) with $A \in$ FFS is a modification of the nondecreasing sequence $1 - [A]_1, 1 - [A]_2, \ldots, 1 - [A]_n, 1 - [A]_{n+1}$ containing the negations of all membership degrees in A. If a value on that list repeats itself, we leave its first copy, and the remaining copies are replaced with zeros. For instance, DBF(A) from Example 9.3 is of the form

$$\mathrm{DBF}(A) = (1 - [A]_1, 0, 1 - [A]_3, 1 - [A]_4, 1 - [A]_5, 0, 0, 1 - [A]_8, 1 - [A]_9).$$

Finally, we like to present a few general properties of DBFCounts which are consequences of (9.8) and (9.9):

- DBF(A) is always normal since DBF(A)(n) = 1;
- DBF(A) is not convex, unless A does not involve repetitions of membership degrees from (0, 1);
- DBF(A)(k) = 0 for $k < m$ and $k > n$;
- DBF(A) is strictly increasing on its support;
- DBF(A) = $1/n$ for $A \in$ FCS;
- DBF(A) = DBF(B) iff $[A]_k = [B]_k$ for each $k \in \mathbb{N}$.

In the next subsection, we will present two types of fuzzy cardinalities which are some derivatives of BFCounts and DBFCounts.

9.1.2 FGCounts and FLCounts

That BF(A) and DBF(A) with $A \in$ FFS are not generally convex is sometimes a troublesome property. Examples 9.1 and 9.3 suggest a simple way of making them convex. Speaking technically, this can be done by replacing the condition "$|A_t| = k$" in (9.4) with "$|A_t| \geq k$", and by replacing "$|A^{1-t}| = k$" in (9.9) with "$|A^{1-t}| \leq k$". The resulting fuzzy cardinalities are called the *FGCount* and *FLCount of A*, and will be denoted by FG(A) and FL(A), respectively. More precisely, we define

$$\begin{aligned} \mathrm{FG}(A) &= \sum_{k\in\mathbf{N}} \bigvee\{t \in (0,1] \colon |A_t| \geq k\}/k \\ &= \sum_{k\in\mathbf{N}} [A]_k/k \\ &= \sum_{k=0}^{n} [A]_k/k, \end{aligned} \tag{9.10}$$

i.e.

$$\mathrm{FG}(A)(k) = [A]_k \quad \text{for } k \in \mathbb{N}. \tag{9.11}$$

Further,

$$\begin{aligned} \mathrm{FL}(A) &= \sum_{k\in\mathbf{N}} \bigvee\{t \in (0,1] \colon |A^{1-t}| \leq k\}/k \\ &= \sum_{k\in\mathbf{N}} \bigvee\{t \in (0,1] \colon [A]_{k+1} \leq 1 - t\}/k \\ &= \sum_{k\in\mathbf{N}} (1 - [A]_{k+1})/k, \end{aligned} \tag{9.12}$$

which means that

$$\mathrm{FL}(A)(k) = 1 - [A]_{k+1} \quad \text{for } k \in \mathbb{N}. \tag{9.13}$$

First, we like to look a bit closer at FGCounts whose idea was introduced in BLANCHARD (1981, 1982) and ZADEH (1981b). In the vector notation, (9.10) can be rewritten as

$$\mathrm{FG}(A) = (1, [A]_1, [A]_2, \dots, [A]_n). \tag{9.14}$$

To give a better insight into the nature of FGCounts, it is very instructive to present them in the language of Łukasiewicz logic (see Subsection 2.1.3). By (2.10), (5.26), (2.9), (2.7) and (2.3), we get

$$\begin{aligned}[\![\exists_{\mathrm{m}} \boldsymbol{D} \in \boldsymbol{P}_k\colon 1_{\boldsymbol{D}} \subset_{\mathrm{m}} A]\!] &= \bigvee_{\boldsymbol{D} \in \boldsymbol{P}_k} [\![1_{\boldsymbol{D}} \subset_{\mathrm{m}} A]\!] \\ &= \bigvee_{\boldsymbol{D} \in \boldsymbol{P}_k} \bigwedge_{x \in \boldsymbol{U}} 1_{\boldsymbol{D}}(x) \to_{\text{Ł}} A(x) \\ &= \bigvee_{\boldsymbol{D} \in \boldsymbol{P}_k} \bigwedge_{x \in \boldsymbol{U}} A(x) \\ &= [A]_k\end{aligned} \tag{9.15}$$

with $k \in \mathbb{N}$ and $\boldsymbol{P}_k$ denoting the family of all k-element subsets of $\boldsymbol{U}$. One can thus say that $\mathrm{FG}(A)(k)$ is a degree to which A contains (in the many-valued sense) a k-element set or, in other words, is a degree to which A has *at least k* elements. This justifies the acronymic name "FGCount" introduced by Lotfi A. Zadeh in which "F" and "G" stand for "fuzzy" and "greater than or equal to", respectively. For completeness, let us notice that the Łukasiewicz implication operator $\to_{\text{Ł}}$ in (9.15) can be replaced with any R-implication, S-implication or QL-implication from Subsection 2.4.5. Main properties of $\mathrm{FG}(A)$ with $A \in \mathrm{FFS}$ are listed below and follow from (9.14) and (9.2):

- $\mathrm{FG}(A)$ is normal and convex;
- $\mathrm{FG}(A)(k) = 1$ for $k \le m$, $\mathrm{FG}(A)(k) = 0$ for $k > n$,
  $\mathrm{FG}(A)(k) \in (0, 1)$ for $m < k \le n$;
- $\mathrm{FG}(A) = (1, 1, \dots, 1)$ with $n+1$ ones if $A \in \mathrm{FCS}$;
- $\mathrm{FG}(A) \subset \mathrm{FG}(B)$ whenever $A \subset B$;
- $\mathrm{FG}(A) = \mathrm{FG}(B)$ iff $[A]_k = [B]_k$ for each $k \in \mathbb{N}$.

Speaking practically, the cost of constructing $\mathrm{FG}(A)$ is equal to the cost of sorting the $A(x)$'s. Second, $\mathrm{FG}(A)$ forms the list of ranking points of elements of $\boldsymbol{U}$ with respect to their membership degree in A. Conversely, after normalization, each list of ranking points of some objects can be viewed as the FGCount of a fuzzy set. FGCounts and ranking lists, a basis for coming to a decision in a lot of situations and dimensions of our life, are thus equivalent constructions in this sense. Moreover, let us notice that

$$(\mathrm{FG}(A))_t = \{0, 1, \ldots, |A_t|\} = \{0, 1, \ldots, \sigma_{f_{1,t}}(A)\} \quad \text{for each } t \in (0, 1]$$

and (9.16)

$$(\mathrm{FG}(A))^t = \{0, 1, \ldots, |A^t|\} = \{0, 1, \ldots, \sigma_{f_{2,t}}(A)\} \quad \text{for each } t \in [0, 1).$$

Hence

$$\sigma_{f_{1,t}}(A) = |A_t| = |(\mathrm{FG}(A))_t| - 1$$

and (9.17)

$$\sigma_{f_{2,t}}(A) = |A^t| = |(\mathrm{FG}(A))^t| - 1.$$

Similarly to the constructions from Subsection 9.1.1, FG(A) thus forms a dynamic and compact piece of information about all possible results of counting by thresholding (sharp or not) in A. FGCounts are, in other words, closely connected with human counting methods under information imprecision. $\mathrm{FG}(A)(k) = t \in (0, 1]$ means that $|A_t|$ contains at least k elements. This forms an additional justification for the acronym "FGCount".

FG(A) is also a convenient basis for determining other types of scalar cardinalities, e.g.

$$\sigma_{id}(A) = \sum_{k=1}^{n} \mathrm{FG}(A)(k),$$

(9.18)

$$\sigma_f(A) = \sum_{k=1}^{|(\mathrm{FG}(A))_t| - 1} \mathrm{FG}(A)(k) \quad \text{for } f = f_{4,\,t,\,1}.$$

On the other hand, FG(A) can be defuzzified by searching for a point k^* of an abrupt fall in $\mathrm{FG}(A)(k)$. Reasonable variants of defining k^* are then

$$k^* \in \operatorname*{argmax}_k ([A]_k - [A]_{k+1}), \tag{9.19}$$

$$k^* \in \operatorname*{argmax}_k (([A]_k - [A]_{k+1})/[A]_k), \tag{9.20}$$

$$k^* \in \operatorname*{argmax}_k ([A]_k \wedge \nu([A]_{k+1})) \tag{9.21}$$

with a strict negation ν (see also Subsection 9.2.3; cf. DUBOIS/PRADE (1990a, b)). Such a defuzzification is frequently used in practice when dealing, say, with a ranking list of some objects and trying to establish a number of acceptable objects, acceptable candidates, etc. (cf. also YAGER (2006)).

As to (9.21), the choice of the smallest or largest k^* satisfying that condition is suggested (see Subsections 9.2.1 and 9.2.3 for details). k^* then collapses to a result of counting by thresholding in A and, thus, forms a sigma f-count.

Example 9.4. Let us return to

$$A = 0.6/x_1 + 1/x_2 + 0.9/x_3 + 0.8/x_4 + 1/x_5 + 0.6/x_6 + 0.6/x_7 + 0.2/x_8$$

from Example 9.1. By (9.14),

$$\mathrm{FG}(A) = (1, 1, 1, 0.9, 0.8, 0.6, 0.6, 0.6, 0.2).$$

This cardinal information about A can be used as a whole in further processing. On the other hand, FG(A) forms a basis for deriving many types of sigma f-counts of A. For instance, using (9.17), we conclude that

$$\sigma_f(A) = |(\mathrm{FG}(A))_t| - 1 = 4 \quad \text{for } f = f_{1,t} \text{ with } t \in (0.6, 0.8]$$

and

$$\sigma_f(A) = |(\mathrm{FG}(A))^t| - 1 = 3 \quad \text{for } f = f_{2,t} \text{ with } t \in [0.8, 0.9).$$

By (9.18),

$$\sigma_{id}(A) = \sum_{k=1}^{8} \mathrm{FG}(A)(k) = 5.7$$

and

$$\sigma_f(A) = \sum_{k=1}^{4} \mathrm{FG}(A)(k) = 3.7 \quad \text{for } f = f_{4,\,0.7,\,1}. \qquad \square$$

Finishing this discussion on FGCounts, let us mention their modification proposed in DUBOIS (1981) (see also DUBOIS/PRADE (1985)). The fuzzy cardinality FG*(A) of $A \in \mathrm{FFS}$ is then defined as

$$\mathrm{FG}^*(A) = [\![\exists_{\mathrm{m}} \boldsymbol{D} \in \boldsymbol{P}_k\, (\boldsymbol{D} \supset \mathrm{core}(A)) \colon 1_D \subset_{\mathrm{m}} A]\!] \quad \text{for } k \in \mathbb{N}. \qquad (9.22)$$

It is easy to check that

$$\mathrm{FG}^*(A) = [A]_k \wedge 1 - [1_{\mathrm{core}(A)}]_{k+1},$$

i.e. (9.23)

$$\mathrm{FG}^*(A) = (0, \ldots, 0, 1, [A]_{m+1}, [A]_{m+2}, \ldots, [A]_n)$$

with m zeros at the beginning of the vector.

We move on to FLCounts, a concept introduced in ZADEH (1983a). By (9.13), one has

$$\mathrm{FL}(A) = (1 - [A]_1, 1 - [A]_2, \ldots, 1 - [A]_n, (1)) \qquad (9.24)$$

for $A \in \mathrm{FFS}$. In this *enriched vector notation*, "(1)" signifies that FL(A)(k) = 1 for each $n \le k \le |\boldsymbol{U}|$ ($\boldsymbol{U}$ - finite) or for each $k \ge n$ ($\boldsymbol{U}$ - infinite). A sequence of transformations analogous to those in (9.15) leads to the following:

$$[\![\exists_{\mathrm{m}} \boldsymbol{D} \in \boldsymbol{P}_k \colon A \subset_{\mathrm{m}} 1_D]\!] = 1 - [A]_{k+1} \quad \text{for each } k \in \mathbb{N}. \qquad (9.25)$$

FL(A)(k) is thus a degree to which A is contained in a k-element set, i.e. is a degree to which A has *at most* k elements. Again, this explains the acronymic name "FLCount": "F" and "L" stand for "fuzzy" and "less than or equal to", respectively. We see that

$$\mathrm{FG}(A)(k) + \mathrm{FL}(A)(k-1) = 1 \quad \text{for each } k \geq 1, \tag{9.26}$$

i.e. the degree to which A has $\geq k$ elements and the degree to which A has $< k$ elements always sum up to 1. FGCounts and FLCounts are thus numerically different, but equivalent constructions. Basic properties of FL(A) are collected below:

- FL(A) is normal and convex;
- $\mathrm{FL}(A)(k) = 0$ for $k < m$, $\mathrm{FL}(A)(k) = 1$ for $k \geq n$,
  $\mathrm{FL}(A)(k) \in (0, 1)$ for $m \leq k < n$;
- $\mathrm{FL}(A) = (0, \dots, 0, (1))$ with n zeros if $A \in \mathrm{FCS}$;
- $\mathrm{FL}(B) \subset \mathrm{FL}(A)$ whenever $A \subset B$;
- $\mathrm{FL}(A) = \mathrm{FL}(B)$ iff $[A]_k = [B]_k$ for each $k \in \mathbb{N}$.

FL(A) forms a nondecreasing list of Łukasiewicz negations of all membership degrees in A, including their repetitions. Let us notice that

$$(\mathrm{FL}(A))_t = \{|A^{1-t}|, |A^{1-t}| + 1, \dots\} \quad \text{for each } t \in (0, 1]. \tag{9.27}$$

In particular, one has

$$(\mathrm{FL}(A))_t = \{|A^{1-t}|, |A^{1-t}| + 1, \dots, |\boldsymbol{U}|\} \quad \text{if } \boldsymbol{U} \text{ is finite.} \tag{9.28}$$

As $|(A')_t| = |\boldsymbol{U}| - |A^{1-t}|$, we then get

$$|(\mathrm{FL}(A))_t| = |(A')_t| + 1$$

and, hence,

$$\boldsymbol{\sigma}_{f_{1,t}}(\boldsymbol{A'}) = |(\mathrm{FL}(A))_t| - 1 \quad \text{for each } t \in (0, 1]. \tag{9.29}$$

If $\boldsymbol{U}$ is finite, FL(A) can thus be viewed as an encoded and compact piece of information about all possible results of counting by thresholding in A'.

Example 9.5. Again, let us refer to A from Example 9.4. By (9.24), we have

$$\mathrm{FL}(A) = (0, 0, 0.1, 0.2, 0.4, 0.4, 0.4, 0.8, (1)).$$

If $\boldsymbol{U}$ is finite, e.g. $|\boldsymbol{U}| = 10$, this collapses to

$$\mathrm{FL}(A) = (0, 0, 0.1, 0.2, 0.4, 0.4, 0.4, 0.8, 1, 1, 1).$$

Applying (9.29), we immediately conclude that, say,

$$\boldsymbol{\sigma}_{f_{1,0.7}}(\boldsymbol{A'}) = 3.$$

□

Similarly to (9.18), other types of scalar cardinality of A' can be derived from FL(A), too, e.g. $\sigma_{id}(A')$.

9.1.3 FECount

The next concept of fuzzy cardinality we like to present, the *FECount of* $A \in \mathrm{FFS}$ denoted by FE(A), forms an aggregation of FG(A) and FL(A). One defines it as FE: FFS $\to [0, 1]^{\mathbb{N}}$ with

$$\mathrm{FE}(A) = \mathrm{FG}(A) \cap \mathrm{FL}(A). \tag{9.30}$$

By (9.10) and (9.12), we get

$$\begin{aligned} \mathrm{FE}(A) &= \sum_{k \in \mathbf{N}} [A]_k \wedge (1 - [A]_{k+1})/k \\ &= \sum_{k=0}^{n} [A]_k \wedge (1 - [A]_{k+1})/k, \end{aligned} \tag{9.31}$$

i.e.

$$\mathrm{FE}(A)(k) = [A]_k \wedge (1 - [A]_{k+1}) \quad \text{for each } k \in \mathbb{N}. \tag{9.32}$$

FECounts are thus always finite fuzzy sets. In terms of Łukasiewicz logic, we have (see (5.28); cf. (9.15))

$$[\![\exists_{\mathrm{m}} \boldsymbol{D} \in \boldsymbol{P}_k \colon A =_{\mathrm{m}} 1_D]\!] = \bigvee_{\boldsymbol{D} \in \boldsymbol{P}_k} [\![A =_{\mathrm{m}} 1_D]\!] = [A]_k \wedge (1 - [A]_{k+1}) \quad \text{for } k \in \mathbb{N}. \tag{9.33}$$

This means that FE(A)(k) can be viewed as a degree to which A has *exactly* k elements, a degree to which A belongs to the equivalence class of k-element subsets of $\boldsymbol{U}$. FECounts were introduced in ZADEH (1983a) and WYGRALAK (1983a) (see also WYGRALAK (1984, 1986)). "F" and "E" in that acronymic name proposed by L. A. Zadeh stand for "fuzzy" and "equal to", respectively. The following list presents basic properties of FE(A) with $A \in \mathrm{FFS}$:

- FE(A) is convex as an intersection of two convex fuzzy sets;
- FE(A) normal iff $A \in \mathrm{FCS}$ and, then, $\mathrm{FE}(A) = 1/n = (0, \ldots, 0, 1)$ with n zeros;
- if $k < m$ or $k > n$, then $\mathrm{FE}(A)(k) = 0$; $\mathrm{FE}(A)(k) > 0$ for $m \le k \le n$;
- $\mathrm{FE}(A) = \mathrm{FE}(B)$ iff $[A]_k = [B]_k$ for each $k \in \mathbb{N}$;
- if $|\boldsymbol{U}| = p < \infty$, then

$$\mathrm{FE}(A')(k) = \mathrm{FE}(A)(p - k) \quad \text{for } 0 \le k \le p. \tag{9.34}$$

As the last property suggests, FE(A') can be easily constructed on the basis of FE(A), and *vice versa*.

Let us describe FE(A) in a more detailed way. By (9.32),

$$\mathrm{FE}(A) = (1 - [A]_1, 1 - [A]_2, \ldots, 1 - [A]_z, [A]_z, [A]_{z+1}, \ldots, [A]_n) \tag{9.35}$$

with $z = \min\{k \in \mathbb{N}: [A]_k + [A]_{k+1} \le 1\}$. Further, one has $\mathrm{FE}(A)(k) \ge 0.5$ at least for one $k \in \mathbb{N}$, and $\mathrm{FE}(A)(k) > 0.5$ at most for one $k \in \mathbb{N}$. If $\mathrm{FE}(A)(k) > 0.5$ for some k, then $\mathrm{FE}(A)(j) < 0.5$ for each $j \ne k$. Put

$$l = |A^{0.5}| \quad \text{and} \quad r = |A_{0.5}|,$$

and consider two cases.

- First, assume A is such that $A(x) \ne 0.5$ for each x, i.e. $l = r$. By virtue of (9.32), we then have

$$\mathrm{FE}(A)(k) = \begin{cases} \mathbf{1 - [A]_{k+1} < 0.5,} & \textbf{if } k < l, \\ \mathbf{[A]_l \wedge (1 - [A]_{l+1}) > 0.5,} & \textbf{if } k = l, \\ \mathbf{[A]_k < 0.5,} & \textbf{otherwise.} \end{cases} \tag{9.36}$$

  And, hence, (9.35) collapses to

$$\mathrm{FE}(A) = (1 - [A]_1, \dots, 1 - [A]_l, [A]_l \wedge (1 - [A]_{l+1}), [A]_{l+1}, \dots, [A]_n) \tag{9.37}$$

  in which

$$0 \le 1 - [A]_1 \le \dots \le 1 - [A]_l < 0.5 < [A]_l \wedge (1 - [A]_{l+1})$$

  and

$$[A]_l \wedge (1 - [A]_{l+1}) > 0.5 > [A]_{l+1} \ge [A]_{l+2} \ge \dots \ge [A]_n > 0. \tag{9.38}$$

- Second, assume that $A(x) = 0.5$ for some x, i.e. $l < r$. Then

$$\mathrm{FE}(A)(k) = \begin{cases} \mathbf{1 - [A]_{k+1} < 0.5,} & \textbf{if } k < l, \\ \mathbf{0.5,} & \textbf{if } l \le k \le r, \\ \mathbf{[A]_k < 0.5,} & \textbf{otherwise.} \end{cases} \tag{9.39}$$

  This time the FECount of A can thus be written in the vector notation as

$$\mathrm{FE}(A) = (1 - [A]_1, \dots, 1 - [A]_l, 0.5, \dots, 0.5, [A]_{r+1}, \dots, [A]_n) \tag{9.40}$$

  with $r - l + 1$ terms 0.5, and

$$0 \le 1 - [A]_1 \le \dots \le 1 - [A]_l < 0.5$$

  and

$$0.5 > [A]_{r+1} \ge [A]_{r+2} \ge \dots \ge [A]_n > 0. \tag{9.41}$$

Example 9.6. Let us illustrate the above discussion and determine the FECounts of A and B, where

$$A = 0.4/x_1 + 1/x_2 + 0.7/x_3 + 0.1/x_4 + 0.8/x_5 + 0.4/x_6 + 0.3/x_7$$

and

$$B = 0.8/y_1 + 0.5/y_2 + 0.3/y_3 + 1/y_4 + 0.7/y_5 + 0.1/y_6 + 0.5/y_7.$$

So,

$$[A]_0 = [A]_1 = 1,\ [A]_2 = 0.8,\ [A]_3 = 0.7,\ [A]_4 = [A]_5 = 0.4,$$
$$[A]_6 = 0.3,\ [A]_7 = 0.1,\ [A]_8 = 0,$$

$$[B]_0 = [B]_1 = 1,\ [B]_2 = 0.8,\ [B]_3 = 0.7,\ [B]_4 = [B]_5 = 0.5,$$
$$[B]_6 = 0.3,\ [B]_7 = 0.1,\ [B]_8 = 0,$$

and

$$l_A = r_A = 3,\ l_B = 3,\ r_B = 5.$$

By (9.37) and (9.40), we get (see also Figure 9.1)

$$\mathrm{FE}(A) = (0, 0.2, 0.3, 0.6, 0.4, 0.4, 0.3, 0.1)$$

and

$$\mathrm{FE}(B) = (0, 0.2, 0.3, 0.5, 0.5, 0.5, 0.3, 0.1).$$

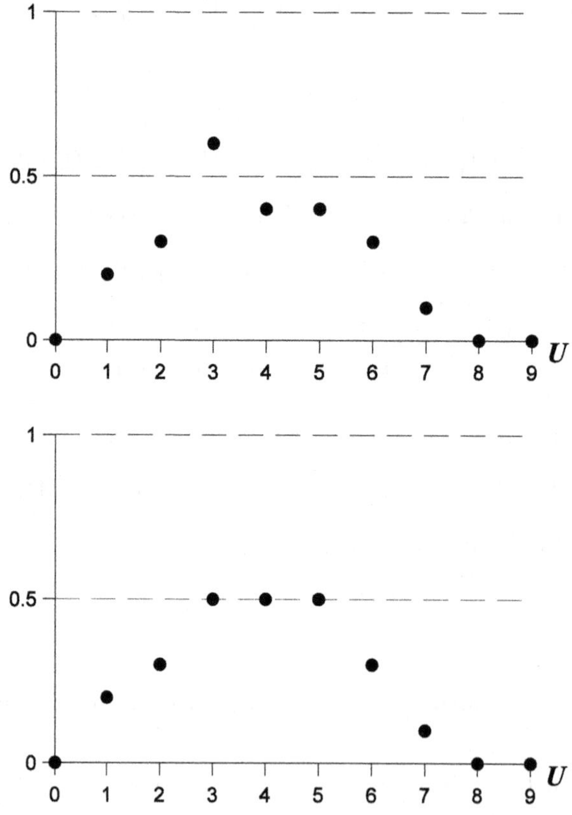

Fig. 9.1 FE(A) (top) and FE(B) (bottom)

Finally, let us look at the FECount of a fuzzy set which is constant on its support, i.e.

$$A = a/x_1 + a/x_2 + \dots + a/x_n \quad \text{with } a \in (0, 1],\ n \geq 1.$$

Applying (9.32), one obtains

$$\text{FE}(A) = (1 - a, a, a, \dots, a) \text{ with } n\ a\text{'s, if } a \in (0, 0.5),$$
$$\text{FE}(A) = (0.5, 0.5, \dots, 0.5) \text{ with } n+1\ 0.5\text{'s, if } a = 0.5,$$
$$\text{FE}(A) = (1 - a, 1 - a, \dots, 1 - a, a) \text{ with } n\ (1 - a)\text{'s, if } a \in (0.5, 1]. \qquad \square$$

As we already pointed out, FE(A)(k) is a degree to which A has exactly k elements. We will try to find another way of interpreting FE(A) with $A \in$ FFS, a way analogous to that offered by (9.17) and (9.29) for FGCounts and FLCounts, respectively. By (9.30), (2.39), (9.16) and (9.27), we get

$$\begin{aligned}(\text{FE}(A))_t &= \{0, 1, \dots, |A_t|\} \cap \{|A^{1-t}|, |A^{1-t}| + 1, \dots\} \\ &= \{|A^{1-t}|, |A^{1-t}| + 1, \dots, |A_t|\},\end{aligned} \tag{9.42}$$

which implies

$$\sigma_{f_{1,t}}(A \cap A') = 0 \vee (|(\text{FE}(A))_t| - 1) \quad \text{for each } t \in (0, 1]. \tag{9.43}$$

Indeed, for $t > 0.5$, this equality is satisfied as $\sigma_f(A \cap A') = 0$ with $f = f_{1,t}$, whereas $|(\text{FE}(A))_t| \leq 1$ (see (9.36) and (9.39)). If $t \leq 0.5$, then $\sigma_f(A \cap A') = |A_t| - |A^{1-t}|$ and, by (9.42), $|(\text{FE}(A))_t| = |A_t| - |A^{1-t}| + 1$, i.e. (9.43) holds true.

FE(A) can thus be viewed as a compact piece of information about all possible results of counting by thresholding in $A \cap A'$. It is a fuzzy set of "embarrassing" elements being simultaneously in A and A', i.e. satisfying to a degree both an imprecise property and its opposite. This suggests a close connection between FECounts and classification issues. We will discuss it in the next subsection.

9.1.4 FECount and Classification

Imagine that basing on a fuzzy set $A \in$ FFS we try to divide the universe $\boldsymbol{U}$ into two disjoint classes, A_+ and A_-. Let us fix an arbitrary $t \in (0, 0.5]$ and formulate the following rules of classification:

- if $A(x) > 1 - t$, we assign x to A_+,
- if $A(x) < t$, x is assigned to A_-,
- if $t \leq A(x) \leq 1 - t$, x is regarded as unclassifiable.

We will refer to the resulting classification of elements of $\boldsymbol{U}$ as the *t-classification* (cf. Subsection 6.3.4). As one sees, each x with $A(x) = 0.5$ remains unclassifiable in each t-classification. For $t \in (0, 0.5]$, the number of unclassifiable elements is equal to

$$|\{x\colon t \leq A(x) \leq 1 - t\}| = |(A \cap A')_t| = \sigma_{f_{1,t}}(A \cap A'). \tag{9.44}$$

Consequently, via (9.43), FE(A) becomes an encoded piece of information about the number of unclassifiable elements in t-classifications (see also Subsection 9.3.2 and Example 9.9).

Example 9.7. Again, let us refer to the eight-bottle example from Subsection 6.3.4 involving a fuzzy set, A, of full bottles described in (6.90) and Figure 6.7. Then

$$\mathrm{FE}(A) = (0, 0, 0.1, 0.2, 0.5, 0.5, 0.3, 0.1).$$

For a fixed $t \in (0, 0.5]$, bottle b_i will be classified as ***full*** whenever $A(b_i) > 1 - t$, and as ***empty*** if $A(b_i) < t$. b_i will be treated as unclassifiable whenever $A(b_i) \in [t, 1 - t]$. It contains too little water to be classified as ***full***, and too much water to be classified as ***empty***. Looking at FE(A), (9.43) and (9.44), one gets the following.

- If $t \in (0.3, 0.5]$, then exactly one bottle, b_4, is unclassifiable as $|(\mathrm{FE}(A))_t| - 1 = 1$. On the other hand, $b_1, b_2, b_6, b_7 \in$ ***full*** and $b_3, b_5, b_8 \in$ ***empty***.
- For $t \in (0.2, 0.3]$, we get $|(\mathrm{FE}(A))_t| - 1 = 2$ and, hence, two bottles are now unclassifiable: b_4 and b_8. Moreover, $b_1, b_2, b_6, b_7 \in$ ***full*** and $b_3, b_5 \in$ ***empty***.
- If $t \in (0.1, 0.2]$, three bottles become unclassifiable: b_2, b_4 and b_8. On the other hand, $b_1, b_6, b_7 \in$ ***full*** and $b_3, b_5 \in$ ***empty***.
- Finally, for $t \in (0, 0.1]$, we have five unclassifiable bottles: b_2, b_4, b_5, b_7 and b_8, whereas $b_1, b_6 \in$ ***full*** and $b_3 \in$ ***empty***. □

The next section will present a study of FECounts and their links with other issues.

9.2 Three Looks at FECounts

In Subsection 9.1.4, we pointed out that there is a close connection between FECounts and classification. Our aim in this section is to reveal other facets of FECounts. We will present them as a result of three different ways of thinking and doing which involve similarity measures, fuzzy rules, and decision making in a fuzzy environment.

9.2.1 Looking through Similarity Measures

Speaking intuitively, a fuzzy set can be more or less similar to different ordinary sets of different cardinalities. The following figure presents an illustrative example.

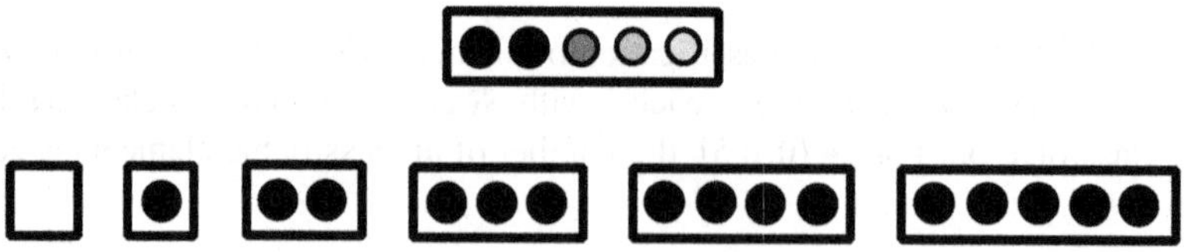

Fig. 9.2 Fuzzy set of *dark* balls (top) and different sets of black balls (bottom)

The above fuzzy set of dark balls seems to show (a) zero similarity to the empty set and a 1-element set, (b) some similarity to a 2-element, 4-element and 5-element set, and (c) strong similarity to a 3-element set of black balls. Let us try to describe those similarities in a formal way. As the reader will see, this task consists in a careful and practical look at (9.33) (see also WYGRALAK (1997b)).

A good candidate for expressing a degree of similarity between a fuzzy set $A \in \mathrm{FFS}$ and a given k-element set $\boldsymbol{D} \subset \boldsymbol{U}$ is the similarity measure R_{eq} from (5.30) forming a $\boldsymbol{t}_{\mathrm{L}}$-similarity (see Subsections 5.1.4 and 5.1.6). By (5.37), we get

$$\begin{aligned} R_{\mathrm{eq}}(A, 1_D) &= [\![A =_{\mathrm{m}} 1_D]\!] \\ &= 1 - \bigvee_{x \in \boldsymbol{U}} |A(x) - 1_D(x)| \\ &= 1 - [\bigvee_{x \in \boldsymbol{D}} (1 - A(x)) \vee \bigvee_{x \notin \boldsymbol{D}} A(x)]. \end{aligned} \tag{9.45}$$

Clearly, we like to find $\boldsymbol{D} \in \boldsymbol{P}_k$ which maximizes $R_{\mathrm{eq}}(A, 1_D)$, i.e. minimizes the subtrahend in (9.45). That maximum value

$$\bigvee_{\boldsymbol{D} \in \boldsymbol{P}_k} R_{\mathrm{eq}}(A, 1_D)$$

will be regarded as the *degree of similarity of A to a k-element set.* $\boldsymbol{P}_0 = \{\varnothing\}$ and, by (9.45), A is thus similar to the empty set to degree $1 - [A]_1 = [A]_0 \wedge (1 - [A]_1)$. If $k > 0$, it is easy to notice that the subtrahend is minimized by $\boldsymbol{D}^* = \{x_1, \ldots, x_k\}$ such that $A(x_i) = [A]_i$ for $i = 1, \ldots, k$. Then

$$\bigvee_{x \in \boldsymbol{D}^*} (1 - A(x)) \vee \bigvee_{x \notin \boldsymbol{D}^*} A(x) = (1 - [A]_k) \vee [A]_{k+1}.$$

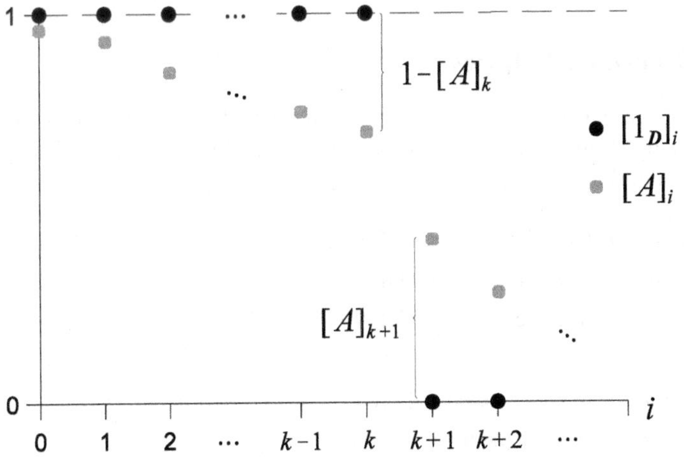

Fig. 9.3 Visualization of $(1 - [A]_k) \vee [A]_{k+1}$

This is the maximum of the deviation of $[A]_k$ from 1 and the deviation of $[A]_{k+1}$ from 0 (see Figure 9.3). $(1 - [A]_k) \vee [A]_{k+1}$ can be treated as a *measure of deviation* of A from the characteristic function of a k-element set. Finally, we get

$$\bigvee_{D \in P_k} R_{eq}(A, 1_D) = 1 - (1 - [A]_k) \vee [A]_{k+1} = [A]_k \wedge (1 - [A]_{k+1}). \qquad (9.46)$$

FE$(A)(k)$ is thus nothing else than the degree of similarity of A to a k-element set, whereas FE(A) forms a vector of similarity degrees for $k = 0, 1, 2, \ldots$.

FE(A) can be used *en bloc* in a further processing. On the other hand, this fuzzy cardinality can also serve as a basis for determining a scalar cardinality of A. It seems that a reasonable and natural way of doing is then to define this scalar cardinality as $k^* \in \mathbb{N}$ maximizing the similarity degree FE$(A)(k)$ to a k-element set (see also Subsection 9.2.2). So,

$$k^* \in \operatorname*{argmax}_k \, ([A]_k \wedge (1 - [A]_{k+1}))$$

and, by (9.36) and (9.39), we have (cf. also WYGRALAK (1997a))

$$k^* \in \{|A^{0.5}|, |A^{0.5}| + 1, \ldots, |A_{0.5}|\}. \qquad (9.47)$$

Especially suitable is the choice of

$$k^* = |A^{0.5}| \quad \text{or} \quad k^* = |A_{0.5}| \qquad (9.48)$$

collapsing to the use of the sigma f-count $\sigma_f(A)$ with $f = f_{2,0.5}$ or $f = f_{1,0.5}$, respectively (cf. also RALESCU (1995)). Indeed, choosing $|A^{0.5}| < k^* < |A_{0.5}|$, we treat the x's with $A(x) = 0.5$ unequally: we "accept" only some of them, which is inconsistent.

9.2.2 Rule-Based Optics

We like to present a fuzzy rule-based approach to FECounts. Let us define the following simple rule base for determining the cardinality $|A|$ of a fuzzy set $A \in$ FFS (see Section 5.3):

IF A *similar* to 0-element set THEN $|A| = 1/0$

IF A *similar* to 1-element set THEN $|A| = 1/1$

IF A *similar* to 2-element set THEN $|A| = 1/2$

⋮

IF A *similar* to $(n-1)$-element set THEN $|A| = 1/(n-1)$

IF A *similar* to n-element set THEN $|A| = 1/n$

By (9.46), the degree of similarity of A to a k-element set equals $[A]_k \wedge (1 - [A]_{k+1})$. Using the Mamdani operator, we get the following results (see (5.58), (5.60)):

- the first rule generates a fuzzy set $\underline{B}^{(0)}$ in $\mathbb{N}$ such that

$$\underline{B}^{(0)}(j) = [A]_0 \wedge (1 - [A]_1) \wedge (1/0)(j) \quad \text{for each } j,$$

i.e.

$$\underline{B}^{(0)} = ([A]_0 \wedge (1 - [A]_1))/0;$$

- the second rule produces $\underline{B}^{(1)}$ with

$$\underline{B}^{(1)}(j) = [A]_1 \wedge (1 - [A]_2) \wedge (1/1)(j) \quad \text{for each } j,$$

i.e.

$$\underline{B}^{(1)} = ([A]_1 \wedge (1 - [A]_2))/1;$$

$$\vdots$$

- the nth rule generates $\underline{B}^{(n-1)}$, where

$$\underline{B}^{(n-1)}(j) = [A]_{n-1} \wedge (1 - [A]_n) \wedge (1/(n-1))(j) \quad \text{for each } j,$$

i.e.

$$\underline{B}^{(n-1)} = ([A]_{n-1} \wedge (1 - [A]_n))/(n-1);$$

- finally, the last rule produces $\underline{B}^{(n)}$ such that

$$\underline{B}^{(n)}(j) = [A]_n \wedge (1 - [A]_{n+1}) \wedge (1/n)(j) \quad \text{for each } j,$$

i.e.

$$\underline{B}^{(n)} = ([A]_n \wedge (1 - [A]_{n+1}))/n.$$

Aggregating these results into one fuzzy set in $\mathbb{N}$, one obtains

$$\underline{B}^{(0)} \cup \underline{B}^{(1)} \cup \ldots \cup \underline{B}^{(n)} = \sum_{k=0}^{n} ([A]_k \wedge (1 - [A]_{k+1}))/k = \mathrm{FE}(A).$$

The last step in the fuzzy rule-based method is a defuzzification of FE(A) which can be performed at least in two ways outlined in Subsection 5.3.3.

- In the maximizing approach, the result of defuzzification is a nonnegative integer k^* from (9.47) or, better, from (9.48). An advantage of this defuzzification procedure is its simplicity. A disadvantage is its insensitivity to the number of elements x with $A(x) \in (0, 0.5)$ which are totally ignored.
- A better approach to defuzzification seems to be the COG method (see (2.122)). Then

$$k^* = \mathrm{COG}_{\mathrm{FE}(A)} = \frac{\sum_{k=0}^{n} k \cdot \mathrm{FE}(A)(k)}{\sum_{k=0}^{n} \mathrm{FE}(A)(k)} \tag{9.49}$$

with the understanding that k^* may be rounded to the nearest integer.

Let us consider an instance with

$$A = 0.4/x_1 + 0.5/x_2 + 1/x_3 + 0.4/x_4 + 0.8/x_5 + 0.4/x_6 + 0.4/x_7 + 0.4/x_8 + 0.4/x_9.$$

Then

$$|A^{0.5}| = 2 \quad \text{and} \quad |A_{0.5}| = 3,$$

whereas

$$\text{FE}(A) = (0, 0.2, 0.5, 0.5, 0.4, 0.4, 0.4, 0.4, 0.4, 0.4)$$

and, hence,

$$\text{COG}_{\text{FE}(A)} = 5.08.$$

The COG method offers a compromise, finds a point of balance between different k's, i.e. between different candidates for the scalar cardinality of A which are acceptable to different degrees.

$\text{COG}_{\text{FE}(A)}$ can be viewed as a new and useful type of scalar cardinality of A derived from FE(A) and being an alternative to sigma f-counts. So, we define

$$\sigma(A) = \text{COG}_{\text{FE}(A)}.$$

This scalar cardinality is generally non-additive:

$$\text{COG}_{\text{FE}(A)} + \text{COG}_{\text{FE}(B)} \neq \text{COG}_{\text{FE}(A \cup B)}$$

for disjoint A and B. Moreover, it seems to be nondecreasing with respect to A, and gives correct results of counting if A collapses to an ordinary set. The use of (9.49) as a scalar cardinality is especially suitable if non-joinable membership degrees are involved (see Subsection 8.1.3; see also Section 10.2).

9.2.3 FECounts as Fuzzy Decisions

Look at an ordinary set $\boldsymbol{D}$ in $\boldsymbol{U}$. Its cardinality is the point of an abrupt decrease of $[1_D]_k$ from 1 to 0. By analogy, we like to determine the cardinality of $A \in \text{FFS}$ as a number $k = k^*$ accompanied by an abrupt decrease of $[A]_k = \text{FG}(A)(k)$. Let us use the Bellman-Zadeh model from Subsection 4.6.1 to solve the problem. To this end, we define

fuzzy goal G: $[A]_k$ should be *large* with $G([A]_k) = [A]_k$,
fuzzy constraint C: $[A]_{k+1}$ should be *small* with $C([A]_{k+1}) = 1 - [A]_{k+1}$.

Using (4.11) with $* = \cap$, the resulting fuzzy decision is

$$D = G * C,$$

i.e.

$$D(k) = [A]_k \wedge (1 - [A]_{k+1}) \quad \text{for } k \in \mathbb{N}, \tag{9.50}$$

which is nothing else than the FECount of A. The two basic defuzzification techniques from (4.13) and (4.14) give k^* from (9.48) and (9.49), respectively:

$$k^* = |A^{0.5}|, \ k^* = |A_{0.5}|, \ \text{or} \ k^* = \mathrm{COG}_{\mathrm{FE}(A)}.$$

Determining the cardinality of A is thus now a decision process in which FE(A) is the fuzzy decision, whereas the above scalar cardinalities k^* are possible crisp decisions.

Finally, we like to present a practical example referring to the crisp decisions $k^* = |A^{0.5}|$ and $k^* = |A_{0.5}|$, and leading to a more general formulation of FECounts.

As we remarked in Subsection 9.1.2, ranking lists and FGCounts are in essence equivalent constructions. Let us focus our attention on a simple instance of a ranking list, namely a nonincreasing list of examination results (points) of a group of students. The examiner uses a predetermined threshold, p, and decides that a student passed the exam if he/she scored (a) over p%, or (b) p% or over. $p = 50$ is then usually applied. After transforming the points into the scale [0, 1], the number of students who passed the exam becomes equal to $k^* = |A^{0.5}|$ or $k^* = |A_{0.5}|$, respectively, with A understood in this case as a fuzzy set of sufficiently good exam papers. k^* is thus a defuzzification of FE(A). Speaking generally, k^* forms a number of objects whose quality is "good enough".

It happens that the examiner uses a threshold differing from 50%, i.e. the number of students who passed the exam equals $k^* = |A^t|$ or $k^* = |A_t|$ with some $t \in (0, 1)$. Then, trying to establish a connection between k^* and a defuzzification of FE(A), one needs a more general definition of FECounts, namely (see also (9.62))

$$\mathrm{FE}_\nu(A)(k) = [A]_k \wedge \nu([A]_{k+1}) \quad \text{for each } k \in \mathbb{N} \tag{9.51}$$

with a strict negation ν having $e(\nu) = t$ as equilibrium point (see Subsection 2.4.1). $\mathrm{FE}_\nu(A)$ will be called the *FECount of* $A \in$ FFS *generated by* ν. As in (9.50), $\mathrm{FE}_\nu(A)$ is still a fuzzy decision provided that $C([A]_{k+1}) = \nu([A]_{k+1})$. Put

$$l = |A^{e(\nu)}| \quad \text{and} \quad r = |A_{e(\nu)}|.$$

It is a routine task to check that (cf. (9.36) and (9.39))

$$\mathrm{FE}_\nu(A)(l) > e(\nu) \quad \text{and} \quad \mathrm{FE}_\nu(A)(k) < e(\nu) \quad \text{for } k \neq l \tag{9.52}$$

whenever $A(x) \neq e(\nu)$ for each $x \in \boldsymbol{U}$. Moreover,

$$\begin{aligned} &\mathrm{FE}_\nu(A)(k) = e(\nu) \quad \text{for } l \le k \le r, \\ \text{and} \quad &\mathrm{FE}_\nu(A)(k) < e(\nu) \quad \text{for } k < l \text{ and } k > r \end{aligned} \tag{9.53}$$

provided that $A(x) = e(\nu)$ for some $x \in \boldsymbol{U}$. Consequently,

$$k^* = |A^{e(\nu)}|, \ k^* = |A_{e(\nu)}|, \ k^* = \mathbf{COG}_{\mathbf{FE}_\nu(A)} \tag{9.54}$$

are reasonable defuzzifications of $\mathrm{FE}_\nu(A)$, and form scalar cardinalities of A derived from $\mathrm{FE}_\nu(A)$.

9.3 Generalizations and Extensions

This subsection presents a triangular norm-based generalization of FG-, FL-, and FECounts of fuzzy sets. Second, these three types of fuzzy cardinalities as well as their generalizations involving t-norms will be extended to IVFSs and IFSs.

9.3.1 Fuzzy Cardinalities with Triangular Norms

At first glance, it seems that FG-, FL-, and FECounts from (9.14), (9.24), and (9.35), respectively, are suitable for fuzzy sets with an arbitrary t-norm $\boldsymbol{t}$. This impression is misleading. Indeed, let us focus our attention on FGCounts for a moment. Since A from FFS is the sum of disjoint singletons composing it, FG(A) should be equal to the sum of FGCounts of those singletons, i.e. we should have

$$\begin{aligned}\mathrm{FG}(A) &= \mathrm{FG}([A]_1/x_1) + \mathrm{FG}([A]_2/x_2) + \ldots + \mathrm{FG}([A]_n/x_n)\\ &= (1, [A]_1) + (1, [A]_2) + \ldots + (1, [A]_n)\end{aligned}$$

with + defined via the $\boldsymbol{t}$-based extension principle (2.103):

$$(P+Q)(k) = \bigvee_{i+j=k} P(i)\,\boldsymbol{t}\,Q(j) \quad \text{for } P, Q \in [0, 1]^{\mathbb{N}}.$$

A routine matter is to check that, however, the above sum of FGCounts equals

$$(1, [A]_1, [A]_1\,\boldsymbol{t}\,[A]_2, \ldots, [A]_1\,\boldsymbol{t}\,[A]_2\,\boldsymbol{t}\ldots\boldsymbol{t}\,[A]_n).$$

This generally differs from FG(A), unless $\boldsymbol{t} = \wedge$. FGCounts are thus suitable only for fuzzy sets with the t-norm minimum. Their appropriate generalization to fuzzy sets with an arbitrary $\boldsymbol{t}$ is just

$$\mathrm{FG}_t(A) = (1, [A]_1, [A]_1\,\boldsymbol{t}\,[A]_2, \ldots, [A]_1\,\boldsymbol{t}\,[A]_2\,\boldsymbol{t}\ldots\boldsymbol{t}\,[A]_n), \tag{9.55}$$

i.e. FG_t: FFS $\to [0, 1]^{\mathbb{N}}$ with

$$\mathrm{FG}_t(A)(k) = [A]_1\,\boldsymbol{t}\,[A]_2\,\boldsymbol{t}\ldots\boldsymbol{t}\,[A]_k \quad \text{for each } k \in \mathbb{N}. \tag{9.56}$$

One says that $\mathrm{FG}_t(A)$ is the *FGCount of A generated by* ***t***. Clearly, $\mathrm{FG}_\wedge(A) = \mathrm{FG}(A)$ (see also (10.39)). Referring to (9.15), $\mathrm{FG}_t(A)(k)$ is a degree to which A has at least i elements for each $i \le k$ (see (2.98)). The following is a list of basic properties satisfied for each A and $\boldsymbol{t}$:

- $\mathrm{FG}_t(A)$ is normal and convex;
- $\mathrm{FG}_t(A)(k) = 1$ for $k \le m$, $\mathrm{FG}_t(A)(k) = 0$ for $k > n$,
  if $\boldsymbol{t}$ is strict or $\boldsymbol{t} = \wedge$, then $\mathrm{FG}_t(A)(k) \in (0, 1)$ for $m < k \le n$, else $\mathrm{FG}_t(A)(k) \in [0, 1)$;

- $\mathrm{FG}_t(A) = (1, 1, \ldots, 1)$ with $n+1$ ones whenever $A \in \mathrm{FCS}$;
- $\mathrm{FG}_t(A)$ is nondecreasing with respect to t; so, $\mathbf{FG}_{t_Ł}(A) \subset \mathbf{FG}_{t_a}(A) \subset \mathrm{FG}(A)$;
- $\mathrm{FG}_t(A) \subset \mathrm{FG}_t(B)$ whenever $A \subset B$.

A way of reasoning analogous to that accompanying the generalization of FGCounts leads to the following generalization of FLCounts:

$$\begin{aligned}\mathrm{FL}_{t,\nu}(A) = (&\nu([A]_1)\,t\,\nu([A]_2)\,t \ldots t\,\nu([A]_n),\\ &\nu([A]_2)\,t\,\nu([A]_3)\,t \ldots t\,\nu([A]_n),\\ &\vdots\\ &\nu([A]_{n-1})\,t\,\nu([A]_n), \nu([A]_n), (1))\end{aligned} \tag{9.57}$$

with a t-norm t and a negation ν. One thus has

$$\begin{aligned}\mathrm{FL}_{t,\nu}(A)(k) &= \nu([A]_{k+1})\,t\,\nu([A]_{k+2})\,t \ldots\\ &= \nu([A]_{k+1})\,t\,\nu([A]_{k+2})\,t \ldots t\,\nu([A]_n)\end{aligned} \tag{9.58}$$

for each $k \in \mathbb{N}$. So, $\mathrm{FL}_{t,\nu}(A) = \mathrm{FL}(A)$ whenever $t = \wedge$ and $\nu = \nu_Ł$. We will say that $\mathrm{FL}_{t,\nu}(A)$ is the *FLCount of A generated by* t *and* ν. $\mathrm{FL}_{t,\nu}(A)(k)$ can be viewed as a degree to which A contains at most i elements for each i greater than or equal to k. In particular, we put

$$\mathrm{FL}_\nu(A) = \mathrm{FL}_{\wedge,\nu}(A), \tag{9.59}$$

i.e.

$$\mathrm{FL}_\nu(A)(k) = \nu([A]_{k+1}) \quad \text{for each } k \in \mathbb{N}.$$

Basic properties of generalized FLCounts of $A \in \mathrm{FFS}$ are listed below:

- $\mathrm{FL}_{t,\nu}(A)$ is normal and convex;
- $\mathrm{FL}_{t,\nu}(A)(k) = 1$ for $k \geq n$, $\mathrm{FL}_{t,\nu}(A)(k) = 0$ for $k < m$,
  if t is strict or $t = \wedge$, and ν is strict, then $\mathrm{FL}_{t,\nu}(A)(k) \in (0, 1)$ for $m \leq k < n$;
- $\mathrm{FL}_{t,\nu}(A) = (0, \ldots, 0, (1))$ with n zeros whenever $A \in \mathrm{FCS}$;
- $\mathrm{FL}_{t,\nu}(A)$ is nondecreasing with respect to t and ν; in particular,
- $\mathbf{FL}_{t_Ł,\nu}(A) \subset \mathbf{FL}_{t_a,\nu}(A) \subset \mathrm{FL}_{\wedge,\nu}(A)$;
- $\mathrm{FL}_{t,\nu}(B) \subset \mathrm{FL}_{t,\nu}(A)$ whenever $A \subset B$.

Finally, we define

$$\mathrm{FE}_{t,\nu}(A) = \mathrm{FG}_t(A) \cap_t \mathrm{FL}_{t,\nu}(A), \tag{9.60}$$

the *FECount of A generated by a t-norm* t *and a negation* ν. Hence

$$\mathrm{FE}_{t,\nu}(A)(k) = \mathrm{FG}_t(A)(k)\,t\,\mathrm{FL}_{t,\nu}(A)(k) \quad \text{for } k \in \mathbb{N}. \tag{9.61}$$

By (9.56) and (9.58), we thus get

$$\mathrm{FE}_{t,\nu}(A)(k) = [A]_1\,t\,[A]_2\,t \ldots t\,[A]_k\,t\,\nu([A]_{k+1})\,t\,\nu([A]_{k+2})\,t \ldots t\,\nu([A]_n) \tag{9.62}$$

for $k \in \mathbb{N}$, and

$$\begin{aligned} \mathrm{FE}_{t,\nu}(A) = (&\nu([A]_1)\, \boldsymbol{t}\, \nu([A]_2)\, \boldsymbol{t} \ldots \boldsymbol{t}\, \nu([A]_n), \\ &[A]_1\, \boldsymbol{t}\, \nu([A]_2)\, \boldsymbol{t}\, \nu([A]_3)\, \boldsymbol{t} \ldots \boldsymbol{t}\, \nu([A]_n), \\ &[A]_1\, \boldsymbol{t}\, [A]_2\, \boldsymbol{t}\, \nu([A]_3)\, \boldsymbol{t} \ldots \boldsymbol{t}\, \nu([A]_n), \\ &\vdots \\ &[A]_1\, \boldsymbol{t}\, [A]_2\, \boldsymbol{t} \ldots \boldsymbol{t}\, [A]_{n-1}\, \boldsymbol{t}\, \nu([A]_n), \\ &[A]_1\, \boldsymbol{t}\, [A]_2\, \boldsymbol{t} \ldots \boldsymbol{t}\, [A]_n). \end{aligned} \tag{9.63}$$

One sees that $\mathrm{FE}_{t,\nu}(A)(k)$ is a degree to which A has at least i elements for each $i \le k$ and, simultaneously, contains at most j elements for each $j \ge k$. Clearly,

$$\mathbf{FE}_{\wedge,\nu_Ł}(\boldsymbol{A}) = \mathrm{FE}(A)$$

and

$$\mathrm{FE}_{\wedge,\nu}(A) = \mathrm{FE}_\nu(A) \quad \text{whenever } \nu \text{ is strict.}$$

Example 9.8. Let $A = 0.5/x_1 + 0.6/x_2 + 1/x_3 + 0.9/x_4 + 1/x_5 + 0.8/x_6 + 0.2/x_7$, i.e.

$$[A]_1 = [A]_2 = 1,\ [A]_3 = 0.9,\ [A]_4 = 0.8,\ [A]_5 = 0.6,\ [A]_6 = 0.5,\ [A]_7 = 0.2,$$

and $[A]_k = 0$ for $k > 7$. Using (9.55), for the three basic t-norms, we get

$$\begin{aligned} &\mathrm{FG}_t(A) = \mathrm{FG}(A) = (1, 1, 1, 0.9, 0.8, 0.6, 0.5, 0.2) \quad \text{for } \boldsymbol{t} = \wedge, \\ &\mathrm{FG}_t(A) = (1, 1, 1, 0.9, 0.72, 0.432, 0.216, 0.0432) \quad \text{for } \boldsymbol{t} = \boldsymbol{t}_a, \\ &\mathrm{FG}_t(A) = (1, 1, 1, 0.9, 0.7, 0.3) \quad \text{for } \boldsymbol{t} = \boldsymbol{t}_Ł. \end{aligned}$$

Put $\nu = \nu_Ł$. By (9.57),

$$\begin{aligned} &\mathrm{FL}_{t,\nu}(A) = \mathrm{FL}(A) = (0, 0, 0.1, 0.2, 0.4, 0.5, 0.8, (1)) \quad \text{for } \boldsymbol{t} = \wedge, \\ &\mathrm{FL}_{t,\nu}(A) = (0, 0, 0.0032, 0.032, 0.16, 0.4, 0.8, (1)) \quad \text{for } \boldsymbol{t} = \boldsymbol{t}_a, \\ &\mathrm{FL}_{t,\nu}(A) = (0, 0, 0, 0, 0, 0.3, 0.8, (1)) \quad \text{for } \boldsymbol{t} = \boldsymbol{t}_Ł. \end{aligned}$$

Finally, applying (9.61) with $\nu = \nu_Ł$, one obtains

$$\begin{aligned} &\mathrm{FE}_{t,\nu}(A) = \mathrm{FE}(A) = (0, 0, 0.1, 0.2, 0.4, 0.5, 0.5, 0.2) \quad \text{for } \boldsymbol{t} = \wedge, \\ &\mathrm{FE}_{t,\nu}(A) = (0, 0, 0.0032, 0.0288, 0.1152, 0.1728, 0.1728, 0.0432) \quad \text{for } \boldsymbol{t} = \boldsymbol{t}_a, \\ &\mathrm{FE}_{t,\nu}(A) = (0, 0, 0, 0, 0, 0.3\, \boldsymbol{t}_Ł\, 0.3) = (0) \quad \text{for } \boldsymbol{t} = \boldsymbol{t}_Ł. \end{aligned}$$

□

As previously, let us list basic properties of $\mathrm{FE}_{t,\nu}(A)$ with $A \in \mathrm{FFS}$, a t-norm $\boldsymbol{t}$ and a negation ν:

- $\mathrm{FE}_{t,\nu}(A)$ is convex; if ν is such that $\nu(a) = 1$ only for $a = 0$, then $\mathrm{FE}_{t,\nu}(A)$ is normal iff $A \in \mathrm{FCS}$;
- $\mathrm{FE}_{t,\nu}(A) = 1/n$ if $A \in \mathrm{FCS}$;

- $\mathrm{FE}_{t,\nu}(A)(k) = 0$ for $k < m$ and $k > n$;
  if t is strict or $t = \wedge$, and ν is strict, then $\mathrm{FE}_{t,\nu}(A)(k) > 0$ for $m \le k \le n$;
- if $|U| = p < \infty$ and ν is strong, then

$$\mathrm{FE}_{t,\nu}(A^{\nu})(k) = \mathrm{FE}_{t,\nu}(A)(p-k) \quad \text{for } 0 \le k \le p; \tag{9.64}$$

- $\mathrm{FE}_{t,\nu}(A)$ is nondecreasing with respect to t and ν; in particular, $\mathbf{FE}_{t_L,\nu}(\boldsymbol{A}) \subset \mathbf{FE}_{t_a,\nu}(\boldsymbol{A}) \subset \mathrm{FE}_{\wedge,\nu}(A)$.

In reference to (9.42), let us notice that

$$(\mathrm{FE}_{t,\nu}(A))_t \subset \{\, |\boldsymbol{A}^{\nu^{-1}(t)}|, \{\, |\boldsymbol{A}^{\nu^{-1}(t)}| + 1, \ldots, |A_t| \,\} \tag{9.65}$$

for each $t \in (0, 1]$, each t-norm t and strict negation ν. If $t \neq \wedge$ and ν is strict, then $\mathrm{FE}_{t,\nu}(A)(k) > e(\nu)$ at most for one $k \in \mathbb{N}$. Furthermore, $\mathrm{FE}_{t,\nu}(A)(k) > e(A)$ implies $\mathrm{FE}_{t,\nu}(A)(j) < e(A)$ for each $j \neq k$ (cf. (9.52)-(9.53)). If t is nilpotent, $\mathrm{FE}_{t,\nu}(A) = (0)$ becomes possible (see Example 9.8). A is then *singular*, totally dissimilar to any set of any cardinality as all the weights assigned to nonnegative integers are equal to zero. A comprehensive study of singular fuzzy sets can be found in DYCZKOWSKI/ WYGRALAK (2001, 2003). Our further discussion about FECounts will be concentrated on the case when $t = \wedge$ or t is strict. The interested reader is referred to WYGRALAK (2001, 2002, 2003a, b) and WYGRALAK/PILARSKI (2000) for further details, properties and proofs concerning FG-, FL-, and FECounts generated by t-norms and negations.

9.3.2 Extensions to IVFSs and IFSs

We will deal with extensions of FG-, FL-, and FECounts to IVFSs and IFSs. Let us start with the case of IFSs.

Assume $\mathscr{E} = (A^+, A^-)$ is a finite IFS, i.e. $(A^-)^{\nu}$ is finite, $A^+ \subset (A^-)^{\nu}$, and ν is a given strong negation (see Subsection 6.3.1). The *FGCount* $\mathrm{FG}(\mathscr{E})$ and the *FLCount* $\mathrm{FL}(\mathscr{E})$ *of* $\mathscr{E}$, respectively, are then basically understood as IFSs in $\mathbb{N}$ defined by (see (9.14) and (9.59))

$$\mathrm{FG}(\mathscr{E}) = (\mathrm{FG}(A^+), (\mathrm{FG}((A^-)^{\nu}))^{\nu})$$

and

$$\mathrm{FL}(\mathscr{E}) = (\mathrm{FL}_{\nu}((A^-)^{\nu}), (\mathrm{FL}_{\nu}(A^+))^{\nu}). \tag{9.66}$$

More convenient in practice, however, seem to be interval forms of these constructions, namely

$$\mathrm{FG}(\mathscr{E}) = (\mathrm{FG}(A^+), \mathrm{FG}((A^-)^{\nu}))$$

and

$$\mathrm{FL}(\mathscr{E}) = (\mathrm{FL}_{\nu}((A^-)^{\nu}), \mathrm{FL}_{\nu}(A^+)) \tag{9.67}$$

being IVFSs in $\mathbb{N}$. Just these forms will be used in our further discussion. For each $k \in \mathbb{N}$, we thus have

$$\mathrm{FG}(\mathscr{E})(k) = [\mathrm{FG}(A^+)(k), \mathrm{FG}((A^-)^\nu)(k))] = [[A^+]_k, [(A^-)^\nu]_k]$$

and (9.68)

$$\mathrm{FL}(\mathscr{E})(k) = [\nu([(A^-)^\nu]_{k+1}), \nu([A^+]_{k+1})].$$

The question of FECounts and their extension to IFSs is more sophisticated and ambiguous. At least two variants of doing have to be taken into account when trying to define the *FECount* $\mathrm{FE}(\mathscr{E})$ *of* $\mathscr{E}$. First, (9.30) suggests that $\mathrm{FE}(\mathscr{E})$ should be constructed via $\mathrm{FG}(\mathscr{E})$ and $\mathrm{FL}(\mathscr{E})$ as

$$\mathrm{FE}(\mathscr{E}) = \mathrm{FG}(\mathscr{E}) \cap \mathrm{FL}(\mathscr{E}). \tag{9.69}$$

By virtue of (9.67) and (6.37), this leads to

$$\mathrm{FE}(\mathscr{E}) = (\mathrm{FG}(A^+) \cap \mathrm{FL}_\nu((A^-)^\nu), \mathrm{FG}((A^-)^\nu) \cap \mathrm{FL}_\nu(A^+)). \tag{9.70}$$

$\mathrm{FE}(\mathscr{E})$ is thus an IVFS in $\mathbb{N}$, but its components are not generally FECounts of fuzzy sets. Nevertheless, if $\mathscr{E}$ collapses to an ordinary fuzzy set, $\mathscr{E} = (A, A^\nu)$, a suitable compatibility is guaranteed as we practically get $\mathrm{FE}(\mathscr{E}) = \mathrm{FE}_\nu(A)$ (see (9.51)).

An alternative extension of FECounts seems to be

$$\mathrm{FE}^*(\mathscr{E}) = (\mathrm{FE}_\nu(A^+), \mathrm{FE}_\nu((A^-)^\nu)). \tag{9.71}$$

Notice that $\mathrm{FE}^*(\mathscr{E})$ does not form an IVFS and, again, $\mathrm{FE}^*((A, A^\nu))$ collapses to $\mathrm{FE}_\nu(A)$. Our further discussion will be focused on (9.69).

Let us look at a particular case of the above extensions corresponding to the standard formulation of IFSs with $\nu = \nu_Ł$. Then

$$\begin{aligned} \mathrm{FG}(\mathscr{E}) &= (\mathrm{FG}(A^+), \mathrm{FG}((A^-)')), \\ \mathrm{FL}(\mathscr{E}) &= (\mathrm{FL}((A^-)'), \mathrm{FL}(A^+)), \\ \mathrm{FE}(\mathscr{E}) &= (\mathrm{FG}(A^+) \cap \mathrm{FL}((A^-)'), \mathrm{FG}((A^-)') \cap \mathrm{FL}(A^+)) \end{aligned} \tag{9.72}$$

and, consequently,

$$\begin{aligned} \mathrm{FG}(\mathscr{E})(k) &= [[A^+]_k, [(A^-)']_k], \\ \mathrm{FL}(\mathscr{E})(k) &= [1 - [(A^-)']_{k+1}, 1 - [A^+]_{k+1}], \\ \mathrm{FE}(\mathscr{E})(k) &= [[A^+]_k \wedge (1 - [(A^-)']_{k+1}), [(A^-)']_k \wedge (1 - [A^+]_{k+1})] \end{aligned} \tag{9.73}$$

for $k \in \mathbb{N}$. By (8.38), (9.17) and (9.29), this leads to

$$\sigma_f(\mathscr{E}) = [|(\mathrm{FG}(A^+))_t| - 1, |(\mathrm{FG}((A^-)'))_t| - 1]$$

and (9.74)

$$\sigma_f(\mathscr{E}') = [|(\mathrm{FL}((A^-)'))_t| - 1, |(\mathrm{FL}(A^+))_t| - 1], \ \boldsymbol{U}\text{ - finite},$$

with a weighting function $f = f_{1,t}$ and $t \in (0, 1]$. $\mathrm{FG}(\mathscr{E})$ thus forms a compact and dynamic piece of information about all possible results of counting by thresholding in an incompletely known fuzzy set A modeled by $\mathscr{E}$. Analogously, FL($\mathscr{E}$) can be viewed as information about all possible results of counting by thresholding in A' represented by $\mathscr{E}'$. The left endpoint of the interval $\sigma_f(\mathscr{E})$ and the right one, respectively, are a minimum and a maximum possible result of counting in A if t is used as the threshold value. The same refers to the interval $\sigma_f(\mathscr{E}')$ and counting in A' (see also Example 9.9).

Unfortunately, as to FE($\mathscr{E}$), there is no simple generalization of (9.43) to IFSs. The interval

$$[0 \vee (|(\mathrm{FG}(A^+) \cap \mathrm{FL}((A^-)'))_t| - 1), 0 \vee (|(\mathrm{FG}((A^-)') \cap \mathrm{FL}(A^+))_t| - 1)]$$

is only a more or less good approximation of $\sigma_f(\mathscr{E} \cap \mathscr{E}')$ with $f = f_{1,t}$ and $t \in (0, 1]$. Nevertheless, $\sigma_f(\mathscr{E} \cap \mathscr{E}')$ can then be interpreted in a way which is a generalization of the interpretation of $\sigma_f(A \cap A')$ from (9.44). More precisely, for each $t \in (0, 0.5]$,

$$\sigma_f(\mathscr{E} \cap \mathscr{E}') = [\sigma_f(A^+ \cap A^-), \sigma_f((A^+)' \cap (A^-)')] \tag{9.75}$$

is an interval (of integers) whose left endpoint (right endpoint, respectively) is the number of elements which are surely (surely or possibly, respectively) unclassifiable whenever the t-classification is used (see Subsection 9.1.4). In other words, those endpoints are a minimum and a maximum possible number of unclassifiable elements. Indeed, we have

$$(A^+ \cap A^-)(x) \geq t \Rightarrow t \leq A(x) \leq 1 - t \Rightarrow ((A^+)' \cap (A^-)')(x) \geq t \tag{9.76}$$

for $t \in (0, 0.5]$. These observations are illustrated below.

Example 9.9. We will refer once more to the eight-bottle example from Subsection 6.3.4 and, more precisely, to the case of $\mathscr{E} = (A^+, A^-)$ modeling an incompletely known fuzzy set A of full bottles from Figure 6.8. By (9.72) and (6.94), we get

$$\begin{aligned}\mathrm{FG}(\mathscr{E}) &= (\mathrm{FG}(A^+), \mathrm{FG}((A^-)')) \\ &= ((1, 1, 0.9, 0.5, 0.4, 0.3, 0.25, 0.1), (1, 1, 1, 1, 1, 0.9, 0.7, 0.2, 0.1)).\end{aligned}$$

This is an encoded and compact piece of information about all possible results of counting by thresholding in A. Indeed, applying (9.74), we conclude that $\sigma_f(\mathscr{E})$ with $f = f_{1,t}$ is equal to

$$\begin{gathered}[1, 4] \text{ for } t \in (0.9, 1], \quad [2, 5] \text{ for } t \in (0.7, 0.9], \quad [2, 6] \text{ for } t \in (0.5, 0.7], \\ [3, 6] \text{ for } t \in (0.4, 0.5], \quad [4, 6] \text{ for } t \in (0.3, 0.4], \quad [5, 6] \text{ for } t \in (0.25, 0.3], \\ [6, 6] \text{ for } t \in (0.2, 0.25], \quad [6, 7] \text{ for } t \in (0.1, 0.2], \quad [7, 8] \text{ for } t \in (0, 0.1].\end{gathered}$$

The left endpoint of each resulting interval of nonnegative integers is the number of bottles being surely full to a degree $\geq t$, whereas the right endpoint is the number of bottles which are surely or possibly full to that degree. For instance, if $t \in (0.4, 0.5]$

is used, three bottles are surely full to a degree $\geq t$, and six bottles are surely or possibly full to that degree. Moreover, those three bottles (b_1, b_2, b_6) contain at least

$$\sigma_{f_{4,t,1}}(A^+) = \sum_{k=1}^{3} \mathrm{FG}(A^+)(k) = 2.4$$

bottles of water in total, whereas the six bottles (b_1, b_2, b_4, b_6, b_7, b_8) contain at most

$$\sigma_{f_{4,t,1}}((A^-)') = \sum_{k=1}^{6} \mathrm{FG}((A^-)')(k) = 5.6$$

bottles of water in total. As we thus see, FG($\mathscr{E}$) is a basis for obtaining various types of information, and is really worth constructing.

Moving on to FLCounts, (9.72) gives

$$\mathrm{FL}(\mathscr{E}) = (\mathrm{FL}((A^-)'), \mathrm{FL}(A^+))$$
$$= ((0, 0, 0, 0, 0.1, 0.3, 0.8, 0.9, 1), (0, 0.1, 0.5, 0.6, 0.7, 0.75, 0.9, 1, 1)).$$

This time, what we get is a compact piece of information about all possible results of counting by thresholding in A' modeled by $\mathscr{E}'$. Using again (9.74) with $f = f_{1,t}$, one gets the following values of $\sigma_f(\mathscr{E}')$:

$$[0, 1] \text{ for } t \in (0.9, 1], \quad [1, 2] \text{ for } t \in (0.8, 0.9], \quad [2, 2] \text{ for } t \in (0.75, 0.8],$$
$$[2, 3] \text{ for } t \in (0.7, 0.75], \quad [2, 4] \text{ for } t \in (0.6, 0.7], \quad [2, 5] \text{ for } t \in (0.5, 0.6],$$
$$[2, 6] \text{ for } t \in (0.3, 0.5], \quad [3, 6] \text{ for } t \in (0.1, 0.3], \quad [4, 7] \text{ for } t \in (0, 0.1].$$

The left endpoint of the resulting interval is now the number of bottles which are surely empty to a degree $\geq t$, and the right endpoint expresses the number of bottles being surely or possibly empty to that degree. So, for instance, if $t \in (0.5, 0.6]$, two bottles (b_3 and b_5) are surely empty to a degree $\geq t$, while five bottles (b_3, b_4, b_5, b_7, b_8) are surely or possibly empty to that degree. The scarcity of water in the two bottles is at least equal to

$$\sigma_{f_{4,t,1}}(A^-) = \sum_{k=6}^{7} \mathrm{FL}((A^-)')(k) = 1.7$$

bottles of water in total. The scarcity in the five bottles is at most

$$\sigma_{f_{4,t,1}}((A^+)') = \sum_{k=3}^{7} \mathrm{FL}(A^+)(k) = 3.95$$

bottles of water in total. Thus, again, FL($\mathscr{E}$) is also a basis for generating various types of information.

Finally, let us refer to the notion of a t-classification, $t \in (0, 0.5]$, introduced in Subsection 9.1.4. By (6.94), we have

$$A^+ \cap A^- = 0.1/b_5 + 0.1/b_7 + 0.3/b_8,$$
$$(A^+)' \cap (A^-)' = 0.1/b_1 + 0.5/b_2 + 0.2/b_3 + 0.75/b_4 + 0.1/b_5 + 0.6/b_7 + 0.7/b_8,$$

and (9.75) with $f = f_{1,t}$ gives

$$\sigma_f(\mathscr{E} \cap \mathscr{E}') = [\sigma_f(A^+ \cap A^-), \sigma_f((A^+)' \cap (A^-)')]$$
$$= \begin{cases} \mathbf{[0, 4],} & \mathbf{if} \ t \in \mathbf{(0.3, 0.5],} \\ \mathbf{[1, 4],} & \mathbf{if} \ t \in \mathbf{(0.2, 0.3],} \\ \mathbf{[1, 5],} & \mathbf{if} \ t \in \mathbf{(0.1, 0.2],} \\ \mathbf{[3, 7],} & \mathbf{if} \ t \in \mathbf{(0, 0.1].} \end{cases}$$

The left endpoint of the resulting interval is here the number of bottles which are surely unclassifiable, whereas the right one expresses the number of surely or possibly unclassifiable bottles whenever the t-classification is used. Let us formulate these conclusions in a more detailed way (cf. Example 9.7).

- If $t \in (0.3, 0.5]$, then no bottle is surely unclassifiable, and four bottles are possibly unclassifiable (b_2, b_4, b_7, b_8). Moreover, $b_1, b_6 \in$ ***full*** and $b_3, b_5 \in$ ***empty***.
- For $t \in (0.2, 0.3]$, one bottle, b_8, is surely unclassifiable, whereas b_2, b_4, b_7 and b_8 are surely or possibly unclassifiable; $b_1, b_6 \in$ ***full*** and $b_3, b_5 \in$ ***empty***.
- If $t \in (0.1, 0.2]$, b_8 is still certainly unclassifiable, while five bottles are certainly or possibly unclassifiable: b_2, b_3, b_4, b_7 and b_8. Moreover, $b_1, b_6 \in$ ***full*** and $b_5 \in$ ***empty***.
- Finally, for $t \in (0, 0.1]$, three bottles are surely unclassifiable (b_5, b_7, b_8), and seven bottles are surely or possibly unclassifiable (all the bottles except b_6). On the other hand, no bottle is now classified as ***empty***, and $b_6 \in$ ***full***. □

Finishing our discussion on FG-, FL- and FECounts of IFSs, one has to mention their further generalization involving a t-norm $\boldsymbol{t}$. Making use of (9.55), (9.57) and (9.63), we define

$$\mathrm{FG}_t(\mathscr{E}) = (\mathrm{FG}_t(A^+), \mathrm{FG}_t((A^-)^\nu)),$$
$$\mathrm{FL}_t(\mathscr{E}) = (\mathrm{FL}_{t,\nu}((A^-)^\nu), \mathrm{FL}_{t,\nu}(A^+)) \qquad (9.77)$$

and

$$\mathrm{FE}_t(\mathscr{E}) = \mathrm{FG}_t(\mathscr{E}) \cap_t \mathrm{FL}_t(\mathscr{E}),$$

which can be called the *FGCount*, *FLCount*, and *FECount of* $\mathscr{E}$ *generated by* $\boldsymbol{t}$, respectively, where $\mathscr{E} = (A^+, A^-)$ with $A^+ \subset (A^-)^\nu$ is finite and ν is a strong negation. For $\mathrm{FE}_t(\mathscr{E})$, the use of a strict $\boldsymbol{t}$ or $\boldsymbol{t} = \wedge$ is recommended (see Subsection 9.3.1). So,

$$\mathrm{FG}_t(\mathscr{E})(k) = [[A^+]_1 \,\boldsymbol{t} \ldots \boldsymbol{t}\, [A^+]_k, [(A^-)^\nu]_1 \,\boldsymbol{t} \ldots \boldsymbol{t}\, [(A^-)^\nu]_k],$$
$$\mathrm{FL}_t(\mathscr{E})(k) = [\nu([(A^-)^\nu]_{k+1}) \,\boldsymbol{t}\, \nu([(A^-)^\nu]_{k+2}) \,\boldsymbol{t} \ldots, \nu([A^+]_{k+1}) \,\boldsymbol{t}\, \nu([A^+]_{k+2}) \,\boldsymbol{t} \ldots], \qquad (9.78)$$
$$\mathrm{FE}_t(\mathscr{E})(k) = [\mathrm{FG}_t(A^+)(k) \,\boldsymbol{t}\, \mathrm{FL}_{t,\nu}((A^-)^\nu)(k), \mathrm{FG}_t((A^-)^\nu)(k) \,\boldsymbol{t}\, \mathrm{FL}_{t,\nu}(A^+)(k)]$$

for each $k \in \mathbb{N}$.

Finally, one should say a few words about extensions of fuzzy cardinalities to IVFSs. They can be easily constructed via a simple modification of (9.67) and (9.69). More precisely, for a finite interval-valued fuzzy set $\mathcal{E} = (A_l, A_u)$ with $A_l \subset A_u$, we define

$$\mathrm{FG}(\mathcal{E}) = (\mathrm{FG}(A_l), \mathrm{FG}(A_u)). \tag{9.79}$$

This *FGCount of* $\mathcal{E}$ is an IVFS in $\mathbb{N}$ with

$$\mathrm{FG}(\mathcal{E})(k) = [\mathrm{FG}(A_l)(k), \mathrm{FG}(A_u)(k)] = [[A_l]_k, [A_u]_k] \tag{9.80}$$

for $k \in \mathbb{N}$. Similarly, the *FLCount of* $\mathcal{E}$ is understood as an IVFS

$$\mathrm{FL}(\mathcal{E}) = (\mathrm{FL}(A_u), \mathrm{FL}(A_l)),$$

i.e.

$$\mathrm{FL}(\mathcal{E})(k) = [1 - [A_u]_{k+1}, 1 - [A_l]_{k+1}]. \tag{9.81}$$

As to the *FECount of* $\mathcal{E}$, one defines it by

$$\mathrm{FE}(\mathcal{E}) = \mathrm{FG}(\mathcal{E}) \cap \mathrm{FL}(\mathcal{E}),$$

i.e.

$$\mathrm{FE}(\mathcal{E})(k) = [[A_l]_k \wedge (1 - [A_u]_{k+1}), [A_u]_k \wedge (1 - [A_l]_{k+1})] \tag{9.82}$$

for $k \in \mathbb{N}$. Further generalizations involving a t-norm and negation are possible, too, using the way of doing for IFSs.

9.4 Equipotency, Comparisons of and Operations on Fuzzy Cardinalities

Comparisons of (finite) sets with respect to their cardinalities as well as arithmetic operations on those cardinalities belong to the most basic mathematical procedures and abilities. Humans make use of them in a lot of areas of activity. As we remember, that two sets, $\boldsymbol{A}$ and $\boldsymbol{B}$, are equipotent (are of the same cardinality, in other words) or that the cardinality of $\boldsymbol{A}$ does not exceed the cardinality of $\boldsymbol{B}$ can be concluded without counting up the elements in $\boldsymbol{A}$ and $\boldsymbol{B}$. One can do it by establishing a 1-1 correspondence between the elements of $\boldsymbol{A}$ and $\boldsymbol{B}$ ($|\boldsymbol{A}| = |\boldsymbol{B}|$) or between the elements of $\boldsymbol{A}$ and of a subset $\boldsymbol{B}^* \subset \boldsymbol{B}$ ($|\boldsymbol{A}| \le |\boldsymbol{B}|$).

As to cardinalities of fuzzy sets, if the scalar approach is used, we do not need any additional definitions and tools as comparisons of and arithmetic operations on those cardinalities collapse to usual comparisons of and operations on nonnegative real numbers (see, however, Subsections 8.2.1 and 8.2.2). The subject of this section is a more sophisticated issue, namely that of comparisons between and arithmetic operations on fuzzy cardinalities of fuzzy sets. Throughout this section, $|E|$ will denote the fuzzy cardinality (of a given type) of a fuzzy set E. Dealing with fuzzy cardinalities generated by a t-norm and a negation, we will limit ourselves to Archimedean t-norms, plus the t-norm minimum, and to strict negations. This restriction seems to be quite sufficient from the viewpoint of applications.

9.4.1 Equipotent Fuzzy Sets

We like to present a practical approach to the issue. Let A, $B \in \mathrm{FFS}$. The equality $|A| = |B|$ of two fuzzy cardinalities will be understood as $|A|(k) = |B|(k)$ for each $k \in \mathbb{N}$. So, $|A| \neq |B|$ means that $|A|(k) \neq |B|(k)$ for some k. If $|A| = |B|$, we will say that A and B are *equipotent* or *are of the same cardinality*. Trivially, = forms an equivalence relation. If convenient, one can also write $A \sim B$ instead of $|A| = |B|$.

Let us formulate a characterization of $|A| = |B|$ in order to make it possible to check whether A and B are equipotent without determining $|A|$ and $|B|$. As our intuition suggests, we should have

$$(\forall k \in \mathbb{N}: [A]_k = [B]_k) \Rightarrow |A| = |B|. \tag{9.83}$$

It is easy to verify that this implication is fulfilled by each type of fuzzy cardinality discussed so far in the current chapter. It is also satisfied by the scalar cardinalities from Chapter 8. Two fuzzy sets are thus equipotent whenever they are identical up to the permutation of their positive membership degrees, i.e. whenever A and B attain the same values, including their repetitions, but possibly at different points of $\boldsymbol{U}$. Worth noticing in this context is that (see e.g. WYGRALAK (2003a))

$$\begin{aligned} \forall k \in \mathbb{N}: [A]_k = [B]_k &\Leftrightarrow \forall t \in (0, 1]: |A_t| = |B_t| \\ &\Leftrightarrow \forall t \in [0, 1): |A^t| = |B^t|. \end{aligned} \tag{9.84}$$

So, two fuzzy sets are of the same cardinality if their corresponding t-cuts (sharp or not) are always equipotent sets.

As already pointed out in Chapter 8, the inverse implication in (9.83) does not generally hold for scalar cardinalities. A question one has to ask is when (9.83) can be strengthened by replacing ⇒ with ⇔ if fuzzy cardinalities are involved. The answer is positive, and we have

$$|A| = |B| \Leftrightarrow \forall k \in \mathbb{N}: [A]_k = [B]_k, \tag{9.85}$$

whenever one uses BFCounts ($|E| = \mathrm{BF}(E)$), DBFCounts ($|E| = \mathrm{DBF}(E)$), FGCounts generated by $\boldsymbol{t} = \wedge$ or a strict t-norm $\boldsymbol{t}$ ($|E| = \mathrm{FG}(E)$ or $|E| = \mathrm{FG}_t(E)$), FLCounts and FECounts generated by $\boldsymbol{t} = \wedge$ or a strict $\boldsymbol{t}$, and by a strict negation ν ($|E| = \mathrm{FL}(E)$, $|E| = \mathrm{FE}(E)$, $|E| = \mathrm{FL}_{t,\nu}(E)$, or $|E| = \mathrm{FE}_{t,\nu}(E)$).

Let us formulate two remarks as to the fuzzy cardinalities fulfilling (9.85). First, equipotent fuzzy sets then have equipotent cores and equipotent supports:

$$|\mathrm{core}(A)| = |\mathrm{core}(B)| \text{ and } |\mathrm{supp}(A)| = |\mathrm{supp}(B)| \text{ whenever } |A| = |B|. \tag{9.86}$$

Second, those fuzzy cardinalities, although numerically different, are thus all equivalent as they generate the same division of FFS into equivalence classes of equipotent fuzzy sets.

In a sense, a disadvantage of the fuzzy cardinalities satisfying (9.85) is that they are "too sensitive" and "too rigorous". Even a very small difference between very small (and, hence, possibly inessential) membership degrees makes A and B non-equipotent, which may be counterintuitive and inconvenient in some applications. This is why we like to mention two more liberal criteria of equipotency in which

$$|A| = |B| \Leftrightarrow [A]_k = [B]_k \text{ for some } k\text{'s}.$$

First, let us look at FGCounts from (9.56) generated by a nilpotent t-norm $\boldsymbol{t}$. One then has

$$|A| = |B| \Leftrightarrow \pi(A) = \pi(B) \ \& \ \forall k \le \pi(A)\colon [A]_k = [B]_k, \tag{9.87}$$

where

$$\pi(A) = \bigvee\{k \in \mathbb{N}\colon [A]_1 \boldsymbol{t} [A]_2 \boldsymbol{t} \ldots \boldsymbol{t} [A]_k > 0\}. \tag{9.88}$$

This time, only the identity of the corresponding top membership degrees in A and B is necessary for having $|A| = |B|$. The bottom membership degrees are not essential (see Example 9.10). It is easy to notice that (9.87) implies

$$|\mathrm{core}(A)| = |\mathrm{core}(B)| \text{ whenever } |A| = |B|. \tag{9.89}$$

Searching for an analogue of (9.84), we get the following:

$$\begin{aligned} &\pi(A) = \pi(B) \ \& \ \forall k \le \pi(A)\colon [A]_k = [B]_k \\ &\Leftrightarrow \pi(A) = \pi(B) \ \& \ [A]_{\pi(A)} = [B]_{\pi(B)} \ \& \ \forall t \in ([A]_{\pi(A)}, 1]\colon |A_t| = |B_t| \\ &\Leftrightarrow \pi(A) = \pi(B) \ \& \ [A]_{\pi(A)} = [B]_{\pi(B)} \ \& \ \forall t \in [[A]_{\pi(A)}, 1)\colon |A^t| = |B^t|. \end{aligned} \tag{9.90}$$

As we thus see, in contrast to the case of strict t-norms enriched by $\boldsymbol{t} = \wedge$, that two fuzzy sets are equipotent/nonequipotent generally depends on the choice of $\boldsymbol{t}$ within the class of nilpotent t-norms if FGCounts generated by $\boldsymbol{t}$ are used.

Second, let us move on to FLCounts from (9.58) generated by a nilpotent $\boldsymbol{t}$ and a strict negation ν. The situation is now quite different in comparison with (9.87). One has for those FLCounts

$$|A| = |B| \Leftrightarrow \pi^*(A) = \pi^*(B) \ \& \ \forall k > \pi^*(A)\colon [A]_k = [B]_k \tag{9.91}$$

with

$$\pi^*(A) = \bigwedge\{k \in \mathbb{N}\colon \nu([A]_{k+1}) \boldsymbol{t} \nu([A]_{k+2}) \boldsymbol{t} \ldots \boldsymbol{t} \nu([A]_n) > 0\}. \tag{9.92}$$

Consequently,

$$|\mathrm{supp}(A)| = |\mathrm{supp}(B)| \text{ whenever } |A| = |B|. \tag{9.93}$$

Identity of the corresponding bottom membership degrees in A and B is now essential for having A and B equipotent. The top membership degrees are insignificant. It seems to be astonishing that just the smallest membership values in A are responsible for its equipotency with another fuzzy set. However, these membership values are the top membership values in A' and A^{ν}, whereas the concept of FLCounts is based just on counting in A' (see Subsections 9.1.1 and 9.1.2). Anyway, our further investigations will be restricted to FLCounts generated by $\wedge$ or a strict t-norm.

Example 9.10. Take

$$A = 0.5/x_1 + 1/x_2 + 0.9/x_3 + 0.5/x_4 + 0.2/x_5,$$
$$B = 0.9/x_4 + 0.5/x_5 + 1/x_6 + 0.2/x_8 + 0.5/x_9,$$
$$C = 0.9/x_1 + 0.5/x_4 + 1/x_8 + 0.3/x_9,$$
$$D = 0.8/x_1 + 0.7/x_2 + 0.2/x_3 + 0.5/x_6 + 0.6/x_8.$$

Since $[A]_k = [B]_k$ for each $k \in \mathbb{N}$, we conclude that $|A| = |B|$ whichever type of fuzzy cardinality we presented is used. For instance,

$$\mathrm{FG}(A) = \mathrm{FG}(B) = (1, 1, 0.9, 0.5, 0.5, 0.2).$$

On the other hand, $|A| \neq |C|$ whenever a fuzzy cardinality satisfying (9.85) is applied as, say, $[A]_4 \neq [C]_4$. Notice that $\pi(A) = \pi(C) = 3$ and $[A]_k = [C]_k$ for each $k < 4$ if $\boldsymbol{t} = t_Ł$. A and C are thus equipotent if FGCounts generated by the Łukasiewicz t-norm are involved. Then

$$\mathrm{FG}_t(A) = \mathrm{FG}_t(C) = (1, 1, 0.9, 0.4).$$

Use again $\boldsymbol{t} = t_Ł$ with $\nu = \nu_Ł$. We have $\pi^*(A) = \pi^*(D) = 3$ and $[A]_k = [D]_k$ for each $k > 3$. So, A and D become equipotent if one employs FLCounts generated by $\boldsymbol{t}$ and ν, namely

$$\mathrm{FL}_{t,\nu}(A) = \mathrm{FL}_{t,\nu}(D) = (0, 0, 0, 0.3, 0.8, (1)).$$ □

It is clear that (9.85), (9.87) and (9.91) can be used in another way. They make it possible to reason about the $[A]_k$'s and $[B]_k$'s when $|A|$ and $|B|$ are identical fuzzy cardinalities.

Finally, the interested reader is referred to WYGRALAK (2003a) for proofs and more theoretical details about equipotent fuzzy sets (see also WYGRALAK (2005b)).

9.4.2 Inequalities

The relationship $|A| \leq |B|$ requires a suitable definition when dealing with fuzzy cardinalities $|A|$ and $|B|$ of $A, B \in \mathrm{FFS}$. Anyway, speaking intuitively, one should have

$$(\forall k \in \mathbb{N}\colon [A]_k \leq [B]_k) \Rightarrow |A| \leq |B|. \tag{9.94}$$

A good and convenient choice seems to be the following classical-like

Definition 9.11. We will say that the (fuzzy) cardinality of A is *less than or equal to* the cardinality of B, and we will write $|A| \le |B|$, if there exists a fuzzy set $B^* \subset B$ equipotent to A. If $|A| \le |B|$ and $|A| \ne |B|$, we will say that the cardinality of A is *less than* that of B, and we then write $|A| < |B|$.

Consequently,

$$\begin{gathered} |A| \le |B| \Leftrightarrow \exists B^* \subset B\colon |A| = |B^*|, \\ A \subset B \Rightarrow |A| \le |B|, \quad |A| < |B| \Rightarrow |A| \le |B|. \end{gathered} \tag{9.95}$$

Assume $\boldsymbol{t} = \wedge$ or $\boldsymbol{t}$ is a strict t-norm, and ν denotes a strict negation. Let us look at the cases of FGCounts generated by $\boldsymbol{t}$ ($|E| = \mathrm{FG}_t(E)$), FLCounts and FECounts generated by $\boldsymbol{t}$ and ν ($|E| = \mathrm{FL}_{t,\nu}(E)$ or $|E| = \mathrm{FE}_{t,\nu}(E)$). Definition 9.11 then collapses to the following:

$$\begin{aligned} |A| \le |B| &\Leftrightarrow \forall k \in \mathbb{N}\colon [A]_k \le [B]_k \\ &\Leftrightarrow \forall t \in (0,1]\colon |A_t| \le |B_t| \\ &\Leftrightarrow \forall t \in [0,1)\colon |A^t| \le |B^t|. \end{aligned} \tag{9.96}$$

Hence

$$|A| < |B| \Leftrightarrow \forall k \in \mathbb{N}\colon [A]_k \le [B]_k \;\&\; \exists i \in \mathbb{N}\colon [A]_i < [B]_i \tag{9.97}$$

and

$$|A| \le |B| \Rightarrow |\mathrm{core}(A)| \le |\mathrm{core}(B)| \;\&\; |\mathrm{supp}(A)| \le |\mathrm{supp}(B)|. \tag{9.98}$$

If FGCounts generated by a nilpotent $\boldsymbol{t}$ are used, Definition 9.11 implies

$$|A| \le |B| \Leftrightarrow \pi(A) \le \pi(B) \;\&\; \forall k \le \pi(A)\colon [A]_k \le [B]_k \tag{9.99}$$

and

$$|A| \le |B| \Rightarrow |\mathrm{core}(A)| \le |\mathrm{core}(B)|. \tag{9.100}$$

In (9.96) and (9.99), $\le$ is only a partial order relation, i.e. there exist fuzzy sets being incomparable with respect to their fuzzy cardinalities (see Example 9.12). Moreover, we then have

$$|A| = |B| \Leftrightarrow |A| \le |B| \;\&\; |B| \le |A|. \tag{9.101}$$

A simple consequence of (9.96) and (9.14) is that

$$\mathrm{FG}(A) \le \mathrm{FG}(B) \Leftrightarrow \mathrm{FG}(A) \subset \mathrm{FG}(B), \tag{9.102}$$

whereas

$$\mathrm{FG}_t(A) \le \mathrm{FG}_t(B) \Rightarrow \mathrm{FG}_t(A) \subset \mathrm{FG}_t(B) \tag{9.103}$$

for an Archimedean $\boldsymbol{t}$.

Again, let ν denote a strict negation. We easily get

$$\mathrm{FL}_\nu(A) \le \mathrm{FL}_\nu(B) \Leftrightarrow \mathrm{FL}_\nu(B) \subset \mathrm{FL}_\nu(A). \tag{9.104}$$

If a strict t-norm $\boldsymbol{t}$ is involved,

$$\mathrm{FL}_{t,\nu}(A) \le \mathrm{FL}_{t,\nu}(B) \Rightarrow \mathrm{FL}_{t,\nu}(B) \subset \mathrm{FL}_{t,\nu}(A). \tag{9.105}$$

By (9.96),

$$\mathrm{FE}_{t,\nu}(A) \le \mathrm{FE}_{t,\nu}(B) \Leftrightarrow \mathrm{FG}_t(A) \le \mathrm{FG}_t(B) \Leftrightarrow \mathrm{FL}_{t,\nu}(A) \le \mathrm{FL}_{t,\nu}(B) \tag{9.106}$$

whenever $\boldsymbol{t} = \wedge$ or $\boldsymbol{t}$ is strict.

Finally, we like to look once more at FGCounts generated by a t-norm $\boldsymbol{t}$ ($\boldsymbol{t} = \wedge$ or $\boldsymbol{t}$ Archimedean), and at FLCounts and FECounts generated by $\boldsymbol{t}$ and ν ($\boldsymbol{t} = \wedge$ or $\boldsymbol{t}$ strict). Assume α and β are fuzzy cardinalities (of the same type specified above) of two fuzzy sets. Their comparisons can be done using the following definitions:

$$\alpha \le \beta \Leftrightarrow \exists A, B \in \mathrm{FFS}\colon |A| = \alpha \ \&\ |B| = \beta \ \&\ |A| \le |B|$$

and

$$\alpha < \beta \Leftrightarrow \exists A, B \in \mathrm{FFS}\colon |A| = \alpha \ \&\ |B| = \beta \ \&\ |A| < |B|. \tag{9.107}$$

Both $\alpha \le \beta$ and $\alpha < \beta$ are well-defined as they do not depend on the choice of fuzzy sets such that $|A| = \alpha$ and $|B| = \beta$. Moreover,

$$\alpha = \beta \Leftrightarrow \alpha \le \beta \ \&\ \beta \le \alpha,$$

$$\alpha < \beta \Leftrightarrow \alpha \le \beta \ \&\ \alpha \ne \beta,$$

$$\alpha < \beta \Rightarrow \alpha \le \beta,$$

$$\alpha \le \beta \ \&\ \beta < \gamma \Rightarrow \alpha < \gamma,$$

$$\alpha < \beta \ \&\ \beta \le \gamma \Rightarrow \alpha < \gamma.$$

Example 9.12. Let us begin with a simple instance of fuzzy sets being incomparable with respect to their cardinalities. Take

$$A = 1/x_1 + 0.9/x_2 + 0.3/x_3 \quad \text{and} \quad B = 1/x_3 + 0.7/x_4 + 0.6/x_5.$$

Then

$$[A]_1 = 1,\ [A]_2 = 0.9,\ [A]_3 = 0.3,\ [B]_1 = 1,\ [B]_2 = 0.7,\ [B]_3 = 0.6,$$

and $[A]_k = [B]_k = 0$ for $k > 3$. Since $[A]_2 > [B]_2$ and $[A]_3 < [B]_3$, (9.96) and (9.99) imply that neither $|A| \le |B|$ nor $|B| \le |A|$ if, say, we use FGCounts, FLCounts, FECounts, or FGCounts generated by the Łukasiewicz t-norm as $\pi(A) = \pi(B) = 3$ for $\boldsymbol{t} = t_Ł$. Generally, if fuzzy cardinalities generated by a t-norm $\boldsymbol{t}$ are involved, the result of comparing $|A|$ and $|B|$ depends on the choice of $\boldsymbol{t}$. Put

$$A = 1/x_1 + 0.6/x_2 + 0.3/x_3 + 0.1/x_4, \quad B = 1/x_3 + 0.6/x_4 + 0.4/x_5 + 0.2/x_6.$$

By (9.97), we get $\mathrm{FG}_t(A) < \mathrm{FG}_t(B)$ for $t = \wedge$ and each strict t, whereas $t = t_Ł$ gives $\mathrm{FG}_t(A) = \mathrm{FG}_t(B) = (1, 1, 0.6)$.

Our last remark relates to (9.107). Take two usual FECounts of some fuzzy sets $(t = \wedge, \nu = \nu_Ł)$, e.g.

$$\alpha = (0.1, 0.2, 0.8, 0.2, 0.2) \quad \text{and} \quad \beta = (0.1, 0.3, 0.7, 0.1).$$

So, $\alpha = |A|$ with A such that

$$[A]_1 = 0.9,\ [A]_2 = 0.8,\ [A]_3 = [A]_4 = 0.2,\ [A]_k = 0 \quad \text{for } k > 4,$$

and $\beta = |B|$ with

$$[B]_1 = 0.9,\ [B]_2 = 0.7,\ [B]_3 = 0.1,\ [B]_k = 0 \quad \text{for } k > 3.$$

By (9.97), we conclude that $\beta < \alpha$. □

One should also mention another possible way of comparing two FECounts α and β generated by t and ν. Drawing inspiration from (2.120), we can define

$$\alpha = \beta \Leftrightarrow \mathrm{COG}_\alpha = \mathrm{COG}_\beta, \quad \alpha < \beta \Leftrightarrow \mathrm{COG}_\alpha < \mathrm{COG}_\beta. \tag{9.108}$$

Again, the interested reader is referred to WYGRALAK (2003a) for further theoretical details about inequalities between fuzzy cardinalities of fuzzy sets.

9.4.3 Operations on Fuzzy Cardinalities

The subject of this subsection are basic arithmetic operations on fuzzy cardinalities. Our main interest, quite understandably, will be in the operations of addition and multiplication. Some remarks about subtraction and division are placed at the end of the subsection.

Let α and β denote two fuzzy cardinalities of the same given type, e.g. two BFCounts or FECounts of fuzzy sets from FFS. The simplest and most natural way of introducing the *sum* $\alpha + \beta$ and the *product* $\alpha \cdot \beta$ of α and β seem to be the following classical-like definitions:

- $\alpha + \beta = |A \cup B|$, (9.109)

  where $A, B \in \mathrm{FFS}$ are arbitrary disjoint fuzzy sets such that $\alpha = |A|$ and $\beta = |B|$;

- $\alpha \cdot \beta = |A \times B|$ (9.110)

  with $A, B \in \mathrm{FFS}$ being arbitrary fuzzy sets such that $\alpha = |A|$ and $\beta = |B|$.

According to the classical arithmetic notation, we shall write $\alpha\beta$ instead of $\alpha \cdot \beta$. (9.109) and (9.110) guarantee that both $\alpha + \beta$ and $\alpha\beta$ are always fuzzy cardinalities of the same type as α and β.

First of all, one has to ask if $\alpha+\beta$ and $\alpha\beta$ are well-defined. We ask, in other words, if they do not depend on the choice of A and B in (9.109) and (9.110), i.e. if

$$|A| = |A^*| \;\&\; |B| = |B^*| \;\&\; A \cap B = A^* \cap B^* = 1_\emptyset \;\Rightarrow\; |A \cup B| = |A^* \cup B^*|$$

and (9.111)

$$|A| = |A^*| \;\&\; |B| = |B^*| \;\Rightarrow\; |A \times B| = |A^* \times B^*|.$$

The answer is affirmative at least for the main sorts of fuzzy cardinalities discussed in this book, e.g. whenever α and β are two BFCounts, two FGCounts generated by $\boldsymbol{t}$ ($\boldsymbol{t} = \wedge$ or $\boldsymbol{t}$ Archimedean), or two FLCounts or FECounts generated by $\boldsymbol{t}$ and ν ($\boldsymbol{t} = \wedge$ or $\boldsymbol{t}$ strict, ν strict). The following shows that (9.109) and (9.110) are quite convenient in practice.

Example 9.13. (a) Assume we like to add two FGCounts, namely

$$\alpha = (1, 1, 0.9, 0.72, 0.432) \quad \text{and} \quad \beta = (1, 0.8, 0.24).$$

Treating them as fuzzy cardinalities of disjoint fuzzy sets, one gets $\alpha = \mathrm{FG}(A)$ and $\beta = \mathrm{FG}(B)$ with

$$A = 1/x_1 + 0.9/x_2 + 0.72/x_3 + 0.432/x_4 \quad \text{and} \quad B = 0.8/y_1 + 0.24/y_2.$$

By (9.109), we obtain

$$\alpha+\beta = \mathrm{FG}(A \cup B) = (1, 1, 0.9, 0.8, 0.72, 0.432, 0.24).$$

(b) Let α and β from (a) be two FGCounts generated by $\boldsymbol{t} = \boldsymbol{t}_{\mathrm{L}}$. Thus, $\alpha = \mathrm{FG}_t(A)$ and $\beta = \mathrm{FG}_t(B)$ with (see (9.55))

$$A = 1/x_1 + 0.9/x_2 + 0.82/x_3 + 0.712/x_4 \quad \text{and} \quad B = 0.8/y_1 + 0.44/y_2,$$

i.e.

$$\alpha+\beta = \mathrm{FG}_t(A \cup B) = \mathrm{FG}_t(1/x_1 + 0.9/x_2 + 0.82/x_3 + 0.8/y_1 + 0.712/x_4 + 0.44/y_2)$$
$$= (1, 1, 0.9, 0.72, 0.52, 0.232).$$

(c) Let us add two FECounts generated by $\boldsymbol{t} = \boldsymbol{t}_{\mathrm{a}}$ and $\nu = \nu_{\mathrm{L}}$ of fuzzy sets

$$A = 1/x_1 + 0.9/x_2 + 0.2/x_3 \quad \text{and} \quad B = 1/x_2 + 0.7/x_4.$$

Since A and B are not disjoint, we have to "part" them and to use $B^* = 1/y_1 + 0.7/y_2$ which is disjoint with A and equipotent to B (see (9.85)). By (9.62),

$$\mathrm{FE}_{t,\nu}(A) + \mathrm{FE}_{t,\nu}(B) = \mathrm{FE}_{t,\nu}(A) + \mathrm{FE}_{t,\nu}(B^*)$$
$$= \mathrm{FE}_{t,\nu}(1/x_1 + 1/y_1 + 0.9/x_2 + 0.7/y_2 + 0.2/x_3)$$
$$= (0, 0, 0.024, 0.216, 0.504, 0.126).$$

(d) Finally, we like to multiply two FECounts

$$\alpha = (0, 0.1, 0.8, 0.2) \quad \text{and} \quad \beta = (0, 0.2, 0.8).$$

It is easy to check that $\alpha = \mathrm{FE}(A)$ and $\beta = \mathrm{FE}(B)$ with

$$A = 1/x_1 + 0.9/x_2 + 0.2/x_3 \quad \text{and} \quad B = 1/y_1 + 0.8/y_2.$$

Hence

$$\alpha\beta = \mathrm{FE}(A \times B) = (0, 0.1, 0.2, 0.2, 0.8, 0.2, 0.2). \qquad \square$$

Referring to operations on FECounts generated by $\boldsymbol{t}$ and v ($\boldsymbol{t} = \wedge$ or $\boldsymbol{t}$ strict, v strict), we see that they can be expressed in terms of operations on the corresponding FGCounts and FLCounts. Indeed, by virtue of (9.60), we have

$$\begin{aligned} \mathrm{FE}_{t,v}(A) + \mathrm{FE}_{t,v}(B) &= \mathrm{FG}_t(A \cup B) \cap_t \mathrm{FL}_{t,v}(A \cup B) \\ &= (\mathrm{FG}_t(A) + \mathrm{FG}_t(B)) \cap_t (\mathrm{FE}_{t,v}(A) + \mathrm{FE}_{t,v}(B)) \end{aligned} \tag{9.112}$$

whenever A and B are disjoint, and

$$\begin{aligned} \mathrm{FE}_{t,v}(A) \cdot \mathrm{FE}_{t,v}(B) &= \mathrm{FG}_t(A \times B) \cap_t \mathrm{FL}_{t,v}(A \times B) \\ &= (\mathrm{FG}_t(A) \cdot \mathrm{FG}_t(B)) \cap_t (\mathrm{FL}_{t,v}(A) \cdot \mathrm{FL}_{t,v}(B)) \end{aligned} \tag{9.113}$$

for each $A, B \in \mathrm{FFS}$.

Another question we should ask is whether the operations of addition and multiplication of fuzzy cardinalities α and β can be equivalently performed via the extension principle (2.103) and (2.105), respectively, or its modification. Let us begin with the case of addition. One shows that $\alpha + \beta$ from (9.109) can be presented as

$$(\alpha + \beta)(k) = \bigvee\{\alpha(i)\, \boldsymbol{t}\, \beta(j) \colon i + j = k\} \quad \text{for each } k \in \mathbb{N} \tag{9.114}$$

whenever α and β are (1) FGCounts generated by $\boldsymbol{t} = \wedge$ or an Archimedean $\boldsymbol{t}$, (2) FLCounts or FECounts generated by $\boldsymbol{t}$ and v ($\boldsymbol{t} = \wedge$ or $\boldsymbol{t}$ strict, v strict). Worth emphasizing is that (9.114) does not hold for BFCounts from (9.4)-(9.5); clearly, $\boldsymbol{t} = \wedge$ should be put in that case. For $k \in \mathbb{N}$, we generally have

$$(\mathrm{BF}(A) + \mathrm{BF}(B))(k) \neq \bigvee\{\mathrm{BF}(A)(i) \wedge \mathrm{BF}(B)(j) \colon i + j = k\}. \tag{9.115}$$

Example 9.14. Take $A = 1/x_1 + 0.8/x_2 + 0.8/x_3$ and $B = 1/y_1 + 0.9/y_2 + 0.9/y_3 + 0.4/y_4$. (9.109) leads to

$$\begin{aligned} \mathrm{BF}(A) + \mathrm{BF}(B) &= \mathrm{BF}(1/x_1 + 1/y_1 + 0.9/y_2 + 0.9/y_3 + 0.8/x_2 + 0.8/x_3 + 0.4/y_4) \\ &= (0, 0, 1, 0, 0.9, 0, 0.8, 0.4). \end{aligned}$$

On the other hand, $\mathrm{BF}(A) = (0, 1, 0, 0.8)$ and $\mathrm{BF}(B) = (0, 1, 0, 0.9, 0.4)$. Computing $\bigvee\{\mathrm{BF}(A)(i) \wedge \mathrm{BF}(B)(j)\colon i+j=k\}$ with $k \in \mathbb{N}$ we thus obtain the following numerical results written in the vector notation:

$$(0, 0, 1, 0, 0.9, 0.4, 0.8, 0.4) \neq \mathrm{BF}(A)+\mathrm{BF}(B).$$

What is worse, we see that the resulting vector does not form the BFCount of any fuzzy set. □

The case of multiplication is more sophisticated. For α and β understood as two FGCounts generated by $\boldsymbol{t} = \wedge$ or an Archimedean $\boldsymbol{t}$, let us ask if

$$(\alpha\beta)(k) = \bigvee\{\alpha(i)\,\boldsymbol{t}\,\beta(j)\colon ij \geq k\} \quad \text{for each } k \in \mathbb{N}. \tag{9.116}$$

The original triangular norm-based extension principle is here modified. α and β are nonincreasing and, moreover, some k's are prime or have a small number of possible decompositions into two factors. The presence of "$ij \geq k$" instead of "$ij = k$" thus guarantees that the right side in (9.116) produces nonincreasing fuzzy sets in $\mathbb{N}$. However, (9.116) is satisfied iff $\boldsymbol{t} = \wedge$. Indeed, it is a routine matter to check that (9.116) is true for $\boldsymbol{t} = \wedge$. If $\boldsymbol{t}$ is Archimedean, the right side of (9.116) creates a fuzzy set F in $\mathbb{N}$ which is not generally an FGCount generated by $\boldsymbol{t}$. A simple counterexample is

$$\alpha = \beta = (1, 1, a)$$

with any $a \in (0, 1)$ such that $a\boldsymbol{t}a > 0$. Then $F = (1, 1, a, a\boldsymbol{t}a, a\boldsymbol{t}a)$. If F were the FGCount of a $D \in \mathrm{FFS}$, $F = \mathrm{FG}_t(D)$, we should have $[D]_1 = 1$, $[D]_2 = [D]_3 = a < 1$, and $[D]_4 = 1$, which forms a contradiction (see (9.55)). Concluding, (9.116) cannot be generally satisfied by Archimedean t-norms.

Assume α and β are now two FLCounts generated by $\boldsymbol{t} = \wedge$ or a strict $\boldsymbol{t}$, and by a strict negation ν. We ask whether

$$(\alpha\beta)(k) = \bigvee\{\alpha(i)\,\boldsymbol{t}\,\beta(j)\colon ij \leq k\} \quad \text{for each } k \in \mathbb{N}. \tag{9.117}$$

The answer is affirmative only for $\boldsymbol{t} = \wedge$. If $\boldsymbol{t}$ is strict, the right side produces a fuzzy set which differs from αβ and does not generally form the FLCount (generated by $\boldsymbol{t}$ and ν) of any fuzzy set from FFS. We see that another type of modification of the original extension principle involving "$ij \leq k$" is now used as what we deal with in this case are nondecreasing fuzzy sets in $\mathbb{N}$. An immediate consequence of this discussion is that

$$\begin{aligned}(\mathrm{FE}_{\wedge,\nu}(A)\cdot\mathrm{FE}_{\wedge,\nu}(B))(k) = {} & \bigvee\{\mathrm{FG}(A)(i) \wedge \mathrm{FG}(B)(j)\colon ij \geq k\} \wedge \\ & \bigvee\{\mathrm{FL}_{\wedge,\nu}(A)(i) \wedge \mathrm{FL}_{\wedge,\nu}(B)(j)\colon ij \leq k\}\end{aligned} \tag{9.118}$$

for each $k \in \mathbb{N}$ and each strict negation ν.

Let us move on to laws and properties of the operations of addition and multiplication of fuzzy cardinalities. Our investigations will be focused on the three cases of cardinalities which - due to their features - already attracted our attention earlier on in this section. We mean FGCounts generated by $\boldsymbol{t}$ ($\boldsymbol{t} = \wedge$ or $\boldsymbol{t}$ Archimedean) and FLCounts as well as FECounts generated by $\boldsymbol{t}$ and a strict negation ν ($\boldsymbol{t} = \wedge$ or $\boldsymbol{t}$ strict). Throughout, lowercase Greek letters α, β, γ and δ, and $|D|$ with $D \in \mathrm{FFS}$ will denote fuzzy cardinalities of any of the three types specified above.

We like to begin with some well-known elementary properties of cardinalities of sets which, generally, cannot be extended to fuzzy sets. First, let us draw attention to the valuation property of sets, and ask whether

$$|A \cap_t B| + |A \cup_s B| = |A| + |B| \quad \text{for each } A, B \in \mathrm{FFS}, \tag{9.119}$$

where $\boldsymbol{s}$ denotes a t-conorm; $\boldsymbol{s} = \vee$ or (1) $\boldsymbol{s}$ is Archimedean if FGCounts are used, (2) $\boldsymbol{s}$ is strict whenever FLCounts or FECounts are involved. For each of the three types of fuzzy cardinalities under discussion, the *valuation property* (9.119) is satisfied iff $\boldsymbol{t} = \wedge$ and $\boldsymbol{s} = \vee$ (see WYGRALAK (2003a); see also WYGRALAK (1999a)). So, it is a rare property.

Second, recollect that nonnegative integers fulfil the following: if $i < k$, there exists a unique number j such that $i + j = k$. Unfortunately, this *compensation property* cannot be extended to the fuzzy cardinalities under discussion. In other words, if we ask whether $\alpha < \gamma$ implies the existence of a unique β such that $\alpha + \beta = \gamma$, the answer is generally negative: it happens that the β does not exist or is not unique. This failure of the compensation property forms one of the most essential differences between the arithmetic of fuzzy cardinalities and the usual arithmetic of nonnegative integers.

Before formulating the properties of sums and products, we have to introduce a new symbol. $\boldsymbol{k}$ will denote a counterpart of $k \in \mathbb{N}$ in the world of fuzzy cardinalities. So, **0**, **1**, **2**, ... are counterparts of 0, 1, 2, ..., respectively. Looking at the results from Subsection 9.3.1, we get

$$\boldsymbol{k} = 1_{\{0, 1, \ldots, k\}} = (1, 1, \ldots, 1) \quad \text{with } k+1 \text{ ones}$$

if FGCounts generated by $\boldsymbol{t}$ are used,

$$\boldsymbol{k} = 1_{\{k, k+1, \ldots\}} = (0, \ldots, 0, (1)) \quad \text{with } k \text{ zeros}$$

whenever FLCounts generated by $\boldsymbol{t}$ and ν are involved, and

$$\boldsymbol{k} = 1_{\{k\}} = (0, \ldots, 0, 1) \quad \text{with } k \text{ zeros}$$

if we deal with FECounts generated by $\boldsymbol{t}$ and ν. $\boldsymbol{k}$ is in each case the fuzzy cardinality of a k-element set viewed as a fuzzy set, and

$$\boldsymbol{k}\alpha = \alpha + \alpha + \ldots + \alpha$$

with k alphas on the right side.

The following list presents classical-like laws and properties which are consequences of (9.109)-(9.110) and hold true for fuzzy cardinalities of any of the three types under discussion. Some of these properties are preceded by $(*)$. This signifies that whenever FGCounts generated by $\boldsymbol{t}$ are used, the property does not hold if $\boldsymbol{t}$ is nilpotent and, thus, $\boldsymbol{t}$ must be strict or $\boldsymbol{t} = \wedge$.

- *Commutativity and associativity*

$$\alpha+\beta = \beta+\alpha, \quad \alpha+(\beta+\gamma) = (\alpha+\beta)+\gamma,$$
$$\alpha\beta = \beta\alpha, \quad \alpha(\beta\gamma) = (\alpha\beta)\gamma. \tag{9.120}$$

- *Neutral elements*

$$\alpha+\mathbf{0} = \alpha, \quad \alpha\mathbf{1} = \alpha. \tag{9.121}$$

- $\alpha, \beta \le \alpha+\beta,$

$(*)$ if $\beta > \mathbf{0}$, then $\alpha < \alpha+\beta$,

if $\beta \ge \mathbf{1}$, then $\alpha \le \alpha\beta$, (9.122)

$(*)$ if $\alpha \ge \mathbf{1}$ and $\beta > \mathbf{1}$, then $\alpha < \alpha\beta$.

- $\alpha+\beta = \mathbf{0}$ iff $\alpha = \beta = \mathbf{0}$,

$\alpha\beta = \mathbf{1}$ iff $\alpha = \beta = \mathbf{1}$, (9.123)

$\alpha\beta = \mathbf{0}$ iff $\alpha = \mathbf{0}$ or $\beta = \mathbf{0}$.

- *Distributivity law*

$$\alpha(\beta+\gamma) = \alpha\beta+\alpha\gamma. \tag{9.124}$$

- *Cancellation laws for addition and multiplication*

$(*)$ If $\alpha+\beta = \alpha+\gamma$, then $\beta = \gamma$,

$(*)$ if $\alpha+\beta < \alpha+\gamma$, then $\beta < \gamma$, (9.125)

$(*)$ if $\alpha+\beta \le \alpha+\gamma$, then $\beta \le \gamma$.

$(*)$ If $\alpha\beta = \alpha\gamma$ and $\alpha \ge \mathbf{1}$, then $\beta = \gamma$,

$(*)$ if $\alpha\beta < \alpha\gamma$ and $\alpha \ge \mathbf{1}$, then $\beta < \gamma$, (9.126)

$(*)$ if $\alpha\beta \le \alpha\gamma$ and $\alpha \ge \mathbf{1}$, then $\beta \le \gamma$.

- *Side-by-side addition and multiplication of inequalities*

If $\alpha \le \beta$ and $\gamma \le \delta$, then $\alpha+\gamma \le \beta+\delta$,

if $\alpha \le \beta$, then $\alpha+\gamma \le \beta+\gamma$, (9.127)

$(*)$ if $\alpha < \beta$ and $\gamma \le \delta$, then $\alpha+\gamma < \beta+\delta$,

$(*)$ if $\alpha < \beta$, then $\alpha+\gamma < \beta+\gamma$.

$$\begin{aligned}
&\text{If } \alpha \le \beta \text{ and } \gamma \le \delta, \text{ then } \alpha\gamma \le \beta\delta,\\
&\text{if } \alpha \le \beta, \text{ then } \alpha\gamma \le \beta\gamma,\\
(*)\ &\text{if } \alpha < \beta \text{ and } \gamma \le \delta \text{ and } \delta \ge \mathbf{1}, \text{ then } \alpha\gamma < \beta\delta,\\
(*)\ &\text{if } \alpha < \beta \text{ and } \gamma \ge \mathbf{1}, \text{ then } \alpha\gamma < \beta\gamma.
\end{aligned} \tag{9.128}$$

- *Comparisons of sums and products*

$$\begin{aligned}
&\text{If } \alpha, \beta \ge \mathbf{2}, \text{ then } \alpha+\beta \le \alpha\beta,\\
&\text{if } \alpha \ge \mathbf{2} \text{ and } \beta \ge \mathbf{3}, \text{ then } \alpha+\beta < \alpha\beta.
\end{aligned} \tag{9.129}$$

The interested reader is referred to WYGRALAK (2003a) for related proofs, further properties and theoretical details as well as for a study of other operations on fuzzy cardinalities, namely subtraction, division, and exponentiation (see also WYGRALAK (1999a, 2001)).

We like to give in this book only a few general remarks about subtraction and division of fuzzy cardinalities. If one tries to define the *difference* $\gamma - \alpha$ of γ and α, at least two ways of doing are possible. First, we can apply the extension principle from (2.104). Then $\gamma - \alpha$ always exists, but $\alpha + (\gamma - \alpha) \neq \gamma$ in general. The other way consists in defining $\gamma - \alpha$ in the classical-like manner via addition. We then say that the difference $\gamma - \alpha$ exists and equals β whenever β is a unique fuzzy cardinality such that $\alpha + \beta = \gamma$. So, by definition, $\alpha + (\gamma - \alpha) = \gamma$ provided that $\gamma - \alpha$ exists. In many cases, however, $\gamma - \alpha$ does not exist as the β is not unique or does not exist.

Similarly, the *quotient* α/β of α and $\beta > \mathbf{0}$ can be defined using the extension principle (2.106). Although α/β always exists in this approach, $(\alpha/\beta)\cdot\beta \neq \alpha$ in general. Making use of the classical-like way of defining via multiplication, one says that the quotient α/β exists and equals δ if δ is a unique fuzzy cardinality such that $\alpha = \beta\delta$. Thus, $(\alpha/\beta)\cdot\beta = \alpha$, but α/β does not necessarily exist.

9.5 Dealing with Cardinalities of IFSs and IVFSs

FGCounts, FLCounts and FECounts, possibly involving a t-norm and a negation, have been extended in Subsection 9.3.2 to IFSs and IVFSs. For those extensions, one can establish a suitable equipotency relation and define their sums and products. We will present very briefly the case of IFSs. That of IVFSs is quite analogous.

Assume $\mathscr{E} = (A^+, A^-)$ and $\mathscr{F} = (B^+, B^-)$ are two finite IFSs, where $A^+ \subset (A^-)^\nu$ and $B^+ \subset (B^-)^\nu$ with a strong negation ν. By virtue of (9.77), we have

$$\begin{aligned}
&\mathrm{FG}_t(\mathscr{E}) = \mathrm{FG}_t(\mathscr{F}) \Leftrightarrow \mathrm{FG}_t(A^+) = \mathrm{FG}_t(B^+) \;\&\; \mathrm{FG}_t((A^-)^\nu) = \mathrm{FG}_t((B^-)^\nu),\\
&\mathrm{FL}_t(\mathscr{E}) = \mathrm{FL}_t(\mathscr{F}) \Leftrightarrow \mathrm{FL}_{t,\nu}((A^-)^\nu) = \mathrm{FL}_{t,\nu}((B^-)^\nu) \;\&\; \mathrm{FL}_{t,\nu}(A^+) = \mathrm{FL}_{t,\nu}(B^+).
\end{aligned} \tag{9.130}$$

For FGCounts, t denotes here $\wedge$ or an Archimedean t-norm, whereas $t = \wedge$ or t is strict if FLCounts are used (see Subsection 9.4.1). Dealing with these two types of cardinalities, equipotency of $\mathcal{E}$ and $\mathcal{F}$ can thus be defined via simultaneous equipotency of fuzzy sets A^+ and B^+ and, on the other hand, of $(A^-)^\nu$ and $(B^-)^\nu$:

$$|\mathcal{E}| = |\mathcal{F}| \Leftrightarrow |A^+| = |B^+| \ \& \ |(A^-)^\nu| = |(B^-)^\nu|. \tag{9.131}$$

For FECounts from (9.77) with $\boldsymbol{t} = \wedge$ or a strict $\boldsymbol{t}$, we get

$$\mathrm{FE}_t(\mathcal{E}) = \mathrm{FE}_t(\mathcal{F}) \Leftrightarrow \mathrm{FG}_t(\mathcal{E}) = \mathrm{FG}_t(\mathcal{F}) \Leftrightarrow \mathrm{FL}_t(\mathcal{E}) = \mathrm{FL}_t(\mathcal{F}). \tag{9.132}$$

Finally, let α and β denote two FGCounts, FLCounts or FECounts from (9.77) with $\boldsymbol{t}$ restricted as in (9.130) and (9.132). Assume $\alpha = |\mathcal{E}|$ and $\beta = |\mathcal{F}|$ for some IFSs $\mathcal{E}$ and $\mathcal{F}$. Then (see also WYGRALAK (2005a, 2008a))

$$\alpha + \beta = |\mathcal{E} \cup \mathcal{F}| \ \text{ whenever } \mathcal{E} \text{ and } \mathcal{F} \text{ are disjoint}$$

and (9.133)

$$\alpha \cdot \beta = |\mathcal{E} \times \mathcal{F}|$$

define the *sum* and *product* of α and β.

Chapter 10

Selected Applications

In the previous two chapters, we have shown that both scalar and fuzzy cardinalities of fuzzy sets formalize and reflect human counting procedures under imprecision of information about the objects of counting. Cardinalities of IVFSs and IFSs were presented from the same angle adding the factor of information incompleteness. Moreover, a relationship between some fuzzy cardinalities and issues like classification was emphasized.

This chapter is devoted to other examples of applications of intelligent counting in various areas of intelligent systems and decision support. In particular, we mean similarity measures and derivative constructions like distance and fuzziness measures. Further, we will deal with applications to time series analysis, linguistic quantification, decision making in a fuzzy environment and, finally, to algorithms of group decision making. The last section contains bibliographical references encompassing these and other areas of applications.

10.1 Cardinality-Based Measures for Fuzzy Sets

We like to return to the coefficients from Subsection 5.1.5 measuring similarities and relationships between sets. We will extend them to fuzzy sets. Throughout, one assumes that U is a finite universe and $A, B \in \mathrm{FFS}$, and one puts $\frac{0}{0} = 1$. We begin with a very important and useful cardinality-based coefficient which, however, in not a similarity measure in the sense of Definition 5.3.

(A) Inclusion coefficient. For fuzzy sets, it is defined as

$$R_{\subset, f}(A, B) = \frac{\sigma_f(A \cap B)}{\sigma_f(A)} \tag{10.1}$$

with a weighting function f (see Subsection 8.1.1). Its particular case

$$R_{\subset, id}(A, B) = \frac{\sum_{x \in U} A(x) \wedge B(x)}{\sum_{x \in U} A(x)} \tag{10.2}$$

was first introduced in SANCHEZ (1977). Let us mention two other useful instances:

M. Wygralak: *Intelligent Counting Under Information Imprecision*, STUDFUZZ 292, pp. 231–260.
DOI: 10.1007/978-3-642-34685-9_10

$$R_{\subset,f}(A,B) = \frac{|A_t \cap B_t|}{|A_t|} \quad \text{for } f = f_{1,t}$$

and (10.3)

$$R_{\subset,f}(A,B) = \frac{\sum_{x \in (A\cap B)_t} A(x) \wedge B(x)}{\sum_{x \in A_t} A(x)} \quad \text{for } f = f_{4,t,1}.$$

It is easy to notice that

$$R_{\subset,f}(A,B) = \sigma_f(B|A)$$

and

$$R_{\subset,f}(A,B) \neq R_{\subset,f}(B',A')$$

in general. Trivially, $R_{\subset,f}(A,B) = 1$ whenever $A \subset B$. $R_{\subset,f}$ is neither symmetrical nor t-transitive even if $t = t_d$, i.e. it does not form a similarity measure. Finally, one should mention a triangular norm-based generalization of (10.1), namely

$$R_{\subset,f,t}(A,B) = \frac{\sigma_f(A \cap_t B)}{\sigma_f(A)} \tag{10.4}$$

with a t-norm t.

(B) Overlap coefficient. One defines it by (cf. (5.24))

$$R_{o,f}(A,B) = R_{\subset,f}(A,B) \vee R_{\subset,f}(B,A) = \frac{\sigma_f(A \cap B)}{\min(\sigma_f(A), \sigma_f(B))}. \tag{10.5}$$

$R_{o,f}$ is always a resemblance relation, $R_{o,f}(A,B) = 1$ whenever $A \subset B$ or $B \subset A$, and $R_{o,f}(A,B) \neq R_{o,f}(A',B')$. Again, the overlap coefficient can be generalized as

$$R_{o,f,t,s}(A,B) = R_{\subset,f,t}(A,B)\, s\, R_{\subset,f,t}(B,A) \tag{10.6}$$

with a t-norm t and t-conorm s.

(C) Jaccard coefficient. For fuzzy sets, its general definition is (cf. (5.22))

$$R_{J,f,t}(A,B) = \frac{\sigma_f(A \cap_t B)}{\sigma_f(A \cup_s B)} \tag{10.7}$$

with a weighting function f, t-norm t, and $s = t^*$. We will focus on two important particular cases. First, let

$$R_{J,t}(A,B) = R_{J,id,t}(A,B) = \frac{\sigma_{id}(A \cap_t B)}{\sigma_{id}(A \cup_s B)} = \frac{\sigma_{id}(A \cap_t B)}{\sigma_{id}(A) + \sigma_{id}(B) - \sigma_{id}(A \cap_t B)}, \tag{10.8}$$

where t and $s = t^*$ are from the Frank families of t-norms and t-conorms, respectively (see Subsection 2.4.3 and Example 8.3(b)). So, $t = t_{F,\lambda}$ and $s = s_{F,\lambda}$ with $\lambda \in [0, \infty]$. As shown in DE BAETS/DE MEYER (2005a), $R_{\mathrm{J},t}$ is always $t_{\mathrm{Ł}}$-transitive. In particular, $R_{\mathrm{J},t}$ is thus $t_{\mathrm{Ł}}$-transitive if $t = \wedge, t_{\mathrm{a}}, t_{\mathrm{Ł}}$. It is also symmetrical for each $\lambda \in [0, \infty]$, i.e. $R_{\mathrm{J},t}(A, B) = R_{\mathrm{J},t}(B, A)$. However, $R_{\mathrm{J},t}$ becomes reflexive only if $t = \wedge$ and, consequently, only $R_{\mathrm{J},\wedge}$ forms a $t_{\mathrm{Ł}}$-similarity.

All this suggests that worth considering is also another special case of (10.7), namely

$$R_{\mathrm{J},f}(A, B) = R_{\mathrm{J},f,\wedge}(A, B) = \frac{\sigma_f(A \cap B)}{\sigma_f(A \cup B)} = \frac{\sigma_f(A \cap B)}{\sigma_f(A) + \sigma_f(B) - \sigma_f(A \cap B)}. \tag{10.9}$$

In particular, we get

$$R_{\mathrm{J},id}(A, B) = \frac{\sum_{x \in U} A(x) \wedge B(x)}{\sum_{x \in U} A(x) \vee B(x)} = \frac{\sum_{x \in U} A(x) \wedge B(x)}{\sum_{x \in U} A(x) + \sum_{x \in U} B(x) - \sum_{x \in U} A(x) \wedge B(x)}$$

and (10.10)

$$R_{\mathrm{J},f}(A, B) = \frac{|A_t \cap B_t|}{|A_t \cup B_t|} = \frac{|A_t \cap B_t|}{|A_t| + |B_t| - |A_t \cap B_t|} \quad \text{for } f = f_{1,t}.$$

Notice that $R_{\mathrm{J},f}$ is a $t_{\mathrm{Ł}}$-similarity for each weighting function f as $R_{\mathrm{J},f}(A, B)$ equals $R_{\mathrm{J},\wedge}(f \circ A, f \circ B)$. Further properties are listed below.

- $R_{\mathrm{J},f}(A, B) \neq R_{\mathrm{J},f}(A', B')$.
- If f is strictly increasing, then (see Subsection 5.1.4)

  $$d_{\mathrm{J},f}(A, B) = 1 - R_{\mathrm{J},f}(A, B) \tag{10.11}$$

  forms a normed metric (*Jaccard metric* for fuzzy sets). It can be treated as a dissimilarity relation in FFS. In particular, a normed metric is thus

  $$d_{\mathrm{J},id}(A, B) = 1 - \frac{\sum_{x \in U} A(x) \wedge B(x)}{\sum_{x \in U} A(x) \vee B(x)}. \tag{10.12}$$

- If f is strictly increasing, then (see (2.53))

  $$\mathrm{Fuzz}_{\mathrm{J},f}(A) = 1 - d_{\mathrm{J},f}(A, A') = \frac{\sigma_f(A \cap A')}{\sigma_f(A \cup A')} \tag{10.13}$$

  becomes a normed fuzziness measure of A (*Jaccard fuzziness measure*). A simple example is

  $$\mathrm{Fuzz}_{\mathrm{J},id}(A) = \frac{\sum_{x \in U} A(x) \wedge (1 - A(x))}{\sum_{x \in U} A(x) \vee (1 - A(x))}. \tag{10.14}$$

(D) Matching coefficient. For $A, B \in \mathrm{FFS}$, it is given by (cf. (5.23))

$$R_{\mathrm{m},f,t}(A, B) = 1 - \frac{\sigma_f(A) + \sigma_f(B) - 2\sigma_f(A \cap_t B)}{|U|} \tag{10.15}$$

with a weighting function f and a t-norm $\boldsymbol{t}$. Again, we like to focus on two basic special cases. The first one is

$$R_{\mathrm{m},\boldsymbol{t}}(A, B) = 1 - \frac{\sigma_{id}(A) + \sigma_{id}(B) - 2\sigma_{id}(A \cap_{\boldsymbol{t}} B)}{|\boldsymbol{U}|}, \tag{10.16}$$

where $\boldsymbol{t}$ belongs to the Frank family of t-norms, i.e. $\boldsymbol{t} = \boldsymbol{t}_{\mathrm{F},\lambda}$ with $\lambda \in [0, \infty]$. It is shown in DE BAETS/DE MEYER (2005a) that $R_{\mathrm{m},\boldsymbol{t}}$ is then $\boldsymbol{t}_{\mathrm{Ł}}$-transitive. So, $R_{\mathrm{m},\boldsymbol{t}}$ is $\boldsymbol{t}_{\mathrm{Ł}}$-transitive for $\boldsymbol{t} = \wedge, \boldsymbol{t}_{\mathrm{a}}, \boldsymbol{t}_{\mathrm{Ł}}$ in particular. As one sees, $R_{\mathrm{m},\boldsymbol{t}}$ remains symmetrical for each λ, whereas the reflexivity property is preserved only if $\boldsymbol{t} = \wedge$. And just this case will be the subject of our further discussion. Let

$$R_{\mathrm{m},f}(A, B) = 1 - \frac{\sigma_f(A) + \sigma_f(B) - 2\sigma_f(A \cap B)}{|\boldsymbol{U}|} \tag{10.17}$$

with a weighting function f. This form of the matching coefficient is a $\boldsymbol{t}_{\mathrm{Ł}}$-similarity for each f as $R_{\mathrm{m},f}(A, B) = R_{\mathrm{m},\wedge}(f \circ A, f \circ B)$. Routine transformations lead to a more explicit formula:

$$R_{\mathrm{m},f}(A, B) = 1 - \frac{1}{|\boldsymbol{U}|} \sum_{x \in \boldsymbol{U}} |f(A(x)) - f(B(x))|. \tag{10.18}$$

In particular, $\boldsymbol{t}_{\mathrm{Ł}}$-similarities are thus (cf. (5.44)):

$$R_{\mathrm{m},id}(A, B) = 1 - \frac{1}{|\boldsymbol{U}|} \sum_{x \in \boldsymbol{U}} |A(x) - B(x)| = 1 - d_1(A, B) = R_1(A, B) \tag{10.19}$$

with the normed Hamming distance $d_1(A, B)$ from (2.54),

$$R_{\mathrm{m},f}(A, B) = 1 - \frac{1}{|\boldsymbol{U}|}(|A_t| + |B_t| - 2|A_t \cap B_t|) \quad \text{with } f = f_{1,t} \tag{10.20}$$

and

$$R_{\mathrm{m},f}(A, B) = 1 - \frac{1}{|\boldsymbol{U}|} \sum_{x \in \boldsymbol{U}} |A^2(x) - B^2(x)| \quad \text{for } f = f_{3,2}. \tag{10.21}$$

Let us list some basic properties of the matching coefficient (10.17).

- $R_{\mathrm{m},f}(A, B) = R_{\mathrm{m},f}(A', B')$ if the complementarity rule $\sigma_f(A) + \sigma_f(A') = |\boldsymbol{U}|$ is satisfied (see Example 8.5). So, one has $R_{\mathrm{m},id}(A, B) = R_{\mathrm{m},id}(A', B')$.
- If f is strictly increasing, then

$$d_{\mathrm{m},f}(A, B) = 1 - R_{\mathrm{m},f}(A, B) \tag{10.22}$$

  is a normed metric. $d_{\mathrm{m},f}(A, B)$ can be viewed as a dissimilarity measure in FFS.
- If f is strictly increasing, then

$$\begin{aligned} \mathrm{Fuzz}_{\mathrm{m},f}(A) = 1 - d_{\mathrm{m},f}(A, A') &= 1 - \frac{\sigma_f(A) + \sigma_f(A') - 2\sigma_f(A \cap A')}{|\boldsymbol{U}|} \\ &= 1 - \frac{1}{|\boldsymbol{U}|} \sum_{x \in \boldsymbol{U}} |f(A(x)) - f(A'(x))| \end{aligned} \tag{10.23}$$

  is a fuzziness measure of A. In particular, we obtain (see (2.56))

$$\mathrm{Fuzz}_{\mathrm{m},\,id}(A) = R_{\mathrm{m},\,id}(A, A') = \mathrm{Fuzz}_1(A).$$

Concluding, we see that $R_{\mathrm{in}}(A, B)$ from (5.31) is based on a maximum violation of the condition $A(x) \le B(x)$, on a maximum depth of that violation. All violations for all x's are taken into account by $R_{\subset,f}(A, B)$ and $R_{\mathrm{in}*}(A, B)$ in (10.1) and (5.44), respectively. Similarly, $R_{\mathrm{eq}}(A, B)$ from (5.37) is expressed via a maximum violation of $A(x) = B(x)$, whereas $R_{\mathrm{J},f}(A, B)$, $R_{\mathrm{m},f}(A, B)$ as well as $R_{\mathrm{eq}*}(A, B)$ are constructions in which violations for all x's are incorporated.

The four coefficients introduced in this section are based on sigma f-counts of fuzzy sets. Alternatively, they can involve scalar cardinalities derived from FECounts by means of the COG method (see (9.49)). Then

$$R_{\subset,\,\mathrm{FE}}(A, B) = \frac{\mathrm{COG}_{\mathrm{FE}(A\cap B)}}{\mathrm{COG}_{\mathrm{FE}(A)}}, \quad R_{\mathrm{o},\,\mathrm{FE}}(A, B) = R_{\subset,\,\mathrm{FE}}(A, B) \vee R_{\subset,\,\mathrm{FE}}(B, A),$$

$$R_{\mathrm{J},\,\mathrm{FE}}(A, B) = \frac{\mathrm{COG}_{\mathrm{FE}(A\cap B)}}{\mathrm{COG}_{\mathrm{FE}(A\cup B)}}, \tag{10.24}$$

$$R_{\mathrm{m},\,\mathrm{FE}}(A, B) = 1 - \frac{1}{|U|}(\mathrm{COG}_{\mathrm{FE}(A)} + \mathrm{COG}_{\mathrm{FE}(B)} - 2\mathrm{COG}_{\mathrm{FE}(A\cap B)}).$$

Example 10.1. (a) Let

$$A = 0.8/x_2 + 0.9/x_3 + 0.2/x_4 + 0.5/x_5 + 1/x_6$$

and

$$B = 1/x_2 + 0.9/x_3 + 0.1/x_4 + 0.2/x_5 + 1/x_6 + 0.4/x_7.$$

By (10.2) and (10.3), we have

$$R_{\subset,\,id}(A, B) = \frac{0.8+0.9+0.1+0.2+1}{0.8+0.9+0.2+0.5+1} = 0.88,$$

$$R_{\subset,f}(A, B) = \frac{|A_{0.3}\cap B_{0.3}|}{|A_{0.3}|} = 0.75 \quad \text{for } f = f_{1,\,0.3},$$

$$R_{\subset,f}(A, B) = \frac{|A_{0.6}\cap B_{0.6}|}{|A_{0.6}|} = 1 \quad \text{for } f = f_{1,\,0.6},$$

whereas

$$R_{\subset,f}(A, B) = \frac{0.8+0.9+1}{0.8+0.9+0.5+1} = 0.84 \quad \text{for } f = f_{4,\,0.3,\,1},$$

$$R_{\subset,f}(A, B) = 1 \quad \text{for } f = f_{4,\,0.6,\,1}.$$

This correctly reflects the fact that $A(x) \le B(x)$ does not generally hold, but is preserved by larger membership degrees, say, those greater than or equal to 0.6. Using (10.24), one gets

$$R_{\subset,\,\mathrm{FE}}(A, B) = 0.9$$

as

$$\mathrm{FE}(A) = (0, 0.1, 0.2, 0.5, 0.5, 0.2), \quad \mathrm{COG}_{\mathrm{FE}(A)} = 3\tfrac{1}{3},$$

$$\mathrm{FE}(A\cap B) = (0, 0.1, 0.2, 0.8, 0.2, 0.1), \quad \mathrm{COG}_{\mathrm{FE}(A\cap B)} = 3.$$

On the other hand, $R_{\subset, id}(B, A) = 0.83$ and, hence, $R_{o, id}(A, B) = 0.88 \vee 0.83 = 0.88$. Further, (10.10) and (10.24) lead to

$$R_{J, id}(A, B) = \frac{0.8+0.9+0.1+0.2+1}{1+0.9+0.2+0.5+1+0.4} = 0.75, \quad R_{J, FE}(A, B) = \frac{3}{4.06} = 0.74,$$

$$R_{J, f}(A, B) = 0.6 \text{ for } f = f_{1, 0.3}, \quad R_{J, f}(A, B) = 1 \text{ for } f = f_{1, 0.6}.$$

If, say, U is a 10-element set, (10.19), (10.20) and (10.24) give

$$R_{m, id}(A, B) = 1 - \frac{1}{10}(0.2+0.1+0.3+0.4) = 0.9,$$

$$R_{m, f}(A, B) = 1 - \frac{1}{10}(4+4-2\cdot 3) = 0.8 \text{ if } f = f_{1, 0.3}, \quad R_{m, f}(A, B) = 1 \text{ if } f = f_{1, 0.6},$$

$$R_{m, FE}(A, B) = 1 - \frac{1}{10}(3.33+3.71-2\cdot 3) = 0.9.$$

(b) The coefficients studied in this section can also be used to measure similarities between, say, arbitrary records of numerical data. To this end, each record should be transformed via normalization into a fuzzy set, i.e. each data item x ranging in a real interval $[a, b]$ has to be transformed into a membership degree $(x - a)/(b - a)$. Let us look at the following example in which a file contains triples of data about age, weight and height of individuals in a population. Assume the values of data fluctuate between 20 and 35 years (age), 65 and 100 kg (weight), and 170 and 200 cm (height). Let us focus on data concerning two persons:

	age	weight	height
John	28	75	180
Mark	31	95	185

The normalization process gives the following results:

John ((28 − 20)/(35 − 20), (75 − 65)/(100 − 65), (180 − 170)/(200 − 170)) = (0.53, 0.29, 0.33),

Mark (0.73, 0.86, 0.50).

What we deal with are now two fuzzy sets in a 3-element universe $\{a_1, a_2, a_3\}$, namely

$$A = 0.53/a_1 + 0.29/a_2 + 0.33/a_3 \quad \text{and} \quad B = 0.73/a_1 + 0.86/a_2 + 0.50/a_3.$$

Use $f = id$ for simplicity. We get

$$\sigma_{id}(A) = \sigma_{id}(A \cap B) = 1.15, \quad \sigma_{id}(B) = \sigma_{id}(A \cup B) = 2.09,$$
$$\mathrm{FE}(A) = \mathrm{FE}(A \cap B) = (0.47, 0.53, 0.33, 0.29),$$
$$\mathrm{FE}(B) = \mathrm{FE}(A \cup B) = (0.14, 0.27, 0.50, 0.50),$$
$$\mathrm{COG}_{\mathrm{FE}(A)} = \mathrm{COG}_{\mathrm{FE}(A \cap B)} = 1.27, \quad \mathrm{COG}_{\mathrm{FE}(B)} = \mathrm{COG}_{\mathrm{FE}(A \cup B)} = 1.96.$$

Consequently,

$$R_{\subset, id}(A, B) = R_{o, id}(A, B) = 1, \quad R_{J, id}(A, B) = 0.55, \quad R_{m, id}(A, B) = 0.69,$$
$$R_{\subset, FE}(A, B) = R_{o, FE}(A, B) = 1, \quad R_{J, FE}(A, B) = 0.65, \quad R_{m, FE}(A, B) = 0.77.$$

These coefficients rightly reflect that John and Mark are quite different (see Jaccard and matching coefficients), and that John is younger, lighter, and shorter than Mark (see inclusion coefficients). □

Finally, let us emphasize that all the coefficients defined in this section can be extended to IVFSs and IFSs. The resulting extensions are understood as pairs or aggregations of coefficients for the corresponding components. For instance, the Jaccard coefficient for IFSs $\mathscr{E} = (A^+, A^-)$ and $\mathscr{F} = (B^+, B^-)$ can be defined as

$$R_{\mathrm{J},f}(\mathscr{E}, \mathscr{F}) = (R_{\mathrm{J},f}(A^+, B^+), R_{\mathrm{J},f}(A^-, B^-))$$

or

$$R_{\mathrm{J},f,Aggr}(\mathscr{E}, \mathscr{F}) = Aggr(R_{\mathrm{J},f}(A^+, B^+), R_{\mathrm{J},f}(A^-, B^-)), \tag{10.25}$$

e.g.

$$R_{\mathrm{J},f,\wedge}(\mathscr{E}, \mathscr{F}) = R_{\mathrm{J},f}(A^+, B^+) \wedge R_{\mathrm{J},f}(A^-, B^-).$$

The reader is also referred to SZMIDT/KACPRZYK (2007) for another approach to this issue.

10.2 Time Series Analysis

We will outline applications of intelligent counting in analysis of time series, sequences of numerical observations that are made at equidistant time moments. Assume $Y = \{y_i\colon i = 1, \ldots, k\}$ and $Z = \{z_i\colon i = 1, \ldots, k\}$ are two time series subjected to a prior normalization, i.e. $y_i, z_i \in [0, 1]$. So, Y and Z can be treated as fuzzy sets. Let us construct the corresponding time series $D(Y, Z)$ of differences:

$$D(Y, Z) = \{d_i\colon i = 1, \ldots, k\} \quad \text{with} \quad d_i = |y_i - z_i|.$$

The scalar cardinality $\mathrm{COG}_{\mathrm{FE}(D(Y,Z))}$, the Jaccard coefficient $R_{\mathrm{J},f}(Y, Z)$ and the matching coefficient $R_{\mathrm{m},f}(Y, Z)$ are then pieces of information about how much Y and Z differ from each other. At least two applications of this information seem to be worth mentioning.

• **Selection of a suitable probabilistic model.** Imagine Y is a normalized time series of observed demand for a product in recent k months. Moreover, assume $Z1$ and $Z2$ are two normalized time series of demand generated, say, by autoregressive moving average models ARMA(p_1, q_1) and ARMA(p_2, q_2) with different pairs of parameters; the reader is referred to BOX/JENKINS (1994) for related details. Our task is to indicate which of these two models generates a time series matching Y better than the other and, thus, which of them will possibly generate a better forecast of future demand. A reasonable decision is then to choose a model corresponding to

$$R_{\mathrm{J},f}(Y, Z1) \vee R_{\mathrm{J},f}(Y, Z2)$$

or

$$R_{\mathrm{m},f}(Y, Z1) \vee R_{\mathrm{m},f}(Y, Z2).$$

An alternative way of doing is to choose a model corresponding to

$$\min(\mathrm{COG}_{\mathrm{FE}(D(Y,Z1))}, \mathrm{COG}_{\mathrm{FE}(D(Y,Z2))}),$$

i.e. a model leading to a smaller fuzzy set of differences.

• **Evaluation of forecast accuracy.** Again, assume Y is a normalized time series of actual demand for a product recorded in recent k months, whereas Z denotes a time series of demand forecasted for that time. $R_{\mathrm{J},f}(Y,Z)$, $R_{\mathrm{m},f}(Y,Z)$ and $\mathrm{COG}_{\mathrm{FE}(D(Y,Z))}$ are now different reasonable evaluations of accuracy of forecast Z. Tests run on real data in a global pharmaceutical corporation have shown that all these evaluations behave in a satisfactory way (KACZMAREK (2010)). $\mathrm{COG}_{\mathrm{FE}(D(Y,Z))}$ forms an especially useful evaluation in this context as it shows a very high degree of consistency with evaluations formulated by market experts.

10.3 Linguistic Quantification

The language of mathematics and logic, including many-valued logic, employs only two types of quantifiers: $\forall x$ (for all x's) and the existential quantifier $\exists x$ (at least for one x). However, in any natural language, humans use a much reacher collection of quantifiers called *linguistic quantifiers* and exemplified by *most*, *a few*, *about* 20, etc. The subject literature offers different computational approaches to these quantifiers. One should mention here Zadeh's approach from ZADEH (1983a) (cf. also ZADEH (1981a)), the substitution calculus proposed in YAGER (1983), and the approach via OWA operators from YAGER (1994) (see KACPRZYK (1997) for a review; cf. also GLÖCKNER (2004, 2006) and DIAZ-HERMIDA *et al.* (2003)).

We like to focus on Zadeh's approach. Linguistic quantifiers are then divided into two classes:

- *absolute quantifiers*, e.g. *about* 100, *much more than* 50,
- *relative quantifiers*, e.g. *most*, *a few*, *about half*.

This division is not disjoint as some linguistic quantifiers (say, *a few*) may be viewed in a context-dependent way as absolute or relative ones.

Looking formally, an absolute quantifier Q is a fuzzy number $Q\colon \mathbb{N} \to [0,1]$. For instance, *about* 100 can be understood as triangular fuzzy number (100, 10, 10) (see Subsection 2.5.1). $Q(k)$ forms a degree to which k satisfies the quantitative requirements specified by the quantifier. Q can be treated in a more general way as a function $Q\colon [0,\infty) \to [0,1]$ whenever a counting procedure giving non-integer results is used.

A relative quantifier is formally a fuzzy number $Q\colon [0,1] \to [0,1]$. $Q(p)$ is now a degree to which proportion p of objects fulfilling a given property to all objects under discussion satisfies the requirements defined by the quantifier. For instance, *about half* can be modeled as a triangular fuzzy number (0.5, 0.2, 0.2), whereas *most* can be represented by s-shaped fuzzy number $(0.8, 0.5)_{s}$, i.e. $most(x) = 2x - 0.6$ for $x \in [0.3, 0.8]$, $most(x) = 1$ for $x > 0.8$, and $most(x) = 0$ for $x < 0.3$.

From the viewpoint of applications, relative quantifiers seem to be generally more important than absolute ones. The most essential class of relative quantifiers is that of non-decreasing quantifiers ("the more the better") exemplified by *most*, *huge majority*, *much more than half*.

We move on to the question of determining the truth degree of a linguistically quantified statement in Zadeh's approach. Its original version is restricted to the use of sigma counts. We will extend it using an arbitrary scalar cardinality σ understood as a sigma *f*-count or another kind of that cardinality (see (9.49), (10.39)). Basically, fuzzy cardinalities could also be employed even in formulae like (10.36) requiring a division of cardinalities (see the end of Subsection 9.4.3). However, the use of scalar cardinalities is here more convenient and simpler in every respect, and seems to be quite sufficient for applications. In particular, we will refer to scalar cardinalities derived from fuzzy ones.

Let us assume that A and B are two imprecise properties identified with fuzzy sets from FFS. Two types of statements on elements $x \in \boldsymbol{U}$ are considered within Zadeh's approach.

- **Type-I statements** of the form

$$Qx\text{'s are } A \tag{10.26}$$

exemplified by

"*Most* (Q) employees (x's) are *experienced* (A)"

and

"*About* 100 (Q) companies (x's) suffered *a big loss* (A)".

If Q is absolute, one defines

$$[\![Qx\text{'s are } A]\!] = Q(\sigma(A)). \tag{10.27}$$

For instance,

$$[\![Qx\text{'s are } A]\!] = Q(\sigma_f(A)) \tag{10.28}$$

with a weighting function f, or

$$[\![Qx\text{'s are } A]\!] = Q(\mathrm{COG}_{\mathrm{FE}(A)}). \tag{10.29}$$

If Q is relative and the universe $\boldsymbol{U}$ is finite, one puts (see (8.34) and (10.1))

$$\begin{aligned}[\![Qx\text{'s are } A]\!] &= Q(\sigma_f(A \mid 1_U)) \\ &= Q(\frac{1}{|\boldsymbol{U}|} \sum_{x \in \mathrm{supp}(A)} f(A(x))).\end{aligned} \tag{10.30}$$

A counterpart of (10.29) is now

$$[\![Qx\text{'s are } A]\!] = Q(\frac{1}{|\boldsymbol{U}|}\mathrm{COG}_{\mathrm{FE}(A)}). \tag{10.31}$$

Example 10.2. Again, let us refer to the fuzzy set A of full bottles discussed in Subsections 6.3.4 and 8.1.3, where (see (8.23))

$$A(b_1) = 1, \quad A(b_2) = 0.8, \quad A(b_3) = 0, \quad A(b_4) = 0.5,$$
$$A(b_5) = 0.1, \quad A(b_6) = 1, \quad A(b_7) = 0.9, \quad A(b_8) = 0.3.$$

Define $about\_4 = (4, 2, 2)$ and put $f = id$. By (10.28), we get

$$[\![about\_4 \text{ bottles are } full]\!] = about\_4(4.6) = 0.7.$$

Alternatively, (10.29) gives

$$[\![about\_4 \text{ bottles are } full]\!] = about\_4(\mathrm{COG}_{\mathrm{FE}(A)}) = about\_4(4.59) = 0.71.$$

Defining $about\_half = (0.5, 0.2, 0.2)$, $most = (0.8, 0.5)_s$, and using again $f = id$, one obtains (see (10.30))

$$[\![about\_half \text{ bottles are } full]\!] = about\_half(\tfrac{4.6}{8}) = 0.625$$

and

$$[\![most \text{ bottles are } full]\!] = most(0.575) = 0.55.$$

On the other hand, by (10.31),

$$[\![about\_half \text{ bottles are } full]\!] = about\_half(\tfrac{1}{8}\mathrm{COG}_{\mathrm{FE}(A)}) = 0.632$$

and

$$[\![most \text{ bottles are } full]\!] = most(\tfrac{1}{8}\mathrm{COG}_{\mathrm{FE}(A)}) = 0.547. \qquad \square$$

- **Type-II statements** of the form

$$Q\,B\,x\text{'s are } A \tag{10.32}$$

with B understood as a *qualifier*. Simple examples are

"*Most* (Q) of the *expensive* (B) cars (x's) are *safe* (A)"

and

"*About* 100 (Q) *small* (B) companies (x's) suffered *a big loss* (A)".

If Q is an absolute quantifier, one defines

$$[\![Q\,B\,x\text{'s are } A]\!] = Q(\sigma(A \cap B)). \tag{10.33}$$

Using a sigma f-count, we thus have

$$[\![Q\,B\,x\text{'s are } A]\!] = Q(\sigma_f(A \cap B)), \tag{10.34}$$

whereas

$$[\![Q\,B\,x\text{'s are } A]\!] = Q(\mathrm{COG}_{\mathrm{FE}(A \cap B)}) \tag{10.35}$$

for scalar cardinalities derived from FECounts.

If Q is relative, we put (cf. (8.32))

$$[\![Q\,B\,x\text{'s are } A]\!] = Q(\sigma(A \mid B)) = Q(\tfrac{\sigma(A \cap B)}{\sigma(B)}). \tag{10.36}$$

Applying sigma *f*-counts, this collapses to

$$[\![Q B\, x\text{'s are } A]\!] = Q(\sigma_f(A\,|\,B)) = Q(R_{\subset, f}(B, A)) = Q(\frac{\sigma(A \cap B)}{\sigma(B)}). \quad (10.37)$$

If scalar cardinalities derived from FECounts are involved, then

$$[\![Q B\, x\text{'s are } A]\!] = Q(R_{\subset, \mathrm{FE}}(B, A)) = Q(\frac{\mathrm{COG}_{\mathrm{FE}(A \cap B)}}{\mathrm{COG}_{\mathrm{FE}(B)}}). \quad (10.38)$$

Example 10.3. Let $U = \{c_1, c_2, c_3, c_4, c_5\}$ be a set of car models characterized by means of the following fuzzy sets with respect to price and safety:

$$expensive = 0.7/c_1 + 0.8/c_2 + 0.9/c_4 + 0.2/c_5,$$
$$safe = 0.8/c_1 + 0.2/c_2 + 0.3/c_3 + 1/c_4 + 0.5/c_5.$$

Using (10.37) with $f = id$ and $most = (0.8, 0.5)_s$,

$$[\![most \text{ of the } expensive \text{ cars are } safe]\!] = most\left(\frac{\sigma_{id}(safe \cap expensive)}{\sigma_{id}(expensive)}\right)$$
$$= most\left(\frac{2}{2.6}\right) = 0.94.$$

Alternatively, (10.38) gives

$$[\![most \text{ of the } expensive \text{ cars are } safe]\!] = most\left(\frac{\mathrm{COG}_{\mathrm{FE}(safe \cap expensive)}}{\mathrm{COG}_{\mathrm{FE}(expensive)}}\right)$$
$$= most\left(\frac{2.07}{2.47}\right) = 1. \qquad \square$$

We like to formulate some generalizations of the methods of determining the truth degree of a linguistically quantified statement presented so far.

First, in (10.33)-(10.38), $A \cap B$ can be replaced with $A \cap_t B$ involving a triangular norm t. Second, the scalar cardinality in (10.29), (10.31), (10.35) and (10.38) derived from FECounts can be replaced with a scalar cardinality derived from FGCounts generated by t (see (9.55)), namely

$$\sigma_{t,f}(A) = \sum_{k=1}^{n} f([A]_1)\, t \ldots t\, f([A]_k)$$

or (10.39)

$$\sigma_{f,t}(A) = \sum_{k=1}^{n} f([A]_1\, t \ldots t\, [A]_k)$$

with a weighting function *f*. Both of them are non-additive for $t \neq \wedge$, whereas $\wedge$ gives the sigma *f*-count from Section 8.1. Their properties and applications to linguistic quantification and linguistic summarization are studied in PILARSKI (2010) (see also further discussion).

Third, worth mentioning are generalizations to IVFSs and IFSs. We mean the case in which (at least) some of fuzzy sets Q, A and B in a linguistically quantified statement (10.26) or (10.32) are replaced with IVFSs or IFSs. A study of all variants of that replacement (when the usual sigma count is used) is placed in NIEWIADOMSKI (2008). In the following discussion we like to focus on three variants involving, however, an arbitrary scalar cardinality σ. We will assume that Q is a non-decreasing function.

- **Variant 1:** A in (10.26) is replaced with a finite IVFS $\mathcal{A} = (A_l, A_u)$ or a finite IFS $A = (A^+, A^-)$ with $A^+ \subset (A^-)^\nu$ and a strong negation ν (see Sections 6.2 and 6.3).

The truth degree of "Qx's are $\mathcal{A}$" is then defined as an interval in $[0, 1]$. If Q is absolute, one puts

$$[\![Qx\text{'s are }\mathcal{A}]\!] = [Q(\sigma(A_l)), Q(\sigma(A_u))]. \tag{10.40}$$

For a relative Q, we have (U finite; cf. (8.52))

$$[\![Qx\text{'s are }\mathcal{A}]\!] = \left[Q\left(\frac{\sigma(A_l)}{|U|}\right), Q\left(\frac{\sigma(A_u)}{|U|}\right)\right]. \tag{10.41}$$

Similarly,

$$[\![Qx\text{'s are }A]\!] = [Q(\sigma(A^+)), Q(\sigma((A^-)^\nu))], \tag{10.42}$$

if Q is absolute, and

$$[\![Qx\text{'s are }A]\!] = \left[Q\left(\frac{\sigma(A^+)}{|U|}\right), Q\left(\frac{\sigma((A^-)^\nu)}{|U|}\right)\right] \tag{10.43}$$

whenever it is relative.

- **Variant 2:** A and B in (10.32) are replaced with finite IVFSs $\mathcal{A} = (A_l, A_u)$ and $\mathcal{B} = (B_l, B_u)$, respectively, or finite IFSs $A = (A^+, A^-)$ and $B = (B^+, B^-)$.

If Q is relative, one puts (cf. (8.54))

$$[\![Q\mathcal{B}x\text{'s are }\mathcal{A}]\!] = \left[Q\left(\frac{\sigma(A_l \cap B_l)}{\sigma(B_l)} \wedge \frac{\sigma(A_l \cap B_u)}{\sigma(B_u)}\right), Q\left(\frac{\sigma(A_u \cap B_l)}{\sigma(B_l)} \vee \frac{\sigma(A_u \cap B_u)}{\sigma(B_u)}\right)\right]. \tag{10.44}$$

The truth degree $[\![QBx$'s are $A]\!]$ is defined by means of the same formula in which A_l, B_l, A_u and B_u, respectively, should be replaced with A^+, B^+, $(A^-)^\nu$ and $(B^-)^\nu$, respectively.

- **Variant 3:** the linguistic quantifier, say, in (10.26) is viewed as an IVFS $\mathcal{Q} = (Q_l, Q_u)$ or IFS $Q = (Q^+, Q^-)$.

Then

$$[\![\mathcal{Q}x\text{'s are }A]\!] = [Q_l(\sigma(A)), Q_u(\sigma(A))] \tag{10.45}$$

whenever $\mathcal{Q}$ forms an absolute quantifier, and

$$[\![\mathcal{Q}x\text{'s are }A]\!] = \left[Q_l\left(\frac{\sigma(A)}{|U|}\right), Q_u\left(\frac{\sigma(A)}{|U|}\right)\right] \tag{10.46}$$

if it is relative. Similar formulae can be given for the case of $Q = (Q^+, Q^-)$.

Finally, one should emphasize that linguistic quantifiers combined with Zadeh's computational approach to them open the door to applications of intelligent counting in a lot of areas of intelligent systems (with special reference to fuzzy systems) and decision support. Let us mention the areas of control and decision making, databases, text categorization, group decision making, etc. (see Sections 10.4-10.6). The most direct and spectacular application of this kind is possibly that to linguistic database summaries in the sense proposed in YAGER (1982). They are also known as the *linguistic quantifier driven aggregation of data* (see KACPRZYK/YAGER (2001)). A *linguistic summary* is then meant as a linguistically quantified (type-I or type-II) statement that subsumes the very essence of a set of numerical data from a certain point of view. That set, as one assumes, is large and, thus, not comprehensible to human beings in its original form. Examples are the following linguistic summaries for a computer retailer:

"*Much* sales on Saturday is *about noon*",

"*About* 1/3 of sales of computers is in *rainy days* to individual customers".

As we see, linguistic summaries form a useful transition from a set of individual data to a synthetic and laconic knowledge expressed in natural language. This knowledge is easy to assimilate by humans and forms a good basis for decision making. Clearly, the user expects that the system of summarization will provide a collection of linguistic summaries whose quality is sufficiently high, including a sufficiently high truth degree. We like to mention two examples of successful implementations of summarization systems: FQUERY for Access (see KACPRZYK/ZADROŻNY (2001a, 2005a)) and QUANTIRIUS (see PILARSKI (2010)). The reader is referred to these papers for details about the summarization process.

10.4 Decision Making in a Fuzzy Environment through Counting

Let us return to decision making in a fuzzy environment presented in Section 4.6. We like to propose an alternative formulation of the Bellman-Zadeh model in which, speaking generally, aggregation as a basis for the decision process is replaced with counting up.

First, we propose to abandon the idea of specifying fuzzy goals and fuzzy constraints separately as this distinction coming from mathematical optimization does not seem to be necessary and would be troublesome in our approach. Both fuzzy goals and constraints will be treated as fuzzy constraints. Referring to (4.12), we thus deal with a set of $p = j + k$ fuzzy constraints $C_1, C_2, \ldots, C_p$: $\boldsymbol{U} \rightarrow [0, 1]$.

Second, those fuzzy constraints will be understood as targets. Each decision alternative x will be treated as a shooter shooting at consecutive targets $C_1, \ldots, C_p$. $C_i(x)$ is now interpreted as a satisfaction level of shooter x after firing a shot at C_i, a degree of nearness of the hit to the center of target, $i = 1, \ldots, p$. $C_i(x) = 1$ signifies a hit at the very center of C_i. The universe of decision alternatives, $\boldsymbol{U}$, is thus viewed as a team of shooters, and we try to search for the best shooter by counting up the hits. We try to search, speaking more formally, for a shooter x^* such that

$$x^* \in \operatorname*{argmax}_x \sigma(\mathit{Hits}(x)), \tag{10.47}$$

where

$$\mathit{Hits}(x) = C_1(x)/1 + C_2(x)/2 + \ldots + C_p(x)/p \tag{10.48}$$

is a fuzzy set of hits scored by x. σ denotes a scalar cardinality, say, a sigma f-count or a scalar cardinality derived from a fuzzy cardinality. In particular, one can use

$$x^* \in \operatorname*{argmax}_x \mathrm{COG}_{\mathrm{FE}(\mathit{Hits}(x))}. \tag{10.49}$$

An alternative to (10.49) may be the choice of (see (9.48))

$$x^* \in \operatorname*{argmax}_x |(\mathit{Hits}(x))_{0.5}|. \tag{10.50}$$

Example 10.4. Let us refer to the first example from Subsection 4.6.2 in which one tries to choose the best TV model from a set $\boldsymbol{U} = \{m_1, m_2, m_3, m_4\}$ of four models. The fuzzy goal G and constraints C_1, C_2 and C_3 are now treated as a set of four constraints:

C_1 - *reliability*,

C_2 - *high* picture and sound quality,

C_3 - *energy-saving*,

C_4 - *good* guarantee conditions.

According to Subsection 4.6.2, the following assessments are given by experts:

$$C_1 = 0.7/m_1 + 0.3/m_2 + 0.7/m_3 + 0.9/m_4,$$
$$C_2 = 0.9/m_1 + 0.7/m_2 + 0.6/m_3 + 0.3/m_4,$$
$$C_3 = 0.7/m_1 + 0.5/m_2 + 0.9/m_3 + 0.7/m_4,$$
$$C_4 = 0.3/m_1 + 0.4/m_2 + 0.6/m_3 + 0.9/m_4.$$

So, by (10.48),

$$\mathit{Hits}(m_1) = 0.7/1 + 0.9/2 + 0.7/3 + 0.3/4,$$
$$\mathit{Hits}(m_2) = 0.3/1 + 0.7/2 + 0.5/3 + 0.4/4,$$
$$\mathit{Hits}(m_3) = 0.7/1 + 0.6/2 + 0.9/3 + 0.6/4,$$
$$\mathit{Hits}(m_4) = 0.9/1 + 0.3/2 + 0.7/3 + 0.9/4.$$

If the membership degrees in $\mathit{Hits}(m_i)$ are viewed as joinable (see Subsection 8.1.3), the usual sigma count can be used to count up in $\mathit{Hits}(m_i)$. Then

$$\sigma_{id}(Hits(m_1)) = 2.6, \quad \sigma_{id}(Hits(m_2)) = 1.9,$$
$$\sigma_{id}(Hits(m_3)) = 2.8, \quad \sigma_{id}(Hits(m_4)) = 2.8$$

and, by (10.47), the choice of m_3 or m_4 is suggested. If those degrees are considered to be non-joinable, a good idea is to use scalar cardinalities derived from FECounts. One gets

$$\mathrm{FE}(Hits(m_1)) = (0.1, 0.3, 0.3, 0.7, 0.3),$$
$$\mathrm{FE}(Hits(m_2)) = (0.3, 0.5, 0.5, 0.4, 0.3),$$
$$\mathrm{FE}(Hits(m_3)) = (0.1, 0.3, 0.4, 0.4, 0.6),$$
$$\mathrm{FE}(Hits(m_4)) = (0.1, 0.1, 0.3, 0.7, 0.3)$$

and

$$\mathbf{COG}_{\mathbf{FE}(\boldsymbol{Hits}(\boldsymbol{m}_1))} = 2.47, \quad \mathbf{COG}_{\mathbf{FE}(\boldsymbol{Hits}(\boldsymbol{m}_2))} = 1.95,$$
$$\mathbf{COG}_{\mathbf{FE}(\boldsymbol{Hits}(\boldsymbol{m}_3))} = 2.61, \quad \mathbf{COG}_{\mathbf{FE}(\boldsymbol{Hits}(\boldsymbol{m}_4))} = 2.67.$$

The choice of m_4 is thus now recommended. Alternatively, applying (10.50), we have

$$|(Hits(m_1))_{0.5}| = 3, \quad |(Hits(m_2))_{0.5}| = 2,$$
$$|(Hits(m_3))_{0.5}| = 4, \quad |(Hits(m_4))_{0.5}| = 3,$$

i.e. m_3 is suggested as a proper choice. □

As one sees, the alternative formulation based on counting leads in the above example to results and conclusions which are in accord with those obtained in Subsection 4.6.2 by means of the original Bellman-Zadeh model. However, this is not always the case: (10.47) may generate essentially different results. An instance is pricing a new product from Subsection 4.6.2 (see also RUTKOWSKI (2008)). One can say that the approach through counting proposed in this section seems to be a bit less universal than the original one, but is useful in many situations and, thus, worth remembering.

10.5 Group Decision Making

The problem of group decision making, GDM, consists in finding an option, decision alternative that "best" suits to individual preferences formulated by the members of a group of decision makers (individuals, experts, agents, organizations, etc.). The main related difficulty is discrepancy between and imprecision (fuzziness) of those preferences, and possible incompleteness of information that is necessary to specify their intensities. One should emphasize that the issue od GDM does not relate to political elections. It refers to decision making by a small group $\boldsymbol{P} = \{p_1, p_2, \ldots, p_m\}$ of decision makers, $m \geq 1$, having to make a choice in a small set $\boldsymbol{S} = \{s_1, s_2, \ldots, s_n\}$ of possible decision alternatives, $n \geq 2$. In other words, one assumes that $\boldsymbol{P}$ and $\boldsymbol{S}$ contain at most a dozen of elements or so.

We will focus on models of GDM involving fuzzy preferences combined with the concept of soft, linguistic majority. Looking historically, it seems that the first model of this kind was proposed in KACPRZYK (1986) by developing and enriching the ideas of GDM with fuzzy preferences from NURMI (1981). That model was then extended in SZMIDT/KACPRZYK (1998) by applying the bipolar methodology of standard I-fuzzy sets and adding the hesitation factor defined via (6.55). Later on, in PANKOWSKA/WYGRALAK (2006) a model and algorithms of GDM inspired by SZMIDT/KACPRZYK (1998) and involving arbitrary sigma f-counts as well as triangular norm-based hesitation degrees (6.51) were proposed. What we like to present in this section are new, improved versions of those algorithms. We will focus on the use of nilpotent triangular norms.

Moving on to details, let us assume that the individual preferences over $\boldsymbol{S}$ of each decision maker p_k, $k = 1, \ldots, m$, are expressed in the form of a binary fuzzy relation $\boldsymbol{S} \times \boldsymbol{S} \to [0. 1]$ (see Section 5.1). It can be presented as an $n \times n$ *preference matrix*

$$R_k = [r^k_{ij}] \quad \text{with } i, j = 1, \ldots, n.$$

The number $r^k_{ij} \in [0, 1]$ is understood as a *degree to which* p_k *prefers* s_i *to* s_j. By convention, $r^k_{ii} = 0$ for each i. We assume that all matrices R_k are complete, i.e. none of their elements is unknown or missing. Although the case of incomplete preference matrices is interesting from the viewpoint of applications, it goes beyond the scope of this discussion. The reader is referred to PANKOWSKA (2007, 2008) for a detailed analysis of that case.

What is worth emphasizing, we do not demand any kind of transitivity from preferences formulated by decision makers as empirical studies show that human preferences are generally non-transitive (see e.g. ŚWITALSKI (2001)). Nevertheless, we will assume that there is a kind of coherence between preference degrees r^k_{ij} and r^k_{ji}, namely

$$r^k_{ij} \le \nu(r^k_{ji}) \quad \text{for each } i, j = 1, \ldots, n, \tag{10.51}$$

where ν denotes a strong negation with equilibrium point $e(\nu)$ (see Subsection 2.4.1). The most important case seems to be that with $\nu = \nu_{Ł}$ and, then, (10.51) collapses to

$$r^k_{ij} + r^k_{ji} \le 1. \tag{10.52}$$

More generally, if $\nu(a) = (1 - a^p)^{1/p}$ with $p > 0$ is used, i.e. $\nu = \nu_t$ with $\boldsymbol{t} = \boldsymbol{t}_{S,p}$, then (10.51) is of the form

$$(r^k_{ij})^p + (r^k_{ji})^p \le 1. \tag{10.53}$$

For brevity, in these formulae as well as in further ones, we usually skip the specification of the range of indices, and we implicitly assume that $k = 1, \ldots, m$ and $i, j = 1, \ldots, n$. A consequence of (6.49) is that

$$r^k_{ij} \le e(\nu) \quad \text{or/and} \quad r^k_{ji} \le e(\nu). \tag{10.54}$$

Condition (10.51) suggests that one has to distinguish between two important classes of preferences.

- All R_k's are *reciprocal*, i.e.

$$r_{ij}^k = \nu(r_{ji}^k) \quad \text{whenever } i \neq j. \tag{10.55}$$

For $\nu = \nu_Ł$, we get

$$r_{ij}^k + r_{ji}^k = 1 \quad \text{whenever } i \neq j. \tag{10.56}$$

One thus assumes that each decision maker p_k has sufficient information for specifying the intensity of preferences as to options s_i and s_j.

- At least some preference matrices R_k are *non-reciprocal*, i.e. for some indices k and $i \neq j$ one has

$$r_{ij}^k < \nu(r_{ji}^k). \tag{10.57}$$

So, if $\nu = \nu_Ł$, this collapses to

$$r_{ij}^k + r_{ji}^k < 1. \tag{10.58}$$

Now, because of incompleteness of information about s_i and s_j, p_k is unable to specify precisely the intensity of preferences as to s_i and s_j.

r_{ij}^k with $i \neq j$ can be viewed as a lower bound on a real degree ρ_{ij}^k to which p_k would prefer s_i to s_j if information about s_i and s_j were complete. More precisely, for $i \neq j$, we have (cf. (6.48))

$$r_{ij}^k \leq \rho_{ij}^k \leq \nu(r_{ji}^k). \tag{10.59}$$

So, the preferences of decision maker p_k as to s_i and s_j are generally accompanied by a *hesitation margin* (cf. (6.51))

$$h_{ij}^k = \nu(r_{ij}^k)\, \boldsymbol{t}\, \nu(r_{ji}^k) = \nu(r_{ij}^k\, \boldsymbol{t}^\nu\, r_{ji}^k) \tag{10.60}$$

with a t-norm $\boldsymbol{t}$. Thus, $h_{ij}^k = h_{ji}^k$. Our attention, as already mentioned, will be focused on the case of nilpotent $\boldsymbol{t}$ and, moreover, we will put $\nu = \nu_t$. By Theorem 6.3, h_{ij}^k is then a "pure" size of ignorance as to ρ_{ij}^k. And just this case seems to be especially interesting from the viewpoint of applications in GDM.

Throughout this section, Q will denote a *most*-type relative linguistic quantifier (see Section 10.3). Q: $[0, 1] \to [0, 1]$ is thus nondecreasing, $Q(0) = 0$ and $Q(a) = 1$ for each $a > t$ with a predefined value $t < 1$. According to KACPRZYK (1986), we like to discuss two general approaches to GDM.

- **Direct approach.** Our task is then to find a *solution* understood as a fuzzy set S_Q of options such that a soft majority, Q decision makers, is not against them. This can be presented schematically as

$$\{R_1, \ldots, R_m\} \to S_Q.$$

- **Indirect approach.** The individual preference matrices $R_1, \ldots, R_m$ are now first aggregated into an $n \times n$ matrix $R = [r_{ij}]$ of *social (group) preferences*. R

represents common preferences of the whole group $\boldsymbol{P}$: r_{ij} forms a *degree to which the group prefers s_i over s_j*. Further, on the basis of R, one constructs a *solution* W_Q understood in this case as a fuzzy set of options preferred by the group over Q remaining alternatives from $\boldsymbol{S}$. Schematically, this procedure is thus of the form

$$\{R_1, \dots, R_m\} \to R \to W_Q.$$

These two approaches will be the subject of further discussion. For notational convenience, we will use the following additional symbols:

$$P = 1_{\boldsymbol{P}}, \ \boldsymbol{S}_j = \boldsymbol{S} \setminus \{s_j\}, \ S_j = 1_{\boldsymbol{S}_j}$$

with $j = 1, \dots, n$.

10.5.1 Direct Approach to GDM

Assume $\boldsymbol{t}$ is a nilpotent t-norm and $\nu = \nu_t$. First of all, let us consider the case when all preference matrices R_k are reciprocal. We present the following algorithm of finding a solution S_Q.

- **Step 1.** For each decision maker p_k and option s_j, create a fuzzy set $R_{kj}: \boldsymbol{S}_j \to [0,1]$ of options such that p_k prefers s_j to them, i.e. $R_{kj}(s_i) = r^k_{ji}$ for $i \neq j$. The membership degrees in R_{kj} are thus the jth row of R_k, excluding the diagonal element. Let

$$r_{kj} = \sigma_f(R_{kj} \mid S_j) = \frac{1}{n-1} \sum_{\substack{i=1 \\ i \neq j}}^{n} f(r^k_{ji}) \tag{10.61}$$

with a weighting function f (see (8.34)). Alternatively, we can define

$$r_{kj} = \frac{1}{n-1} \mathrm{COG}_{\mathrm{FE}(R_{kj})}. \tag{10.62}$$

r_{kj} says to what extent p_k is *not against* s_j. All r_{kj}'s can be collected in an $m \times n$ matrix

$$R^{\#} = [r_{kj}].$$

- **Step 2.** For each s_j, construct a fuzzy set I_j of decision makers p_k being not against s_j. So, $I_j: \boldsymbol{P} \to [0, 1]$ with

$$I_j(p_k) = r_{kj}.$$

Compute the relative cardinality

$$d_j = \sigma_{f^1}(I_j \mid P) = \frac{1}{m} \sum_{k=1}^{m} f^1(r_{kj}) \tag{10.63}$$

with a weighting function f^1 possibly differing from f. Alternatively, one can put

$$d_j = \frac{1}{m} \mathrm{COG}_{\mathrm{FE}(I_j)}. \tag{10.64}$$

- **Step 3.** For $j = 1, \ldots, n$, compute $Q(d_j)$, a degree to which most decision makers are not against s_j. Finally, create

$$S_Q = Q(d_1)/s_1 + Q(d_2)/s_2 + \ldots + Q(d_n)/s_n, \tag{10.65}$$

a fuzzy set of options such that a soft majority is not against them.

From the viewpoint of practice, the most important case is that with $t = t_Ł$ and $\nu = \nu_t = \nu_Ł$. As to the use of sigma f-counts, the following particular cases of the above algorithm seem to be especially interesting (cf. KACPRZYK (1986)).

- $f = f_{2,\,0.5}$, $f^1 = id$. S_Q is then called the *fuzzy Q-core*.
- More generally, let $f = f_{2,\,\alpha}$ with $0.5 \le \alpha < 1$ and $f^1 = id$. S_Q is now said to be the *fuzzy α/Q-core*.
- $f = f_{7,\,0.5}$ with $f_{7,\,0.5}(a) = 0 \vee (2a - 1)$, $f^1 = id$. S_Q is then called the *fuzzy s/Q-core*. $f_{7,\,0.5}$ makes it possible to describe the strength (intensity) of the domination of s_j over the s_i's with $i \ne j$. Generally, if a negation ν with $e(\nu) \ne 0.5$ is used, one should employ $f = f_{7,\,\alpha}$ with $\alpha = e(\nu)$, where

$$f_{7,\,\alpha}(a) = 0 \vee \left(\frac{1}{1-\alpha}(a - \alpha)\right).$$

Example 10.5. Let $m = n = 4$ and (KACPRZYK (1986))

$$R_1 = \begin{bmatrix} 0 & 0.3 & 0.7 & 0.1 \\ 0.7 & 0 & 0.6 & 0.6 \\ 0.3 & 0.4 & 0 & 0.2 \\ 0.9 & 0.4 & 0.8 & 0 \end{bmatrix}, \quad R_2 = \begin{bmatrix} 0 & 0.4 & 0.6 & 0.2 \\ 0.6 & 0 & 0.7 & 0.4 \\ 0.4 & 0.3 & 0 & 0.1 \\ 0.8 & 0.6 & 0.9 & 0 \end{bmatrix},$$

$$R_3 = \begin{bmatrix} 0 & 0.5 & 0.7 & 0 \\ 0.5 & 0 & 0.8 & 0.4 \\ 0.3 & 0.2 & 0 & 0.2 \\ 1 & 0.6 & 0.8 & 0 \end{bmatrix}, \quad R_4 = \begin{bmatrix} 0 & 0.4 & 0.7 & 0.3 \\ 0.6 & 0 & 0.4 & 0.3 \\ 0.3 & 0.6 & 0 & 0.1 \\ 0.7 & 0.7 & 0.9 & 0 \end{bmatrix}.$$

We will apply $t = t_Ł$ and $\nu = \nu_Ł$, i.e. $e(\nu) = 0.5$, and $Q = (0.8, 0.5)_s$.
(a) First, take $f = f_{2,\,0.5}$ and $f^1 = id$. By (10.61),

$$R^{\#} = \begin{bmatrix} \frac{1}{3} & 1 & 0 & \frac{2}{3} \\ \frac{1}{3} & \frac{2}{3} & 0 & 1 \\ \frac{1}{3} & \frac{1}{3} & 0 & 1 \\ \frac{1}{3} & \frac{1}{3} & \frac{1}{3} & 1 \end{bmatrix}.$$

Using (10.63), one gets

$$d_1 = \frac{1}{3},\ d_2 = \frac{7}{12},\ d_3 = \frac{1}{12},\ d_4 = \frac{11}{12}$$

and, finally, the fuzzy Q-core is

$$S_Q = \frac{2}{30}/s_1 + \frac{17}{30}/s_2 + 1/s_4.$$

The choice of s_4 is thus suggested.

(b) Let $f = f_{2,\,2/3}$ and $f^1 = id$. Now

$$d_1 = \frac{1}{4},\ d_2 = \frac{1}{4},\ d_3 = 0,\ d_4 = \frac{3}{4},$$

i.e.

$$S_Q = 0.9/s_4.$$

(c) Put $f = f_{7,\,0.5}$ and, again, $f^1 = id$. Then $f(r_{ji}^k) = 0 \vee (2r_{ji}^k - 1)$ and

$$R^{\#} = \begin{bmatrix} \frac{4}{30} & \frac{8}{30} & 0 & \frac{14}{30} \\ \frac{2}{30} & \frac{6}{30} & 0 & \frac{16}{30} \\ \frac{4}{30} & \frac{6}{30} & 0 & \frac{18}{30} \\ \frac{4}{30} & \frac{2}{30} & \frac{2}{30} & \frac{16}{30} \end{bmatrix}.$$

Hence

$$d_1 = \frac{7}{60},\ d_2 = \frac{11}{60},\ d_3 = \frac{1}{60},\ d_4 = \frac{32}{60},$$

and

$$S_Q = \frac{7}{15}/s_4$$

forms the fuzzy s/Q-core. Again, s_4 is recommended.

(d) Use (10.62) and (10.64) instead of sigma f-counts. For instance,

$$R_{11} = 0.3/s_2 + 0.7/s_3 + 0.1/s_4,\quad \mathrm{FE}(R_{11}) = (0.3, 0.7, 0.3, 0.1),$$

$$\mathbf{COG}_{\mathbf{FE}(R_{11})} = (0.7 + 0.6 + 0.3)/1.4 = 1.14,\quad r_{11} = \frac{1.14}{3} = 0.38.$$

Similar computations lead to

$$R^{\#} = \begin{bmatrix} 0.38 & 0.59 & 0.36 & 0.67 \\ 0.42 & 0.55 & 0.31 & 0.72 \\ 0.38 & 0.56 & 0.31 & 0.78 \\ 0.48 & 0.45 & 0.36 & 0.71 \end{bmatrix}.$$

Hence, say,

$$I_1 = 0.38/p_1 + 0.42/p_2 + 0.38/p_3 + 0.48/p_4,$$

$$\mathrm{FE}(I_1) = (0.52,\ 0.48,\ 0.42,\ 0.38,\ 0.38)$$

and

$$\mathbf{COG}_{\mathrm{FE}(I_1)} = 1.83, \quad \text{i.e. } d_1 = \frac{1.83}{4} = 0.46.$$

Similarly,

$$d_2 = 0.52, \quad d_3 = 0.41, \quad d_4 = 0.63$$

and the solution is now

$$S_Q = Q(0.46)/s_1 + Q(0.52)/s_2 + Q(0.41)/s_3 + Q(0.63)/s_4$$
$$= 0.32/s_1 + 0.44/s_2 + 0.22/s_3 + 0.66/s_4.$$

As previously, the choice of s_4 is thus recommended. We see that this approach is more complex from the computational viewpoint, but behaves well and relieves us of the necessity of choosing the weighting functions f and f^1. Since $\mathrm{COG}_{\mathrm{FE}(\cdot)}$ is a sensitive measure of cardinality, S_Q is now "flattened" in comparison with (a)-(c).

(e) Let us add a semantics to preference matrices $R_1, \ldots, R_m$. We like to present two examples borrowed from PANKOWSKA (2005).

Environmental management. In 2004, the High Tatra Mountains in Slovakia were struck by destructive, gale-force wind. The storm lasted about four hours. Many square kilometers of Tatra forest were then completely flattened. This calamity started a national discussion: is it necessary to remove and process all fallen trees? How forest management should look like in the future? In the aftermath of that discussion, four options were emerged:

- s_1 - change the boundaries of protected areas, and develop sport and tourist industry,
- s_2 - remove and utilize windfallen trees as soon as possible, and plant new trees,
- s_3 - remove windfallen trees, let the forest regenerate in some areas and afforest the rest,
- s_4 - leave all windfallen trees, let the forest regenerate naturally.

In the related decision process, the following four decision makers were involved:

- p_1 - government,
- p_2 - ecological organizations,
- p_3 - Department of Agriculture and Forestry,
- p_4 - businessmen.

Assume their preferences as to the four options are of the following form:

$$R_1 = \begin{bmatrix} 0 & 0.6 & 0.7 & 0.9 \\ 0.4 & 0 & 0.8 & 0.9 \\ 0.3 & 0.2 & 0 & 0.7 \\ 0.1 & 0.1 & 0.3 & 0 \end{bmatrix}, \quad R_2 = \begin{bmatrix} 0 & 0 & 0 & 0 \\ 1 & 0 & 0.2 & 0.1 \\ 1 & 0.8 & 0 & 0.4 \\ 1 & 0.9 & 0.6 & 0 \end{bmatrix},$$

$$R_3 = \begin{bmatrix} 0 & 0 & 0 & 0.1 \\ 1 & 0 & 0.6 & 1 \\ 1 & 0.4 & 0 & 0.9 \\ 0.9 & 0 & 0.1 & 0 \end{bmatrix}, \quad R_4 = \begin{bmatrix} 0 & 0.7 & 0.9 & 1 \\ 0.3 & 0 & 0.7 & 0.9 \\ 0.1 & 0.3 & 0 & 0.6 \\ 0 & 0.1 & 0.4 & 0 \end{bmatrix}.$$

The reader can easily check that, say, the fuzzy Q-core is now (see (a))

$$S_Q = \frac{6}{15}/s_1 + \frac{11}{15}/s_2 + \frac{6}{15}/s_3 + \frac{1}{15}/s_4,$$

i.e. the choice of s_2 is suggested.

Worker recruitment. This example offers a non-standard look at the set $\boldsymbol{P}$ of decision makers. Assume an employer is searching for a good candidate for SAP Business Analyst position. $\boldsymbol{S}$ is a set of candidates, whereas $\boldsymbol{P}$ will be understood as a set of requirements formulated by the employer. For instance,

p_1 - suitable knowledge and experience in using SAP,

p_2 - knowledge of accounting principles and financial systems, and business knowledge,

p_3 - fluency in German and English,

p_4 - teamworking and cooperation skills.

r^k_{ij} forms a degree to which candidate s_i is preferred to s_j by the employer with respect to requirement p_k. Assume that four candidates are taken into account and the preference matrices look as follows:

$$R_1 = \begin{bmatrix} 0 & 0.7 & 0.6 & 0.2 \\ 0.3 & 0 & 0.3 & 0.2 \\ 0.4 & 0.7 & 0 & 0.2 \\ 0.8 & 0.8 & 0.8 & 0 \end{bmatrix}, \quad R_2 = \begin{bmatrix} 0 & 0.7 & 0.6 & 0.6 \\ 0.3 & 0 & 0 & 0 \\ 0.4 & 1 & 0 & 0.2 \\ 0.4 & 1 & 0.8 & 0 \end{bmatrix},$$

$$R_3 = \begin{bmatrix} 0 & 0.7 & 0.6 & 0.6 \\ 0.3 & 0 & 0 & 0 \\ 0.4 & 1 & 0 & 0.8 \\ 0.4 & 1 & 0.2 & 0 \end{bmatrix}, \quad R_4 = \begin{bmatrix} 0 & 0.6 & 0.5 & 0.3 \\ 0.4 & 0 & 0.2 & 0 \\ 0.5 & 0.8 & 0 & 0 \\ 0.7 & 1 & 1 & 0 \end{bmatrix}.$$

For instance, the s/Q-core is then of the form

$$S_Q = \frac{16}{30}/s_4.$$

The choice of s_4 is thus recommended. □

Let us move on to the case when at least some preference matrices R_k are not reciprocal (see (10.57)). By (10.59), each R_k can now be viewed as a matrix of minimum preferences of p_k. These preferences may increase if decision maker p_k will get more information about the decision alternatives. Speaking more precisely, the degree to which p_k currently prefers s_i to s_j, $i \neq j$, may then increase from r^k_{ij} up to $r^k_{ij}\, \boldsymbol{t}^\circ\, h^k_{ij} = \nu(r^k_{ji})$ (see Theorem 6.3(b)). It may increase, in other words, by the hesitation margin (see (10.60), (3.16))

$$h^k_{ij} = \nu(r^k_{ij})\, \boldsymbol{t}\, \nu(r^k_{ji}) = \nu(r^k_{ij}\, \boldsymbol{t}^\circ\, r^k_{ji}). \tag{10.66}$$

As previously, $\boldsymbol{t}$ is a nilpotent t-norm and $\nu = \nu_t$. In the most important case of $\boldsymbol{t} = \boldsymbol{t}_Ł$ and $\nu = \nu_Ł$, one gets $h^k_{ij} = 1 - r^k_{ij} - r^k_{ji}$, i.e. the preference degree may increase from r^k_{ij} up to $r^k_{ij} + h^k_{ij} = 1 - r^k_{ji}$.

A result of all maximum increases are matrices of maximum preferences

$$M_k = [\, m^k_{ij}] \quad \text{with} \quad m^k_{ij} = \nu(r^k_{ji}) \quad \text{for} \quad i \neq j. \tag{10.67}$$

By convention, $m^k_{ii} = 0$. If R_k is reciprocal, one thus has $M_k = R_k$. We propose the following algorithm of finding a solution.

- **Step 1a.** Perform twice Step 1-Step 2 of the algorithm for reciprocal matrices. First, this procedure should be applied to $R_1, \ldots, R_m$ giving the matrix $R^{\#}$ and vector $(d_1, \ldots, d_n)$ with d_j defined by (10.63) or (10.64). Further, apply that procedure to $M_1, \ldots, M_m$ and create the corresponding matrix $M^{\#}$ and vector $(d_1^*, \ldots, d_n^*)$.
- **Step 2a.** For $j = 1, \ldots, n$, compute $Q(d_j)$ and $Q(d_j^*)$. The solution is now an IVFS

$$S_Q = [Q(d_1), Q(d_1^*)]/s_1 + [Q(d_2), Q(d_2^*)]/s_2 + \ldots + [Q(d_n), Q(d_n^*)]/s_n. \tag{10.68}$$

  $Q(d_j)$ and $Q(d_j^*)$, respectively, are a minimum and a maximum possible degree to which a soft majority (Q decision makers) is not against s_j. $Q(d_j^*)$ refers to the case of a maximum increase of preferences of all decision makers in favour of s_j.

Example 10.6. Again, let $m = n = 4$ and

$$R_1 = \begin{bmatrix} 0 & 0.3 & 0.7 & 0.4 \\ 0.7 & 0 & 0.6 & 0.9 \\ 0.3 & 0.4 & 0 & 0.5 \\ 0.4 & 0 & 0.3 & 0 \end{bmatrix}, \quad R_2 = \begin{bmatrix} 0 & 0.4 & 0.6 & 0.5 \\ 0.6 & 0 & 0.7 & 0.7 \\ 0.4 & 0.3 & 0 & 0.4 \\ 0.3 & 0.1 & 0.3 & 0 \end{bmatrix},$$

$$R_3 = \begin{bmatrix} 0 & 0.5 & 0.7 & 0.3 \\ 0.5 & 0 & 0.8 & 0.7 \\ 0.3 & 0.2 & 0 & 0.5 \\ 0.4 & 0.1 & 0.2 & 0 \end{bmatrix}, \quad R_4 = \begin{bmatrix} 0 & 0.4 & 0.7 & 0.6 \\ 0.6 & 0 & 0.4 & 0.6 \\ 0.3 & 0.6 & 0 & 0.4 \\ 0.1 & 0.1 & 0.1 & 0 \end{bmatrix}.$$

Use $\boldsymbol{t} = t_Ł$, $\nu = \nu_Ł$, $Q = (0.8, 0.5)_s$, $f = f_{2,0.5}$ and $f^1 = id$. So, by (10.61),

$$R^{\#} = \begin{bmatrix} \frac{1}{3} & 1 & 0 & 0 \\ \frac{1}{3} & 1 & 0 & 0 \\ \frac{1}{3} & \frac{2}{3} & 0 & 0 \\ \frac{2}{3} & \frac{2}{3} & \frac{1}{3} & 0 \end{bmatrix}$$

and (10.63) gives

$$d_1 = \frac{5}{12},\ d_2 = \frac{10}{12},\ d_3 = \frac{1}{12},\ d_4 = 0,$$

i.e.

$$Q(d_1) = \frac{7}{30},\ Q(d_2) = 1,\ Q(d_3) = Q(d_4) = 0.$$

Further, we construct the corresponding matrices of maximum possible preferences (see (10.67)):

$$M_1 = \begin{bmatrix} 0 & 0.3 & 0.7 & 0.6 \\ 0.7 & 0 & 0.6 & 1 \\ 0.3 & 0.4 & 0 & 0.7 \\ 0.6 & 0.1 & 0.5 & 0 \end{bmatrix},\quad M_2 = \begin{bmatrix} 0 & 0.4 & 0.6 & 0.7 \\ 0.6 & 0 & 0.7 & 0.9 \\ 0.4 & 0.3 & 0 & 0.7 \\ 0.5 & 0.3 & 0.6 & 0 \end{bmatrix},$$

$$M_3 = \begin{bmatrix} 0 & 0.5 & 0.7 & 0.6 \\ 0.5 & 0 & 0.8 & 0.9 \\ 0.3 & 0.2 & 0 & 0.8 \\ 0.7 & 0.3 & 0.5 & 0 \end{bmatrix},\quad M_4 = \begin{bmatrix} 0 & 0.4 & 0.7 & 0.9 \\ 0.6 & 0 & 0.4 & 0.9 \\ 0.3 & 0.6 & 0 & 0.9 \\ 0.4 & 0.4 & 0.6 & 0 \end{bmatrix}.$$

The ith row of each M_k presents the preferences of decision maker p_k after their maximum possible increase in favour of s_i. So, each row of M_k should be treated autonomously. One sees that $m^k_{ij} + m^k_{ji} > 1$ whenever $r^k_{ij} + r^k_{ji} < 1$. Using (10.61), we create $M^{\#} = [m_{kj}]$ with

$$m_{kj} = \frac{1}{n-1}\sum_{\substack{i=1\\ i\neq j}}^{n} f(m^k_{ji}).$$

Then

$$M^{\#} = \begin{bmatrix} \frac{2}{3} & 1 & \frac{1}{3} & \frac{1}{3} \\ \frac{2}{3} & 1 & \frac{1}{3} & \frac{1}{3} \\ \frac{2}{3} & \frac{2}{3} & \frac{1}{3} & \frac{1}{3} \\ \frac{2}{3} & \frac{2}{3} & \frac{2}{3} & \frac{1}{3} \end{bmatrix}.$$

The jth column of $M^{\#}$ shows to what extent decision makers $p_1, \ldots, p_m$ will not be against s_j after a maximum possible increase of their preferences in favour of s_j. By (10.63), one gets

$$d_1^* = \frac{8}{12},\quad d_2^* = \frac{10}{12},\quad d_3^* = \frac{5}{12},\quad d_4^* = \frac{4}{12}$$

and

$$Q(d_1^*) = \frac{22}{30},\quad Q(d_2^*) = 1,\quad Q(d_3^*) = \frac{7}{30},\quad Q(d_4^*) = \frac{2}{30}.$$

Finally, performing Step 2a, the solution is the following IVFS:

$$S_Q = \left[\frac{7}{30}, \frac{22}{30}\right]/s_1 + [1, 1]/s_2 + \left[0, \frac{7}{30}\right]/s_3 + \left[0, \frac{2}{30}\right]/s_4.$$

s_2 is thus recommended. Analogous computations can be done for other choices of f and f^1 (see Example 10.5 (b-d)). □

10.5.2 Indirect Approach

Let us apply the notation and terminology established in the introductory part of the present section. In the indirect approach, individual preference matrices $R_1, \ldots, R_m$ are first aggregated into a social (group) preference $n \times n$ matrix $R = [r_{ij}]$. That matrix is then a basis for constructing a solution W_Q, a fuzzy set of options preferred by the whole group $\boldsymbol{P}$ over Q remaining options from $\boldsymbol{S}$.

Again, let us assume that $\nu = \nu_t$ with a nilpotent t-norm t, and begin with the case of reciprocal preference matrices R_k. We present the following algorithm of finding W_Q.

- **Step 1b.** Construct fuzzy sets P_{ij} of decision makers preferring s_i to s_j. Speaking precisely, P_{ij}: $\boldsymbol{P} \to [0, 1]$ with $P_{ij}(p_k) = r^k_{ij}$ for $k = 1, \ldots, m$. Put

$$r_{ij} = \sigma_f(P_{ij} \mid P) = \frac{1}{m}\sum_{k=1}^{m} f(r^k_{ij}) \tag{10.69}$$

  with a weighting function f. Further, we create matrix $R = [r_{ij}]$. Each r_{ij} is a degree to which the whole group $\boldsymbol{P}$ prefers s_i to s_j (clearly, $r_{ii} = 0$). One should point out that R is not generally reciprocal.

- **Step 2b.** Let J_i: $\boldsymbol{S}_i \to [0, 1]$ with $J_i(s_j) = r_{ij}$. This ith row of R, excluding the diagonal element, is a fuzzy set of options such that s_i is preferred over them by the group. Its relative cardinality equals

$$b_i = \sigma_{f^1}(J_i \mid S_i) = \frac{1}{n-1}\sum_{\substack{j=1 \\ j \neq i}}^{n} f^1(r_{ij}) \tag{10.70}$$

  with a weighting function f^1 possibly differing from f.

- **Step 3b.** For $j = 1, \dots, n$, compute $Q(b_j)$, a degree to which s_j is preferred by the group over Q remaining options. Finally, the solution is of the form

$$W_Q = Q(b_1)/s_1 + Q(b_2)/s_2 + \dots + Q(b_n)/s_n. \tag{10.71}$$

Worth emphasizing are three particular cases of the above algorithm (cf. KACPRZYK (1986) and SZMIDT/KACPRZYK (1998)).

- $f = f^1 = f_{2,\, e(\nu)}$. W_Q is then called the *fuzzy Q-consensus winner*.
- $f = f_{2,\, e(\nu)}$ and $f^1 = f_{2,\, \alpha}$ with $e(\nu) \le \alpha < 1$. W_Q is said to be the *fuzzy* α/Q*-consensus winner*.
- $f = f_{2,\, e(\nu)}$ and $f^1 = f_{7,\, e(\nu)}$. What we now get as W_Q is called the *fuzzy* s/Q*-consensus winner*.

Similarly to the direct approach, (10.69) and (10.70) can be replaced with

$$r_{ij} = \frac{1}{m}\mathbf{COG}_{\mathbf{FE}(P_{ij})} \quad \text{and} \quad b_i = \frac{1}{n-1}\mathbf{COG}_{\mathbf{FE}(J_i)}, \tag{10.72}$$

respectively.

Example 10.7. Take the preference matrices R_1, R_2, R_3, R_4 from Example 10.5. And again, use $t = t_Ł$ leading to $\nu = \nu_Ł$, i.e. $e(\nu) = 0.5$. Let $f = f_{2,\, 0.5}$ and $Q = (0.8, 0.5)_s$. Performing Step 1b, we obtain

$$R = \begin{bmatrix} 0 & 0 & 1 & 0 \\ \frac{3}{4} & 0 & \frac{3}{4} & \frac{1}{4} \\ 0 & \frac{1}{4} & 0 & 0 \\ 1 & \frac{3}{4} & 1 & 0 \end{bmatrix}.$$

Further, by (10.70), one gets the following results (cf. KACPRZYK (1986)).
(a) If $f^1 = f$, then

$$b_1 = \frac{1}{3}, \quad b_2 = \frac{2}{3}, \quad b_3 = 0, \quad b_4 = 1,$$

i.e. the fuzzy Q-consensus winner is

$$W_Q = \frac{1}{15}/s_1 + \frac{11}{15}/s_2 + 1/s_4.$$

(b) $f^1 = f_{2,\, 0.8}$ gives

$$b_1 = \frac{1}{3}, \quad b_2 = 0, \quad b_3 = 0, \quad b_4 = \frac{2}{3}.$$

So,

$$W_Q = \frac{1}{15}/s_1 + \frac{11}{15}/s_4$$

forms the fuzzy $0.8/Q$-consensus winner.

(c) For $f^1 = f_{7,\,0.5}$, we have

$$b_1 = \frac{1}{3},\;\; b_2 = \frac{1}{3},\;\; b_3 = 0,\;\; b_4 = \frac{5}{6}$$

and, thus, the fuzzy s/Q-consensus winner is

$$W_Q = \frac{1}{15}/s_1 + \frac{1}{15}/s_2 + 1/s_4.$$

Concluding, s_4 is recommended in (a)-(c).

(d) Finally, use (10.72) instead of (10.69) and (10.70). Then

$$R = \begin{bmatrix} 0 & 0.44 & 0.59 & 0.19 \\ 0.56 & 0 & 0.59 & 0.45 \\ 0.41 & 0.41 & 0 & 0.23 \\ 0.81 & 0.55 & 0.77 & 0 \end{bmatrix}.$$

For instance,

$$P_{32} = 0.4/p_1 + 0.3/p_2 + 0.2/p_3 + 0.6/p_4,$$

$$\mathrm{FE}(P_{32}) = (0.4, 0.6, 0.4, 0.3, 0.2),$$

$$r_{32} = \frac{1}{4}\mathbf{COG}_{\mathbf{FE}(P_{32})} = 0.41.$$

Further, we determine the b_i's. Trying to compute, say, b_2 one gets

$$J_2 = 0.56/s_1 + 0.59/s_3 + 0.45/s_4,$$

$$\mathrm{FE}(J_2) = (0.41, 0.44, 0.55, 0.45),$$

$$b_2 = \frac{1}{3}\mathbf{COG}_{\mathbf{FE}(J_2)} = 0.52.$$

Generally, we have

$$b_1 = 0.42,\;\; b_2 = 0.52,\;\; b_3 = 0.39,\;\; b_4 = 0.65,$$

i.e.

$$Q(b_1) = 0.24,\;\; Q(b_2) = 0.44,\;\; Q(b_3) = 0.18,\;\; Q(b_4) = 0.7.$$

The solution is thus

$$W_Q = 0.24/s_1 + 0.44/s_2 + 0.18/s_3 + 0.7/s_4.$$

The choice of s_4 is again suggested although, similarly to Example 10.5, W_Q is now more "flat". □

Finally, it remains to consider the case when at least some of the individual preference matrices R_k are not reciprocal. We will use twice the way of doing described in Step 1b - Step 2b. We propose the following procedure of finding W_Q.

- **Step 1c.** Perform Step 1b - Step 2b for the matrices R_k of individual preferences, and create the vector $(Q(b_1), \dots, Q(b_n))$. Furthermore, for each R_k, construct a matrix $M_k = [m_{ij}^k]$ of maximum possible preferences (see (10.67)). Aggregate $M_1, \dots, M_m$ into a matrix $M = [m_{ij}]$ of group preferences. So,

$$m_{ij} = \frac{1}{m} \sum_{k=1}^{m} f(m_{ij}^k). \tag{10.73}$$

Compute

$$b_i^* = \frac{1}{n-1} \sum_{\substack{j=1 \\ j \neq i}}^{n} f^1(m_{ij}). \tag{10.74}$$

An alternative variant is to use formulae analogous to those from (10.72).

- **Step 2c.** Create the solution which is now an interval-valued fuzzy set

$$W_Q = [Q(b_1), Q(b_1^*)]/s_1 + [Q(b_2), Q(b_2^*)]/s_2 + \dots + [Q(b_n), Q(b_n^*)]/s_n. \tag{10.75}$$

$Q(b_i)$ and $Q(b_i^*)$, respectively, are a minimum and a maximum degree to which s_i is preferred by the group over Q remaining options.

Example 10.8. Let us return to the individual preference matrices from Example 10.6. Again, $t = t_Ł$, $v = v_Ł$, $Q = (0.8, 0.5)_s$, and put $f = f_{2,\,0.5}$. By (10.69) and (10.73),

$$R = \begin{bmatrix} 0 & 0 & 1 & \frac{1}{4} \\ \frac{3}{4} & 0 & \frac{3}{4} & 1 \\ 0 & \frac{1}{4} & 0 & 0 \\ 0 & 0 & 0 & 0 \end{bmatrix}, \quad M = \begin{bmatrix} 0 & 0 & 1 & 1 \\ \frac{3}{4} & 0 & \frac{3}{4} & 1 \\ 0 & \frac{1}{4} & 0 & 1 \\ \frac{2}{4} & 0 & \frac{2}{4} & 0 \end{bmatrix}.$$

(a) Let $f^1 = f$. Applying (10.70) and (10.74), one gets

$$b_1 = \frac{1}{3}, \quad b_2 = 1, \quad b_3 = 0, \quad b_4 = 0$$

and

$$b_1^* = \frac{2}{3}, \quad b_2^* = 1, \quad b_3^* = \frac{1}{3}, \quad b_4^* = 0.$$

The solution, an interval-valued counterpart of Q-consensus winner, is thus

$$W_Q = \left[\frac{1}{15}, \frac{11}{15}\right]/s_1 + [1, 1]/s_2 + \left[0, \frac{1}{15}\right]/s_3.$$

(b) Put $f^1 = f_{7,\,0.5}$. Now

$$b_1 = \tfrac{1}{3},\ \ b_2 = \tfrac{2}{3},\ \ b_3 = 0,\ \ b_4 = 0$$

and

$$b_1^* = \tfrac{2}{3},\ \ b_2^* = \tfrac{2}{3},\ \ b_3^* = \tfrac{1}{3},\ \ b_4^* = 0,$$

i.e. the interval-valued version of s/Q-consensus winner has the form

$$W_Q = \left[\tfrac{1}{15}, \tfrac{11}{15}\right]/s_1 + \left[\tfrac{11}{15}, \tfrac{11}{15}\right]/s_2 + \left[0, \tfrac{1}{15}\right]/s_3. \qquad \square$$

10.6 Bibliographical References to Applications

This final section presents selected bibliographical references involving other applications of intelligent counting in various areas of intelligent systems and decision support. For completeness, the list also includes the areas and references which were already mentioned in the previous sections.

- Computational approach to linguistic quantifiers:

  DELGADO *et al.* (2000), DUBOIS/PRADE (1985, 1990b), GLŐCKNER (2004, 2006), KACPRZYK (1997), LIU/KERRE (1998a, b), RALESCU (1995), ZADEH (1983a).

- Similarity measures:

  DE BAETS *et al.* (2002, 2006), DE BAETS/DE MEYER (2005a, b).

- Flexible querying in databases, information retrieval:

  BOSC/PIVERT (1991, 1994a, b), JAWORSKA *et al.* (2010), KACPRZYK/ZADROŻNY (1997, 2001b), LIETARD/ROCACHER (2007), ROCACHER (2003), ROCACHER/BOSC (2005), WU *et al.* (1994), ZADROŻNY/KACPRZYK (1996).

- Discovery of association rules and sequential patterns:

  DUBOIS/PRADE *et al.* (2005), DYCZKOWSKI (2007), HONG/CHEN *et al.* (2006), HONG/LIN *et al.* (2002, 2006).

- Linguistic summarization of databases:

  BOSC *et al.* (2002), KACPRZYK/YAGER (2001), KACPRZYK/ZADROŻNY (2001a, 2005a, b, 2010b), NIEWIADOMSKI (2008), PILARSKI (2010), YAGER *et al.* (1991).

- Linguistic summarization of time series:

  KACPRZYK/WILBIK (2010), KACPRZYK/WILBIK *et al.* (2006, 2008).

- Text categorization:

  PILARSKI (2010), ZADROŻNY/KACPRZYK (2003, 2006).

- Control and decision making:

  FELIX (1994, 2006), KACPRZYK (1992, 1997), KACPRZYK/YAGER (1984).

- Group decision making:

  KACPRZYK (1986), PANKOWSKA (2005, 2007, 2008), PANKOWSKA/WYGRALAK (2005, 2006), SZMIDT/KACPRZYK (1998).

- Consensus reaching:

  HERRERA *et al.* (1996), KACPRZYK/NURMI *et al.* (1999), KACPRZYK/ZADROŻNY (2010a).

- Expert systems:

  SILER/BUCKLEY (2005), ZADEH (1983c).

- Inductive learning:

  KACPRZYK/IWAŃSKI (1991).

- Recommender systems, social networks and trust propagation:

  DYCZKOWSKI/STACHOWIAK (2012), STACHOWIAK (2009, 2010).

- Demand forecast:

  KACZMAREK (2010).

- Miscellanea

  Formal concept analysis: CERAVOLO *et al.* (2006),
  Banknote validation: IONESCU/RALESCU (2005),
  Probabilities of fuzzy events: CASASNOVAS/ROSELLÓ (2002).

Chapter 11

Cardinalities of Infinite Fuzzy Sets

For completeness, this closing chapter presents a short study of cardinalities of arbitrary, generally infinite fuzzy sets. Especially interesting in that case seem to be fuzzy cardinalities. We like to extend FGCounts, FLCounts, and FECounts to infinite fuzzy sets. This will carry new threads and properties and, consequently, will give the reader a deeper insight into the issue of cardinality under information imprecision. Among other things, questions related to the Continuum Hypothesis and axiomatic approaches to fuzzy cardinalities will be discussed. Finally, cardinalities of infinite interval-valued fuzzy sets and I-fuzzy sets will be defined, too.

11.1 Notation and Introductory Remarks

Throughout this chapter, let $\boldsymbol{U}$ be an infinite universe. The family of all fuzzy sets in $\boldsymbol{U}$, i.e. of all functions $\boldsymbol{U} \to [0, 1]$, will be denoted by FS. We will use finite cardinal numbers (nonnegative integers, in other words) as well as transfinite ones. Both of them are denoted in our study in the same way by means of small letters i, j, k, m, n. Additional special symbols, $\aleph_0$ and $\mathfrak{C}$, denote the two most important transfinite cardinals: $\aleph_0 = |\mathbb{N}|$ and $\mathfrak{C} = |\mathbb{R}|$. Moreover, CN denotes the set of all cardinal numbers less than or equal to $|\boldsymbol{U}|$, i.e. $\mathrm{CN} = \{k\colon k \le |\boldsymbol{U}|\}$.

The definitions of m, n and $[A]_k$ from Section 9.1 can be easily extended to infinite fuzzy sets. So, we still put

$$m = |\mathrm{core}(A)|, \quad n = |\mathrm{supp}(A)|$$

and (11.1)

$$[A]_k = \bigvee\{t \in (0,1]\colon |A_t| \ge k\} \text{ with } k \in \mathrm{CN}$$

for $A \in \mathrm{FS}$. This time, however, $\bigvee$ cannot be generally replaced with max. It is easy to check that the following properties are satisfied:

- $[A]_k$ is nonincreasing with respect to k, and $[A]_k \le [B]_k$ for each $k \in \mathrm{CN}$ whenever $A \subset B$,
- $[A]_k = 1$ for each $k \le m$,
- $[A]_k = 0$ for each $k > n$.

$[A]_k$ collapses to the kth greatest membership degree in A whenever $A \in \mathrm{FFS}$.

M. Wygralak: *Intelligent Counting Under Information Imprecision*, STUDFUZZ 292, pp. 261–270.
DOI: 10.1007/978-3-642-34685-9_11

As to scalar cardinalities of fuzzy sets from FS, it seems that only $|A_t|$ and $|A^t|$, the results of counting by thresholding (sharp or not), can be extended to infinite fuzzy sets without any difficulties and additional preassumptions. The corresponding cardinality of an IVFS $\mathcal{E} = (A_l, A_u)$ with $A_l \subset A_u$ is then

$$|\mathcal{E}| = [\,|(A_l)_t|,\ |(A_u)_t|\,] \quad \text{with } t \in (0, 1]$$

or

$$|\mathcal{E}| = [\,|(A_l)^t|,\ |(A_u)^t|\,] \quad \text{with } t \in [0, 1). \tag{11.2}$$

Dealing with an IFS $\mathscr{E} = (A^+, A^-)$ with $A^+ \subset (A^-)^\nu$ and a strong negation ν, we define

$$|\mathscr{E}| = [\,|(A^+)_t|,\ |((A^-)^\nu)_t|\,]$$

or

$$|\mathscr{E}| = [\,|(A^+)^t|,\ |((A^-)^\nu)^t|\,]. \tag{11.3}$$

Extensions of fuzzy cardinalities to infinite fuzzy sets will be discussed in the next section.

11.2 Fuzzy Cardinalities of Infinite Fuzzy Sets

FGCounts, FLCounts and FECounts of finite fuzzy sets from Section 9 have very natural extensions to $A \in \mathrm{FS}$. FG, FL, FE: FS $\to [0, 1]^{\mathrm{CN}}$ are then defined as follows:

$$\mathrm{FG}(A)(k) = [A]_k$$

and

$$\mathrm{FL}(A)(k) = 1 - [A]_{k^+} \tag{11.4}$$

for each $k \in \mathrm{CN}$, where k^+ denotes the *successor of k*. Speaking precisely, k^+ is the smallest cardinal being greater than k. If the Generalized Continuum Hypothesis is accepted, we thus get $k^+ = 2^k$ for each transfinite k (see Section 11.3). If k is finite, then simply $k^+ = k + 1$. There exist cardinal numbers which are not successors of other cardinals, e.g. 0 and $\aleph_0$. One calls them *limit cardinal numbers*. Further, we define

$$\mathrm{FE}(A) = \mathrm{FG}(A) \cap \mathrm{FL}(A)$$

i.e.

$$\mathrm{FE}(A)(k) = [A]_k \wedge (1 - [A]_{k^+}) \quad \text{for } k \in \mathrm{CN}. \tag{11.5}$$

FG(A), FL(A) and FE(A) are convex fuzzy sets in CN. A comprehensive study of FG-, FL- and FECounts of infinite fuzzy sets can be found in WYGRALAK (1996a); see also WYGRALAK (1991a, 1992, 1993a, b, c, 1994, 1995, 1996b). FGCounts are also investigated in ŠOSTAK (1989). Similarly to Chapter 9, FG(A)(k), FL(A)(k), and FE(A)(k), respectively, form a degree to which $A \in$ FS has at least, at most, and exactly k elements, respectively.

Example 11.1. Put $\boldsymbol{U} = [0, \infty)$ and define $A \in$ FS in the following way:

$$A(x) = \begin{cases} 1, & \text{if } x = 1, 2, 3, \\ 0.8, & \text{if } 0 \le x < 3 \text{ is rational and } x \ne 1, 2, \\ 0.4, & \text{if } x > 3, \\ 0, & \text{otherwise.} \end{cases}$$

Then CN $= \{k\colon k \le \mathfrak{C}\}$, $[A]_0 = [A]_1 = [A]_2 = [A]_3 = 1$, $[A]_k = 0.8$ for $3 < k \le \aleph_0$, and $[A]_k = 0.4$ for $k = \mathfrak{C}$. Let us extend the enriched vector notation used in (9.24) to fuzzy cardinalities of some infinite fuzzy sets. We will write

$$\alpha = (a_0, a_1, \ldots, a_k, (a) \mid b_1, b_2), \tag{11.6}$$

which means that

$$\alpha(i) = a_i \text{ for each } i \le k \ (k \text{ finite}),$$
$$\alpha(i) = a \text{ for each finite } i > k,$$
$$\alpha(\aleph_0) = b_1, \ \alpha(\mathfrak{C}) = b_2$$

whenever the Continuum Hypothesis is accepted (see Section 11.3). By (11.4)-(11.5), one thus gets

$$\mathrm{FG}(A) = (1, 1, 1, 1, (0.8) \mid 0.8, 0.4),$$
$$\mathrm{FL}(A) = (0, 0, 0, (0.2) \mid 0.6, 1),$$
$$\mathrm{FE}(A) = (0, 0, 0, (0.2) \mid 0.6, 0.4).$$

□

In our further discussion, $|A|$ will denote the FGCount, FLCount or FECount of $A \in$ FS. If $|A|(k) = |B|(k)$ for each $k \in$ CN, we will write $|A| = |B|$ and say that A and B are *equipotent* fuzzy sets. That A and B are *non-equipotent* ($|A| \ne |B|$) thus means that $|A|(k) \ne |B|(k)$ for some $k \in$ CN. The following characterization holds true:

$$|A| = |B| \ \Leftrightarrow \ \forall k \in \mathrm{CN}\colon [A]_k = [B]_k \ \Leftrightarrow \ \forall t \in [0, 1)\colon |A^t| = |B^t|. \tag{11.7}$$

Using infinite fuzzy sets, however, sharp t-cuts in these equivalences cannot be replaced with t-cuts (cf. (9.84)).

Example 11.2. A simple counterexample is $\boldsymbol{U} = [0, 1]$ with

$$A(x) = 1 - x \text{ and } B(x) = 1 \text{ for each } x \in \boldsymbol{U}.$$

We then have $\mathrm{CN} = \{k: k \leq \mathfrak{C}\}$,

$$[A]_k = [B]_k = 1 \text{ for each } k \in \mathrm{CN}$$

and

$$|A^t| = |B^t| = \mathfrak{C} \text{ for each } t \in [0, 1),$$

which means that $|A| = |B|$, whereas $|A_1| = 1$ and $|B_1| = \mathfrak{C}$. Moreover, worth emphasizing in this example is also that A and B are equipotent although they look very different. This difference is, however, only apparent: A has uncountably many elements whose membership degrees lie as near to 1 as one likes. □

As to inequalities between fuzzy cardinalities of $A, B \in \mathrm{FS}$, one defines them by

$$|A| \leq |B| \Leftrightarrow \forall k \in \mathrm{CN}: [A]_k \leq [B]_k$$

and

$$|A| < |B| \Leftrightarrow |A| \leq |B| \;\&\; |A| \neq |B|. \tag{11.8}$$

We then say that the (fuzzy) cardinality of A is *less than or equal to* the cardinality of B, and that the cardinality of A is *less than* the cardinality of B, respectively. $\leq$ is only a partial order relation. Fuzzy cardinalities with $\leq$, speaking more precisely, form a lattice structure. We still have

$$|A| < |B| \Rightarrow |A| \leq |B|,$$
$$|A| = |B| \Leftrightarrow |A| \leq |B| \;\&\; |B| \leq |A|, \tag{11.9}$$
$$A \subset B \Rightarrow |A| \leq |B|.$$

Moreover, the following equivalence is satisfied:

$$|A| \leq |B| \Leftrightarrow \forall t \in [0, 1): |A^t| \leq |B^t|. \tag{11.10}$$

However, we have

$$\exists B^* \subset B: |A| = |B^*| \Rightarrow |A| \leq |B|. \tag{11.11}$$

The inverse implication does not generally hold for $A, B \in \mathrm{FS}$, i.e. the definition of $\leq$ in (11.8) cannot be replaced with Definition 9.11. This forms another essential difference between cardinalities of sets and cardinalities of fuzzy sets.

Example 11.3. Let us consider the following counterexample with $\boldsymbol{U} = [0, 1]$ and $\mathrm{CN} = \{k: k \leq \mathfrak{C}\}$:

$$A(x) = \begin{cases} 1, & \text{if } x = 0, 0.1, \\ 0.2, & \text{if } 0.2 \leq x \leq 1, \\ 0, & \text{otherwise,} \end{cases} \qquad B(x) = 1 - x.$$

Then

$$[A]_k = 1 \text{ for each } k \leq 2,$$
$$[A]_k = 0.2 \text{ for each } 2 < k \in \text{CN},$$
$$[B]_k = 1 \text{ for each } k \in \text{CN},$$

i.e. $|A| \leq |B|$ and even $|A| < |B|$. However, trying to construct $B^* \subset B$ such that $|A| = |B^*|$, we should have

$$[B^*]_k = 1 \text{ for } k \leq 2, \text{ and } [B^*]_k = 0.2 \text{ for } k > 2.$$

This means that $B^*(x) = 1$ for two different x's, which excludes $B^* \subset B$.

Moreover, let us point out that replacing (11.8) with Definition 9.11 extended to fuzzy sets from FS we would get into trouble. Indeed, using A and B from this example together with C such that $C(x) = 1$ for each $x \in \boldsymbol{U}$, one has

$$|A| \leq |C| \text{ and } |C| = |B|, \text{ whereas } |A| \leq |B| \text{ does not hold,}$$

i.e. $\leq$ is then no longer transitive, which is difficult to accept. □

Similarly to finite fuzzy sets,

$$\text{FG}(A) \leq \text{FG}(B) \Leftrightarrow \text{FG}(A) \subset \text{FG}(B),$$
$$\text{FL}(A) \leq \text{FL}(B) \Leftrightarrow \text{FL}(B) \subset \text{FL}(A), \tag{11.12}$$
$$\text{FE}(A) \leq \text{FE}(B) \Leftrightarrow \text{FG}(A) \leq \text{FG}(B) \Leftrightarrow \text{FL}(A) \leq \text{FL}(B)$$

for $A, B \in \text{FS}$.

Finally, it is possible to introduce graded (or approximated) equipotencies and inequalities involving arbitrary fuzzy sets. This can be done in various ways (see WYGRALAK (1996a); cf. Subsection 5.1.6). For instance, one defines

$$|A| =^t |B| \Leftrightarrow \forall k \in \text{CN}: |[A]_k - [B]_k| \leq 1 - t \tag{11.13}$$

for $t \in (0, 1]$ and $A, B \in \text{FS}$. If $|A| =^t |B|$, we say that A and B are *equipotent to degree t*.

11.3 References to the Continuum Hypothesis

In Subsection 9.4.3, $\boldsymbol{k}$ was used to denote the fuzzy cardinality of a finite, k-element set viewed as a fuzzy set. Here we like to extend that notation to infinite sets. $\boldsymbol{k}$ will denote the FGCount, FLCount, or FECount of a set of cardinality $k \in \text{CN}$, a counterpart of k in the world of fuzzy cardinalities. So,

$$\boldsymbol{k} = 1_{\{i \in \mathrm{CN}:\, i \le k\}} \text{ for FGCounts,}$$
$$\boldsymbol{k} = 1_{\{i \in \mathrm{CN}:\, i \ge k\}} \text{ for FLCounts,} \tag{11.14}$$
$$\boldsymbol{k} = 1_{\{k\}} \text{ for FECounts.}$$

Clearly, for each $k \in \mathbb{N}$, there exists an intermediate fuzzy cardinality α being the FGCount, FLCount, or FECount of a fuzzy set from FFS such that

$$\boldsymbol{k} < \alpha < \boldsymbol{k}+\mathbf{1}.$$

An instance is

$$\boldsymbol{k} = |1/x_1 + \ldots + 1/x_k|,$$
$$\boldsymbol{k}+\mathbf{1} = |1/x_1 + \ldots + 1/x_k + 1/x_{k+1}|$$

and

$$\alpha = |1/x_1 + \ldots + 1/x_k + 0.5/x_{k+1}|.$$

Let us move on to transfinite cardinal numbers. Recollect that the Generalized Continuum Hypothesis (GCH) is a sentence which is independent on the axioms of set theory and says that $k^+ = 2^k$ for each transfinite k. In particular, the sentence saying that $\aleph_0^{\ +} = \mathfrak{C}$ is called the Continuum Hypothesis (CH). GCH thus states that, for each transfinite k, a set of cardinality greater than k and less than 2^k does not exist and, hence, there is no intermediate cardinal number i such that $k < i < 2^k$. As to CH, a set of cardinality k with $\aleph_0 < k < \mathfrak{C}$ does not exist. Consider in this context the following example involving fuzzy sets. Let $\boldsymbol{U} = [0, \infty)$, $\mathrm{CN} = \{k\colon k \le \mathfrak{C}\}$, $B = 1_{\mathbb{N}}$, $D = 1_{[0,\infty)}$, and

$$C(x) = \begin{cases} \mathbf{1}, & \textbf{if } \boldsymbol{x} \in \mathbf{N}, \\ \mathbf{0.5}, & \textbf{if } \boldsymbol{x} \in \boldsymbol{U} \setminus \mathbf{N}. \end{cases}$$

Since $B \subset C \subset D$, (11.9) implies $|B| \le |C| \le |D|$. More precisely,

$$[B]_k = \begin{cases} \mathbf{1}, & \textbf{if } \boldsymbol{k} \le \aleph_0, \\ \mathbf{0}, & \textbf{otherwise}, \end{cases} \qquad [D]_k = 1 \text{ for each } k \in \mathrm{CN},$$

$$[C]_k = \begin{cases} \mathbf{1}, & \textbf{if } \boldsymbol{k} \le \aleph_0, \\ \mathbf{0.5}, & \textbf{if } \aleph_0 < \boldsymbol{k} \le \mathfrak{C}. \end{cases}$$

Hence

$$\aleph_0 = |B| < |C| < |D| = \mathfrak{C}.$$

With reference to CH, we thus point out that, anyway, there exists a fuzzy set whose cardinality is greater than $\aleph_0$ and less than $\mathfrak{C}$. Analogous examples and conclusions can be easily given for GCH. This builds a strong feeling that all mathematical discussions and questions arisen around independency and acceptability of CH and GCH are essentially conditioned by the two-valued character of classical logic used by mathematical theories. Adding even one new logical value, 0.5, intermediate cardinalities start to exist and are quite easy to find. A more detailed treatment of this issue is presented in WYGRALAK (1993a, 1996a); see also LI *et al.* (1993).

11.4 Arithmetic Operations

Let α and β denote two FGCounts, FLCounts, or FECounts of fuzzy sets from FS. The *sum* $\alpha+\beta$ and *product* $\alpha\beta$ of α and β are then defined as follows:

- $\alpha+\beta = |A\cup B|,$ (11.15)

  where $A, B \in \mathrm{FS}$ are arbitrary disjoint fuzzy sets such that $\alpha = |A|$ and $\beta = |B|$;
- $\alpha\beta = |A\times B|$ (11.16)

  with arbitrary $A, B\in \mathrm{FS}$ such that $\alpha = |A|$ and $\beta = |B|$.

They can be equivalently rewritten by means of the extension principle. Then

$$(\alpha+\beta)(k) = \bigvee\{\alpha(i)\wedge\beta(j)\colon i+j=k\} \quad \text{for each } k\in \mathrm{CN} \tag{11.17}$$

whenever α and β are two FGCounts or FLCounts. If α and β are FECounts, we get

$$\begin{aligned}(\alpha+\beta)(k) = \bigvee\{\mathrm{FG}(A)(i)\wedge \mathrm{FG}(B)(j)\colon i+j=k\} \wedge \\ \bigvee\{\mathrm{FL}(A)(i)\wedge \mathrm{FL}(B)(j)\colon i+j=k\}\end{aligned} \tag{11.18}$$

for $k\in\mathrm{CN}$ provided that A and B are disjoint and such that $\alpha = |A|$ and $\beta = |B|$. As to multiplication, one has (see also (9.116)-(9.118))

$$\text{for FGCounts:} \quad (\alpha\beta)(k) = \bigvee\{\alpha(i)\wedge\beta(j)\colon ij\geq k\} \quad \text{for each } k\in\mathrm{CN}, \tag{11.19}$$

$$\text{for FLCounts:} \quad (\alpha\beta)(k) = \bigvee\{\alpha(i)\wedge\beta(j)\colon ij\leq k\} \quad \text{for each } k\in\mathrm{CN}, \tag{11.20}$$

$$\begin{aligned}\text{for FECounts:} \quad (\alpha\beta)(k) = \bigvee\{\mathrm{FG}(A)(i)\wedge \mathrm{FG}(B)(j)\colon ij\geq k\} \wedge \\ \bigvee\{\mathrm{FL}(A)(i)\wedge \mathrm{FL}(B)(j)\colon ij\leq k\} \quad \text{for each } k\in\mathrm{CN}\end{aligned} \tag{11.21}$$

whenever $A, B\in\mathrm{FS}$ are such that $\alpha = |A|$ and $\beta = |B|$.

Many of the properties formulated in (9.120)-(9.129) can be transferred to FG-, FL-, and FECounts of infinite fuzzy sets. In particular, this is the case of (9.120), (9.121), (9.123), (9.124), and of the valuation property $|A\cap B| + |A\cup B| = |A| + |B|$. Some properties of fuzzy cardinalities of infinite fuzzy sets are peculiar to those fuzzy cardinalities, e.g.

$$\text{if } \alpha, \beta \geq \mathbf{1} \text{ and } \alpha\geq\aleph_0 \text{ or } \beta\geq\aleph_0, \text{ then } \alpha+\beta = \alpha\beta. \tag{11.22}$$

A detailed study, proofs and many other threads concerning FGCouts, FLCounts and FECounts of infinite fuzzy sets, e.g. properties of their infinite sums and products, can be found in WYGRALAK (1996a); see also WYGRALAK(1992, 1993a, b, c, 1994, 1995, 1996b). The reader is also referred to LI *et al.* (1994) presenting a cardinality theory which seems to collapse to a theory of FGCounts, and in which the cardinality of A from FS is defined as the equivalence class of all fuzzy sets being equipotent to A.

11.5 Axiomatic Approaches to Fuzzy Cardinalities

One has to mention that what WYGRALAK (1996a) actually offers is an axiomatic approach to fuzzy cardinalities. We like to present some elements of that approach looking at them in a bit different way tailored to the needs of this book.

First, a fuzzy set $A \in \mathrm{FS}$, possibly incompletely known, is then represented by means of a pair $(f(A), g(A))$ called a *VD-object* (*vaguely defined object*), where f, g: FS → FS are such that

(A1) $f(A) \subset A \subset g(A)$,

(A2) $A \in \{0,1\}^U \Rightarrow f(A), g(A) \in \{0,1\}^U$,

(A3) $A(x) \le B(y) \Rightarrow f(A)(x) \le f(B)(y)$ & $g(A)(x) \le g(B)(y)$,

(A4) $(f, g) \neq (id, id) \Rightarrow f(\mathrm{FS}) \subset \{0,1\}^U$ # $g(\mathrm{FS}) \subset \{0,1\}^U$

for each $A, B \in \mathrm{FS}$ and $x, y \in \boldsymbol{U}$. As previously, *id* denotes the identity function. Let us explain some motivations which are behind these four axioms.

By (A1), $(f(A), g(A))$ is an IVFS (see Section 6.2). However, contrary to general IVFSs, the bounds $f(A)$ (lower) and $g(A)$ (upper) on A cannot be quite arbitrary. (A2) says that both $f(A)$ and $g(A)$ must be sets whenever A is a set. (A3) indicates that the bounds should be results of monotonic transformations of membership grades. By the way, (A3) is stronger than the usual monotonicity condition

$$A \subset B \Rightarrow f(A) \subset f(B) \;\&\; g(A) \subset g(B).$$

As to (A4), it is obvious that $(f, g) = (id, id)$ fulfils (A1)-(A3) and corresponds to the case of a completely known A. Otherwise, we assume that our ignorance about A is more or less total in the following sense:

- $A(x) \ge 0$ is the only unquestionable lower bound for $A(x)$, possibly excluding the $A(x)$'s equal to 1, if they are assumed to be precise,

or/and

- $A(x) \le 1$ is the only unquestionable upper bound for $A(x)$, possibly excluding the $A(x)$'s equal to 0, if they are assumed to be precise.

We thus postulate that at least one of the bounds $f(A)$ and $g(A)$ has to be a set. And this is guaranteed just by (A4).

Put $\mathrm{f1}(A) = 1_{\mathrm{core}(A)}$ and $\mathrm{gs}(A) = 1_{\mathrm{supp}(A)}$. Let us notice that if $(f, g) \neq (id, id)$, then $f = 1_{\emptyset}$ or $f = \mathrm{f1}$ or/and $g = 1_U$ or $g = \mathrm{gs}$. The following eight pairs (f, g) are therefore particularly important:

$$(id, id),$$
$$(\mathrm{f1}, id), (1_{\emptyset}, id),$$
$$(id, \mathrm{gs}), (id, 1_U),$$
$$(1_{\emptyset}, \mathrm{gs}), (\mathrm{f1}, 1_U), (\mathrm{f1}, \mathrm{gs}).$$

If $(f, g) = (1_{\emptyset}, \mathrm{gs}), (\mathrm{f1}, 1_U), (\mathrm{f1}, \mathrm{gs})$, then $(f(A), g(A))$ becomes a partial set (see Section 6.1). For $(f, g) \neq (id, id)$, we have $f(A)(x) = 0$ or/and $g(A)(x) = 1$, i.e. $(f(A), g(A))$ is a twofold fuzzy set (see (6.15)).

The fuzzy cardinality $\alpha = |A|_{f,g}$ of $A \in \mathrm{FS}$ represented by $(f(A), g(A))$ with (f, g) satisfying (A1)-(A4) is defined as a convex fuzzy set in CN such that

$$\alpha(k) = [g(A)]_k \wedge (1 - [f(A)]_{k^+}) \quad \text{for each } k \in \mathrm{CN}. \tag{11.23}$$

We then have ($\beta = |B|_{f,g}$)

$$\alpha = \beta \Leftrightarrow \forall k \in \mathrm{CN}\colon [f(A)]_k = [f(B)]_k \;\&\; [g(A)]_k = [g(B)]_k. \tag{11.24}$$

It is easy to point out that

$$\alpha = \mathrm{FG}(A) \quad \text{for } (f, g) = (1_{\emptyset}, id),$$
$$\alpha = \mathrm{FL}(A) \quad \text{for } (f, g) = (id, 1_U),$$
$$\alpha = \mathrm{FE}(A) \quad \text{for } (f, g) = (id, id).$$

Moreover, if $(f, g) = (\mathrm{f1}, id)$, α becomes the fuzzy cardinality from (9.23) extended to arbitrary fuzzy sets. For $(f, g) = (\mathrm{f1}, 1_U), (1_{\emptyset}, \mathrm{gs}), (\mathrm{f1}, \mathrm{gs})$, α collapses to an interval cardinality defined in KLAUA (1969) for partial sets, namely

$$\alpha = 1_{\{k \in \mathrm{CN}:\, k \geq m\}} \quad \text{if } (f, g) = (\mathrm{f1}, 1_U),$$
$$\alpha = 1_{\{k \in \mathrm{CN}:\, k \leq n\}} \quad \text{if } (f, g) = (1_{\emptyset}, \mathrm{gs}),$$
$$\alpha = 1_{\{k \in \mathrm{CN}:\, m \leq k \leq n\}} \quad \text{if } (f, g) = (\mathrm{f1}, \mathrm{gs}).$$

An exhaustive study of fuzzy cardinalities from (10.23), including FGCounts, FLCounts and FECounts as particular cases, is contained in WYGRALAK (1996a). It also presents a cardinality theory of incompletely known fuzzy sets $A \in \mathrm{FS}$ represented via pairs $(f(A), g(A))$, where $f, g\colon \mathrm{FS} \to \mathrm{FS}$ are transformations satisfying just one general postulate: $f(A)$ must be contained in the core of $g(A)$ (see (6.15); see also WYGRALAK (1998a)). Finally, we refer the interested reader to yet another axiomatic approach to fuzzy cardinalities proposed in CASASNOVAS/TORRENS (2003a) and oriented to finite fuzzy sets (see also HOLČAPEK (2010)).

11.6 Using IVFSs and IFSs

In this closing section we like to introduce extensions of FGCounts, FLCounts and FECounts of arbitrary fuzzy sets to the case of IVFSs and IFSs (cf. Subsection 9.3.2; cf. also KRÁL (2006)).

Let $\mathfrak{E} = (A_l, A_u)$ with $A_l \subset A_u$ be an arbitrary IVFS in $\boldsymbol{U}$. The *FGCount of* $\mathfrak{E}$ is then defined as an IVFS

$$\mathrm{FG}(\mathfrak{E}) = (\mathrm{FG}(A_l), \mathrm{FG}(A_u)) \tag{11.25}$$

in CN, whereas the *FLCount of* $\mathfrak{E}$ is of the form

$$\mathrm{FL}(\mathfrak{E}) = (\mathrm{FL}(A_u), \mathrm{FL}(A_l)) \tag{11.26}$$

with the components specified via (11.4). Further, the *FECount of* $\mathfrak{E}$ is understood as

$$\mathrm{FE}(\mathfrak{E}) = \mathrm{FG}(\mathfrak{E}) \cap \mathrm{FL}(\mathfrak{E}) = (\mathrm{FG}(A_l) \cap \mathrm{FL}(A_u), \mathrm{FG}(A_u) \cap \mathrm{FL}(A_l)). \tag{11.27}$$

Moving on to the corresponding cardinalities of an arbitrary IFS $\mathscr{E} = (A^+, A^-)$ with $A^+ \subset (A^-)'$, we define them as the following IVFSs in CN:

$$\begin{aligned} \mathrm{FG}(\mathscr{E}) &= (\mathrm{FG}(A^+), \mathrm{FG}((A^-)')), \\ \mathrm{FL}(\mathscr{E}) &= (\mathrm{FL}((A^-)'), \mathrm{FL}(A^+)), \\ \mathrm{FE}(\mathscr{E}) = \mathrm{FG}(\mathscr{E}) \cap \mathrm{FL}(\mathscr{E}) &= (\mathrm{FG}(A^+) \cap \mathrm{FL}((A^-)'), \mathrm{FG}((A^-)') \cap \mathrm{FL}(A^+)). \end{aligned} \tag{11.28}$$

This means that

$$\begin{aligned} \mathrm{FG}(\mathscr{E})(k) &= [[A^+]_k, [(A^-)']_k], \\ \mathrm{FL}(\mathscr{E})(k) &= [1 - [(A^-)']_{k^+}, 1 - [A^+]_{k^+}], \\ \mathrm{FE}(\mathscr{E})(k) &= [[A^+]_k \wedge (1 - [(A^-)']_{k^+}), [(A^-)']_k \wedge (1 - [A^+]_{k^+})] \end{aligned} \tag{11.29}$$

for each cardinal number $k \in \mathrm{CN}$.

References

Ahn, E.-Y., Kim, J.-W., Kwak, N.-Y., Han, S.-H.: Emotion-based crowd simulation using fuzzy algorithm. In: Zhang, S., Jarvis, R.A. (eds.) AI 2005. LNCS (LNAI), vol. 3809, pp. 330–338. Springer, Heidelberg (2005)

Ajdukiewicz, K.: On the problem of universals. Przegląd Filozoficzny 38, 219–234 (1935)

Arieli, O., Cornelis, C., Deschrijver, G., Kerre, E.E.: Relating intuitionistic fuzzy sets and interval-valued fuzzy sets through bilattices. In: Ruan, D., D'Hondt, P., De Cock, M., Nachtegael, M., Kerre, E.E. (eds.) Applied Computational Intelligence, pp. 57–64. World Scientific, Singapore (2004)

Atanassov, K.T.: Intuitionistic fuzzy sets. Fuzzy Sets and Systems 20, 87–96 (1986)

Atanassov, K.T.: Intuitionistic Fuzzy Sets. Theory and Applications. Physica-Verlag, Heidelberg (1999)

Atanassov, K.T.: Intuitionistic fuzzy sets – Past, present and future. Proc. 3rd Conf. European Society for Fuzzy Logic and Technology (EUSFLAT'2003), Zittau, Germany, pp. 12-19 (2003)

Atanassov, K.T., Stoeva, S.: Intuitionistic fuzzy sets. In: Proc. Polish Symp. Interval and Fuzzy Mathematics, Poznań, Poland, pp. 23-26 (1983)

Baczyński, M., Jayaram, B.: Fuzzy Implications. Springer, Berlin (2008)

Barrenechea, E., Bustince, H., Pagola, M., Fernandez, J., Sanz, J.: Generalized Atanassov's intuitionistic fuzzy index. Construction methods. In: Proc. 2009 IFSA World Congress and 2009 EUSFLAT Conf., Lisbon, Portugal, pp. 478–482 (2009)

Batyrshin, I., Kacprzyk, J., Sheremetov, L., Zadeh, L.A. (eds.): Perception-Based Data Mining and Decision Making in Economics and Finance. Springer, Heidelberg (2007)

Beliakov, G., Pradera, A., Calvo, T.: Aggregation Functions: A Guide for Practitioners. Springer, Berlin (2007)

Bellman, R.E., Zadeh, L.A.: Decision making in a fuzzy environment. Management Science 17, 141–164 (1970)

Belluce, L.P., Di Nola, A., Sessa, S.: Triangular norms, MV-algebras and bold fuzzy set theory. Math. Japonica 36, 481–487 (1991)

Black, M.: Vagueness: An exercise in logical analysis. Philosophy of Science 4, 427–455 (1934)

Black, M.: Reasoning with loose concepts. Dialogue 2, 1–12 (1963)

Blanchard, N.: Theories Cardinale et Ordinale des Ensembles Flous. PhD Thesis, Université Claude-Bernard, Lyon, France (1981)

Blanchard, N.: Cardinal and ordinal theories about fuzzy sets. In: Gupta, M.M., Sanchez, E. (eds.) Fuzzy Information and Decision Processes, pp. 149–157. North-Holland, Amsterdam (1982)

Boixander, D., Jacas, J., Recasens, J.: Fuzzy equivalence relations: Advanced material. In: Dubois, D., Prade, H. (eds.) Fundamentals of Fuzzy Sets, pp. 261–290. Kluwer, Dordrecht (2000)

Borkowski, L. (ed.): Jan Łukasiewicz – Selected Works, Jan. 1970. North-Holland, Amsterdam and Polish Scientific Publishers, Warsaw (1970)

Bosc, P., Dubois, D., Pivert, O., Prade, H., De Calmès, M.: Fuzzy summarization of data using fuzzy cardinalities. In: Proc. 9th Inter. Conf. on Information Processing and Management of Uncertainty in Knowledge-Based Systems (IPMU2002), Annecy, France, pp. 1553–1559 (2002)

Bosc, P., Kacprzyk, J. (eds.): Fuzziness in Database Management Systems. Physica-Verlag, Heidelberg (1995)

Bosc, P., Kraft, D., Petry, F.: Fuzzy sets in database and information systems: Status and opportunities. Fuzzy Sets and Systems 156, 418–426 (2005)

Bosc, P., Pivert, O.: Some algorithms for evaluating fuzzy relational queries. In: Bouchon-Meunier, B., Zadeh, L.A., Yager, R.R. (eds.) IPMU 1990. LNCS, vol. 521, pp. 431–442. Springer, Heidelberg (1991)

Bosc, P., Pivert, O.: Imprecise data management and flexible querying in databases. In: Yager, R.R., Zadeh, L.A. (eds.) Fuzzy Sets, Neural Networks, and Soft Computing, pp. 368–395. Van Nostrand Reinhold, New York (1994a)

Bosc, P., Pivert, O.: Fuzzy queries and relational databases. In: Proc. ACM Symp. on Applied Computing, Phoenix, pp. 170–174 (1994b)

Bosteels, K., Kerre, E.E.: On a reflexivity-preserving family of cardinality-based fuzzy comparison measures. Information Sciences 179, 2342–2352 (2009)

Box, G., Jenkins, G.: Time Series Analysis: Forecasting and Control. Prentice Hall, Upper Saddle River (1994)

Burillo, P., Frago, N., Fuentes, R.: Inclusion grade and fuzzy implication operator. Fuzzy Sets and Systems 114, 417–429 (2000)

Bustince, H.: Indicator of inclusion grade for interval-valued fuzzy sets. Application to approximate reasoning based on interval-valued fuzzy sets. Inter. Jour. of Approximate Reasoning 23, 137–209 (2000)

Bustince, H., Mohedano, V., Barrenechea, E., Pagola, M.: Image thresholding using intuitionistic fuzzy sets. In: Atanassov, K.T., Kacprzyk, J., Krawczak, M., Szmidt, E. (eds.) Issues in the Representation and Processing of Uncertain and Imprecise Information, pp. 72–82. EXIT Publ., Warsaw (2005)

Bustince, H., Montero, J., Pagola, M., Barrenechea, E., Gómez, D.: A survey of interval-valued fuzzy sets. In: Pedrycz, W., Skowron, A., Kreinovich, V. (eds.) Handbook of Granular Computing, pp. 491–516. Wiley, New York (2008)

Bustince, H., Pagola, M., Barrenechea, E., Fernandez, J., Melo-Pinto, P., Couto, P., Tizhoosh, H.R., Montero, J.: Ignorance functions. An application to the calculation of the threshold in prostate ultrasound images. Fuzzy Sets and Systems 161, 20–36 (2010)

Calvo, T., Mayor, G., Mesiar, R. (eds.): Aggregation Operators. New Trends and Applications. Physica-Verlag, Heidelberg (2002)

Casasnovas, J., Roselló, F.: Probabilities of fuzzy events based on scalar cardinalities. In: Grzegorzewski, P., Hryniewicz, O., Gil, M.Á. (eds.) Soft Methods in Probability, Statistics and Data Analysis, pp. 92–97. Physica-Verlag, Heidelberg (2002)

Casasnovas, J., Torrens, J.: An axiomatic approach to fuzzy cardinalities of finite fuzzy sets. Fuzzy Sets and Systems 133, 193–209 (2003a)

Casasnovas, J., Torrens, J.: Scalar cardinalities of finite fuzzy sets for t-norms and t-conorms. Inter. Jour. of Uncertainty, Fuzziness and Knowledge-Based Systems 11, 599–614 (2003b)

Ceravolo, P., Damiani, E., Viviani, M.: Extending formal concept analysis by fuzzy bags. In: Proc. 11th Inter. Conf. on Information Processing and Management of Uncertainty in Knowledge-Based Systems (IPMU2006), Paris, pp. 2230–2237. Editions EDK, Paris (2006)

Chang, P.T., Lee, E.S.: Fuzzy arithmetic and comparison of fuzzy numbers. In: Delgado, M., Kacprzyk, J. (eds.) Fuzzy Optimization, pp. 69–82. Physica-Verlag, Heidelberg (1994)

Christiansen, H., Hacid, M.-S., Andreasen, T., Larsen, H.L. (eds.): FQAS 2004. LNCS (LNAI), vol. 3055. Springer, Heidelberg (2004)

Cornelis, C., Deschrijver, G., Kerre, E.E.: Implication in intuitionistic and interval-valued fuzzy set theory: construction, classification and application. Int. J. Approx. Reasoning 35, 55–95 (2004)

De Baets, B., De Meyer, H.: Transitivity-preserving fuzzification schemes for cardinality-based similarity measures. Europ. J. Oper. Research 160, 726–740 (2005a)

De Baets, B., De Meyer, H.: Transitivity frameworks for reciprocal relations: cycle-transitivity versus FG-transitivity. Fuzzy Sets and Systems 152, 249–270 (2005b)

De Baets, B., De Meyer, H., Naessens, H.: On rational cardinality-based inclusion measures. Fuzzy Sets and Systems 128, 169–183 (2002)

De Baets, B., Janssens, S., De Meyer, H.: Meta-theorems on inequalities for scalar fuzzy set cardinalities. Fuzzy Sets and Systems 157, 1463–1476 (2006)

De Cock, M., Pinheiro da Silva, P.: A many-valued representation and propagation of trust and distrust. In: Bloch, I., Petrosino, A., Tettamanzi, A.G.B. (eds.) WILF 2005. LNCS (LNAI), vol. 3849, pp. 114–120. Springer, Heidelberg (2006)

De Luca, A., Termini, S.: A definition of a non-probabilistic entropy in the setting of fuzzy sets theory. Inform. and Control 20, 301–312 (1972)

De Luca, A., Termini, S.: Entropy and energy measures of a fuzzy set. In: Gupta, M.M., Ragade, R.K., Yager, R.R. (eds.) Advances in Fuzzy Set Theory and Applications, pp. 321–338. North-Holland, Amsterdam (1979)

De Luca, A., Termini, S.: On some algebraic aspects of the measures of fuzziness. In: Gupta, M.M., Sanchez, E. (eds.) Fuzzy Information and Decision Processes, pp. 17–24. North-Holland, Amsterdam (1982)

Delgado, M., Sánchez, D., Vila, M.A.: Fuzzy cardinality based evaluation of quantified sentences. Int. J. Approx. Reasoning 23, 23–66 (2000)

Deschrijver, G.: Archimedean t-norms in interval-valued fuzzy set theory. In: Proc. 11th Inter. Conf. on Information Processing and Management of Uncertainty in Knowledge-Based Systems (IPMU2006), Paris, pp. 580–586. Editions EDK, Paris (2006a)

Deschrijver, G.: The Archimedean property for t-norms in interval-valued fuzzy set theory. Fuzzy Sets and Systems 157, 2311–2327 (2006b)

Deschrijver, G., Cornelis, C., Kerre, E.E.: On the representation of intuitionistic fuzzy t-norms and t-conorms. IEEE Trans. on Fuzzy Systems 12, 45–61 (2004)

Deschrijver, G., Kerre, E.E.: A generalization of operators on intuitionistic fuzzy sets using triangular norms and conorms. Notes on IFS 8, 19–27 (2002)

Deschrijver, G., Kerre, E.E.: On the relationship between some extensions of fuzzy set theory. Fuzzy Sets and Systems 133, 227–235 (2003)

Deschrijver, G., Král, P.: On the cardinalities of interval-valued fuzzy sets. Fuzzy Sets and Systems 158, 1728–1750 (2007)

Detyniecki, M.: Mathematical aggregation operators and their application to video querying. PhD Thesis, Univ. Paris VI (2000)

Diaz-Hermida, F., Bugarin, A., Barro, S.: Definition and classification of semi-fuzzy quantifiers for the evaluation of fuzzy quantified sentences. Int. J. Approx. Reasoning 34, 49–88 (2003)

Driankov, D., Hellendoorn, H., Reinfrank, M.: An Introduction to Fuzzy Control. Springer, Berlin (1993)

Dubois, D.: A new definition of the fuzzy cardinality of finite fuzzy sets preserving the classical additivity property. Bull Stud. Exch. on Fuzziness and its Appl (BUSEFAL) 8, 65–67 (1981)

Dubois, D.: On degrees of truth, partial ignorance and contradiction. In: Proc. 12th Inter. Conf. on Information Processing and Management of Uncertainty in Knowledge-Based Systems (IPMU2008), Malaga, Spain, pp. 31–38 (2008)

Dubois, D., Gottwald, S., Hájek, P., Kacprzyk, J., Prade, H.: Terminological difficulties in fuzzy set theory – The case of "Intuitionistic Fuzzy Sets". Fuzzy Sets and Systems 156, 485–491 (2005)

Dubois, D., Prade, H.: Operations on fuzzy numbers. Int. J. Systems Sci. 9, 613–626 (1978)

Dubois, D., Prade, H.: Fuzzy cardinality and the modeling of imprecise quantification. Fuzzy Sets and Systems 16, 199–230 (1985)

Dubois, D., Prade, H.: Twofold fuzzy sets and rough sets – Some issues in knowledge representation. Fuzzy Sets and Systems 23, 3–18 (1987a)

Dubois, D., Prade, H.: Fuzzy numbers: An overview. In: Bezdek, J.C. (ed.) Analysis of Fuzzy Information, vol. 1, pp. 3–39. CRC Press, Boca Raton (1987b)

Dubois, D., Prade, H.: Scalar evaluation of fuzzy sets: Overview and applications. Appl. Math. Lettr. 3, 37–42 (1990a)

Dubois, D., Prade, H.: Measuring properties of fuzzy sets: A general technique and its use in fuzzy query evaluation. Fuzzy Sets and Systems 38, 137–152 (1990b)

Dubois, D., Prade, H. (eds.): Fundamentals of Fuzzy Sets. Kluwer, Dordrecht (2000)

Dubois, D., Prade, H.: An overview of the asymmetric bipolar representation of positive and negative information in possibility theory. Fuzzy Sets and Systems 160, 1355–1366 (2009)

Dubois, D., Prade, H., Sudkamp, T.: On the representation, measurement and discovery of fuzzy associations. IEEE Trans. on Fuzzy Systems 13, 250–262 (2005)

Dyczkowski, K.: A less cumulative algorithm of mining linguistic browsing patterns in the World Wide Web. In: Proc. 5th EUSFLAT Conf., Vol. II, Ostrava, Czech Rep, pp. 129–135 (2007)

Dyczkowski, K.: Application of IF-sets to modeling of lip shapes similarities. In: Hüllermeier, E., Kruse, R., Hoffmann, F. (eds.) Information Processing and Management of Uncertainty in Knowledge-Based Systems. Theory and Methods. Communications in Computer and Information Science, vol. 80, pp. 611–617. Springer, Heidelberg (2010)

Dyczkowski, K., Stachowiak, A.: A recommender system with uncertainty on the example of political elections. In: Greco, S., Bouchon-Meunier, B., Coletti, G., Fedrizzi, M. (eds.) Advances in Computational Intelligence. Part II. Communications in Computer and Information Science, vol. 298, pp. 441–449. Springer, Heidelberg (2012)

Dyczkowski, K., Wygralak, M.: On cardinality and singular fuzzy sets. In: Reusch, B. (ed.) Fuzzy Days 2001. LNCS, vol. 2206, pp. 261–268. Springer, Heidelberg (2001)

Dyczkowski, K., Wygralak, M.: On triangular norm-based generalized cardinals and singular fuzzy sets. Fuzzy Sets and Systems 133, 211–226 (2003)

Ebanks, B.R.: On measures of fuzziness and their representations. Jour. Math. Anal. Appl. 94, 24–37 (1983)

Felix, R.: Relationships between goals in multiple attribute decision making. Fuzzy Sets and Systems 67, 47–52 (1994)

Felix, R.: Decision making with interactions between goals. In: Proc. 13th Zittau Fuzzy Coloquium, Zittau, Germany, pp. 3–10 (2006)

Fodor, J.: A new look at fuzzy connectives. Fuzzy Sets and Systems 57, 141–148 (1993)

Fodor, J., Roubens, M.: Fuzzy Preference Modeling and Multicriteria Decision Support. Kluwer, Dordrecht (1994)

Fodor, J., Yager, R.R.: Fuzzy set-theoretic operators and quantifiers. In: Dubois, D., Prade, H. (eds.) Fundamentals of Fuzzy Sets, pp. 3–74. Kluwer, Boston (2000)

Frank, M.J.: On the simultaneous associativity of F(x, y) and x + y − F(x, y). Aequationes Math 19, 194–226 (1979)

Gentilhomme, Y.: Les ensembles flous en linguistique. Cahiers dé Linguistique Theorique et Appliquée, Bucharest 5, 47 (1968)

Giles, R.: Łukasiewicz logic and fuzzy set theory. Inter. Jour. Man-Machine Stud. 8, 313–327 (1976)

Glöckner, I.: Fuzzy Quantifiers in Natural Language Semantics and Computational Models. Der Andere Verlag, Osnabrück (2004)

Glöckner, I.: Fuzzy Quantifiers – A Computational Theory. Springer, Heidelberg (2006)

Goguen, J.A.: L-fuzzy sets. Jour. Math. Anal. Appl. 18, 145–174 (1967)

Gottwald, S.: A note on fuzzy cardinals. Kybernetika 16, 156–158 (1980)

Gottwald, S.: Fuzzy set theory: some aspects of the early development. In: Skala, H.J., Termini, S., Trillas, E. (eds.) Aspects of Vagueness, pp. 13–29. Reidel, Dordrecht (1984)

Gottwald, S.: Many-valued logic and fuzzy set theory. In: Höhle, U., Rodabaugh, S.E. (eds.) Mathematics of Fuzzy Sets – Logic, Topology, and Measure Theory, pp. 5–89. Kluwer, Boston (1999)

Gottwald, S.: A Treatise on Many-Valued Logics. Research Studies Press, Baldock, Hertfordshire (2001)

Grattan-Guiness, I.: Fuzzy membership mapped onto interval and many-valued quantities. Z. Math. Logik Grundl. Math. 22, 149–160 (1975)

Grzegorzewski, P.: Decision Support under Uncertainty. Statistical Methods for Imprecise Data (in Polish). EXIT Publ., Warsaw (2006)

Grzegorzewski, P., Mrówka, E.: Some notes on (Atanassov's) intuitionistic fuzzy sets. Fuzzy Sets and Systems 156, 492–495 (2005)

Hájek, P.: Metamathematics of Fuzzy Logic. Kluwer, Dordrecht (1998)

Herrera, F., Herrera-Viedma, E., Verdegay, J.L.: A model of consensus in group decision making under linguistic assessments. Fuzzy Sets and Systems 78, 73–87 (1996)

Herrera, F., Martinez, L., Sanchez, P.J.: Managing non-homogeneous information in group decision making. Europ. J. Oper. Research 166, 115–132 (2005)

Higashi, M., Klir, G.J.: On measures of fuzziness and fuzzy complements. Int. Jour. General Systems 8, 169–180 (1982)

Holčapek, M.: An axiomatic approach to fuzzy measures like set cardinality for finite fuzzy sets. In: Hüllermeier, E., Kruse, R., Hoffmann, F. (eds.) Information Processing and Management of Uncertainty in Knowledge-Based Systems. Theory and Methods. Communications in Computer and Information Science, vol. 80, pp. 505–514. Springer, Heidelberg (2010)

Hong, T.-P., Chen, C.-H., Wu, Y.-L., Lee, Y.-C.: A GA-based fuzzy mining approach to achieve a trade-off between number of rules and suitability of membership functions. Soft Computing 10, 1091–1101 (2006)

Hong, T.-P., Lin, K.-Y., Wang, S.-L.: Mining linguistic browsing patterns in the world wide web. Soft Computing 6, 329–336 (2002)

Hong, T.-P., Lin, K.-Y., Wang, S.-L.: Mining fuzzy sequential patterns from quantitative transactions. Soft Computing 10, 925–932 (2006)

Ionescu, M., Ralescu, A.: Fuzzy Hamming distance based banknote validator. In: Proc. FUZZ-IEEE Conf., Reno, Nevada, USA, pp. 300–305 (2005)

Jahn, K.U.: Intervall-wertige Mengen. Math. Nachr. 68, 115–132 (1975)

Jaworska, T., Kacprzyk, J., Marin, N., Zadrożny, S.: On dealing with imprecise information in a content based image retrieval system. In: Hüllermeier, E., Kruse, R., Hoffmann, F. (eds.) Computational Intelligence for Knowledge-Based Systems Design. LNCS (LNAI), vol. 6178, pp. 149–158. Springer, Heidelberg (2010)

Kacprzyk, J.: Group decision making with a fuzzy linguistic majority. Fuzzy Sets and Systems 18, 106–118 (1986)

Kacprzyk, J.: Fuzzy logic with linguistic quantifiers in decision making and control. Archives of Control Sciences 37, 127–141 (1992)

Kacprzyk, J.: Multistage Fuzzy Control. A Model-Based Approach to Fuzzy Control and Decision Making. John Wiley, Chichester (1997)

Kacprzyk, J., Iwański, C.: Inductive learning from incomplete and imprecise examples. In: Bouchon-Meunier, B., Zadeh, L.A., Yager, R.R. (eds.) IPMU 1990. LNCS, vol. 521, pp. 424–430. Springer, Heidelberg (1991)

Kacprzyk, J., Nurmi, H., Fedrizzi, M.: Group decision making and a measure of consensus under fuzzy preferences and a fuzzy linguistic majority. In: Zadeh, L.A., Kacprzyk, J. (eds.) Computing with Words in Information/Intelligent Systems, Vol. 2 – Applications, pp. 243–269. Physica-Verlag, Heidelberg (1999)

Kacprzyk, J., Wilbik, A.: Temporal linguistic summaries of time series using fuzzy logic. In: Hüllermeier, E., Kruse, R., Hoffmann, F. (eds.) Information Processing and Management of Uncertainty in Knowledge-Based Systems. Theory and Methods. Communications in Computer and Information Science, vol. 80, pp. 436–445. Springer, Heidelberg (2010)

Kacprzyk, J., Wilbik, A., Zadrożny, S.: Linguistic summarization of trends: A fuzzy logic based approach. In: Proc. 11th Inter. Conf. on Information Processing and Management of Uncertainty in Knowledge-Based Systems (IPMU2006), Paris, pp. 2166–2172. Editions EDK, Paris (2006)

Kacprzyk, J., Wilbik, A., Zadrożny, S.: Linguistic summarization of time series using a fuzzy quantifier driven aggregation. Fuzzy Sets and Systems 159, 1485–1499 (2008)

Kacprzyk, J., Yager, R.R.: "Softer" optimization and control models via fuzzy linguistic quantifiers. Information Sciences 34, 157–178 (1984)

Kacprzyk, J., Yager, R.R.: Linguistic summaries of data using fuzzy logic. Inter. Jour. of General Systems 30, 133–154 (2001)

Kacprzyk, J., Zadrożny, S.: Flexible querying using fuzzy logic: An implementation for Microsoft Access. In: Andreasen, T., Christiansen, H., Larsen, H.L. (eds.) Flexible Query Answering Systems, pp. 247–275. Kluwer, Boston (1997)

Kacprzyk, J., Zadrożny, S.: Data mining via linguistic summaries of databases: an interactive approach. In: Ding, L. (ed.) A New Paradigm of Knowledge Engineering by Soft Computing, pp. 325–345. World Scientific, Singapore (2001a)

Kacprzyk, J., Zadrożny, S.: Computing with words in intelligent database querying: standalone and internet-based applications. Information Sciences 134, 71–109 (2001b)

Kacprzyk, J., Zadrożny, S.: Linguistic database summaries and their protoforms: towards natural language based knowledge discovery tools. Information Sciences 173, 281–304 (2005a)

Kacprzyk, J., Zadrożny, S.: Towards more powerful information technology via computing with words and perceptions: Precisiated natural language, protoforms and linguistic data summaries. In: Nikravesh, M., Zadeh, L.A., Kacprzyk, J. (eds.) Soft Computing for Information Processing and Analysis, pp. 19–33. Springer, Heidelberg (2005b)

Kacprzyk, J., Zadrożny, S.: Soft computing and Web intelligence for supporting consensus reaching. Soft Computing 14, 833–846 (2010a)

Kacprzyk, J., Zadrożny, S.: Computing with words is an implementable paradigm: fuzzy queries, linguistic data summaries and natural language generation. IEEE Trans. on Fuzzy Systems 18, 461–472 (2010b)

Kaczmarek, K.: A Model of Decision Support for Demand Forecast. MSc Thesis, Adam Mickiewicz University, Poznań, Poland, (in Polish) (2010)

Kaufmann, A.: Introduction à la Théorie des Sous-Ensembles Flous, Vol. IV. Masson, Paris (1977)

Klaua, D.: Partiell definierte Mengen. Monatsber. Deut. Akad. Wiss. Berlin 10, 571–578 (1968)

Klaua, D.: Partielle Mengen und Zahlen. Monatsber. Deut. Akad. Wiss. Berlin 11, 585–599 (1969)

Klement, E.P., Mesiar, R., Pap, E.: Triangular Norms. Kluwer, Dordrecht (2000)

Klement, E.P., Mesiar, R., Pap, E.: Triangular norms, Position paper I: Basic analytical and algebraic properties. Fuzzy Sets and Systems 143, 5–26 (2004a)

Klement, E.P., Mesiar, R., Pap, E.: Triangular norms, Position paper II: General constructions and parameterized families. Fuzzy Sets and Systems 145, 411–438 (2004b)

Klement, E.P., Mesiar, R., Pap, E.: Triangular norms, Position paper III: Continuous t-norms. Fuzzy Sets and Systems 145, 439–454 (2004c)

Klir, G.J.: Measures of uncertainty and information. In: Dubois, D., Prade, H. (eds.) Fundamentals of Fuzzy Sets, pp. 439–457. Kluwer, Dordrecht (2000)

Klir, G.J., Yuan, B.: Fuzzy Sets and Fuzzy Logic: Theory and Applications. Prentice Hall, Englewood Cliffs (1995)

Kolesárová, A.: Revision of parametric evaluation of aggregation functions. In: Proc. 13th Zittau Fuzzy Colloquium, Zittau, Germany, pp. 202–211 (2006)

Kotarbiński, T.: Elements of Cognition Theory, Formal Logic, and Methodology of Sciences (in Polish). Ossolineum, Lwów (1929)

Král, P.: On the scalar cardinality of IF sets. Jour. of Electrical Engineering 12, 83–86 (2004)

Král, P.: Cardinalities and new measure of entropy for IF sets. In: Atanassov, K.T., Kacprzyk, J., Krawczak, M., Szmidt, E. (eds.) Issues in the Representation and Processing of Uncertain and Imprecise Information. Fuzzy Sets, Intuitionistic Fuzzy Sets, Generalized Nets, and Related Topics, pp. 209–216. EXIT Publ., Warsaw (2005a)

Král, P.: T-operations and cardinalities of IF sets. In: Hryniewicz, O., Kacprzyk, J., Kuchta, D. (eds.) Issues in Soft Computing – Decisions and Operations Research, pp. 219–228. EXIT Publ., Warsaw (2005b)

Král, P. (Interval-valued) fuzzy cardinality of IF-sets. In: Proc. 11th Inter. Conf. on Information Processing and Management of Uncertainty in Knowledge-Based Systems (IPMU2006), pp. 2367–2374. Editions EDK, Paris (2006)

Larsen, H.L., Kacprzyk, J., Zadrożny, S., Andreasen, T., Christiansen, H. (eds.): Flexible Query Answering Systems. Recent Advances. Physica-Verlag, Heidelberg (2001)

Li, D.-F., Yang, J.-B.: A multiattribute decision making approach using intuitionistic fuzzy sets. In: Proc. 3rd Conf. European Society for Fuzzy Logic and Technology (EUSFLAT2003), Zittau, Germany, pp. 183–186 (2003)

Li, H.-X., Luo, C.-Z., Wang, P.-Z.: The cardinality of fuzzy sets and the continuum hypothesis. Fuzzy Sets and Systems 55, 61–78 (1993)

Li, H.-X., Wang, P.-Z., Lee, E.S., Yen, V.C.: The operations of fuzzy cardinalities. Jour. Math. Anal. Appl. 182, 768–778 (1994)

Lietard, L., Rocacher, D.: Complex quantified statements evaluated using gradual numbers. In: Castillo, O., Melin, P., Ross, O., Cruz, R., Pedrycz, W., Kacprzyk, J. (eds.) Theoretical Advances and Applications of Fuzzy Logic and Soft Computing, pp. 46–53. Springer, Heidelberg (2007)

Ling, C.H.: Representation of associative functions. Publ. Math. Debrecen 12, 189–212 (1965)

Liu, H.-W., Wang, G.-J.: Multi-criteria decision-making methods based on intuitionistic fuzzy sets. Europ. J. Oper. Research 179, 220–233 (2007)

Liu, T., Kerre, E.E.: An overview of fuzzy quantifiers, Part I: Interpretations. Fuzzy Sets and Systems 95, 1–21 (1998a)

Liu, T., Kerre, E.E.: An overview of fuzzy quantifiers, Part II: Reasoning and applications. Fuzzy Sets and Systems 95, 135–146 (1998b)

Lowen, R.: Fuzzy Set Theory. Basic Concepts, Techniques and Bibliography. Kluwer, Dordrecht (1996)

Łukasiewicz, J.: Interpretacja liczbowa teorii zdań (in Polish). Ruch Filozoficzny 7, 92–93 (1922). Translated as 'A numerical interpretation of the theory of propositions' in Borkowski, 129-130 (1970)

Mamdani, E.H.: Advances in the linguistic synthesis of fuzzy controllers. Int. Jour. of Man-Machine Studies 8, 669–678 (1976)

Menger, K.: Statistical metrics. Proc. Nat. Acad. Sci. U. S. A. 28, 535–537 (1942)

Mizumoto, M., Tanaka, K.: Some properties of fuzzy sets of type 2. Inf. and Control 31, 312–340 (1976)

Mizumoto, M., Tanaka, K.: Fuzzy sets of type 2 under algebraic product and algebraic sum. Fuzzy Sets and Systems 5, 277–290 (1981)

Narin'yani, A.S.: Sub-definite sets – New data type for knowledge representation. Memo no. 4-232, Computing Center, Novosibirsk, Russia (in Russian) (1980)

Negoita, C.V., Ralescu, D.A.: Applications of Fuzzy Sets to Systems Analysis. Birkhäuser, Basel (1975)

Nguyen, H.T.: A note on the extension principle of fuzzy sets. J. Math. Anal. Appl. 64, 369–380 (1978)

Nguyen, H.T., Walker, E.A.: A First Course in Fuzzy Logic, 3rd edn. CRC Press, Boca Raton (2005)

Niewiadomski, A.: Methods for the Linguistic Summarization of Data: Applications of Fuzzy Sets and their Extensions. EXIT Publ., Warsaw (2008)

Novák, V.: Are fuzzy sets a reasonable tool for modeling vague phenomena? Fuzzy Sets and Systems 156, 341–348 (2005)

Nurmi, H.: Approaches to collective decision making with fuzzy preference relations. Fuzzy Sets and Systems 6, 249–259 (1981)

Ovchinnikov, S.: An introduction to fuzzy relations. In: Dubois, D., Prade, H. (eds.) Fundamentals of Fuzzy Sets, pp. 233–259. Kluwer, Dordrecht (2000)

Pal, N.R., Bezdek, J.C.: Measuring fuzzy uncertainty. IEEE Trans. on Fuzzy Systems 2, 107–118 (1994)

Pałubicki, W.: Fuzzy Plant Modeling with OpenGL. Novel Approaches in Simulating Phototrophism and Environmental Conditions. VDM Verlag Dr. Müller, Saarbrücken (2007)

Pankowska, A.: Examples of applications of IF-sets with triangular norms to group decision making problems. In: Atanassov, K.T., Kacprzyk, J., Krawczak, M., Szmidt, E. (eds.) Issues in the Representation and Processing of Uncertain and Imprecise Information, pp. 290–306. EXIT Publ., Warsaw (2005)

Pankowska, A.: Dealing with incompleteness of preferences in group decision making problems. In: Proc. 5th Conf. European Society for Fuzzy Logic and Technology (EUSFLAT2007), Vol. II, pp. 317–322. Ostrava, Czech Rep. (2007)

Pankowska, A.: The problem of uncertain and incomplete information in group decision making process. In: Atanassov, K., Bustince, H., Hryniewicz, O., Kacprzyk, J. (eds.) Developments in Fuzzy Sets, Intuitionistic Fuzzy Sets, Generalized Nets and Related Topics. Foundations, pp. 265–274. EXIT Publ., Warsaw (2008)

Pankowska, A., Wygralak, M.: Intuitionistic fuzzy sets – An alternative look. In: Proc. 3rd Conf. European Society for Fuzzy Logic and Technology (EUSFLAT 2003), Zittau, Germany, pp. 135–140 (2003)

Pankowska, A., Wygralak, M.: A general concept of IF-sets with triangular norms. In: Atanassov, K.T., Hryniewicz, O., Kacprzyk, J. (eds.) Soft Computing – Foundations and Theoretical Aspects, pp. 319–335. EXIT Publ., Warsaw (2004a)

Pankowska, A., Wygralak, M.: On hesitation degrees in IF-set theory. In: Rutkowski, L., Siekmann, J.H., Tadeusiewicz, R., Zadeh, L.A. (eds.) ICAISC 2004. LNCS (LNAI), vol. 3070, pp. 338–343. Springer, Heidelberg (2004b)

Pankowska, A., Wygralak, M.: Algorithms of group decision making based on generalized IF-sets. In: Hryniewicz, O., Kacprzyk, J., Koronacki, J., Wierzchoń, S.T. (eds.) Issues in Intelligent Systems – Paradigms, pp. 185–197. EXIT Publ., Warsaw (2005)

Pankowska, A., Wygralak, M.: General IF-sets with triangular norms and their applications to group decision making. Information Sciences 176, 2713–2754 (2006)

Pawlak, Z.: Rough sets. Inter. Jour. Information and Computer Sciences 11, 341–356 (1982)

Pawlak, Z.: Rough Sets – Theoretical Aspects of Reasoning about Data. Kluwer, Dordrecht (1991)

Pedrycz, W., Gomide, F.: Fuzzy Systems Engineering: Towards Human-Centric Computing. Wiley, Hoboken (2007)

Petrovic, D., Xie, Y., Burnham, K.: Fuzzy decision support system for demand forecasting with a learning mechanism. Fuzzy Sets and Systems 157, 1713–1725 (2006)

Piegat, A.: Fuzzy Modeling and Control. Physica-Verlag, Heidelberg (2001)

Pilarski, D.: Generalized relative cardinalities of fuzzy sets. Inter. Jour. of Uncertainty, Fuzziness and Knowledge-Based Systems 13, 1–10 (2005)

Pilarski, D.: Linguistic summarization of databases with Quantirius: a reduction algorithm for generated summaries. Inter. Jour. of Uncertainty, Fuzziness and Knowledge-Based Systems 18, 305–331 (2010)

Ralescu, D.: Cardinality, quantifiers, and the aggregation of fuzzy criteria. Fuzzy Sets and Systems 69, 355–365 (1995)

Rocacher, D.: On fuzzy bags and their applications to flexible querying. Fuzzy Sets and Systems 140, 93–110 (2003)

Rocacher, D., Bosc, P.: The set of fuzzy rational numbers and flexible querying. Fuzzy Sets and Systems 155, 317–339 (2005)

Russell, B.: Vagueness. Australasian Jour. of Psychology and Philosophy 1, 84–92 (1923)

Rutkowski, L.: Computational Intelligence: Methods and Techniques. Springer, Heidelberg (2008)

Sambuc, R.: Fonctions Φ-Floues – Application á l'Aide au Diagnostic en Pathologie Thyroïdienne. Thése de Doctorat en Médecine, Marseille, France (1975)

Sanchez, E.: Inverses of fuzzy relations: applications to possibility distributions and medical diagnosis. In: Proc. IEEE Conf, Decision Control, New Orleans, USA, pp. 1384–1389 (1977)

Schweizer, B., Sklar, A.: Associative functions and statistical triangle inequalities. Publ. Math. Debrecen 8, 169–186 (1961)

Schweizer, B., Sklar, A.: Probabilistic Metric Spaces. North-Holland, New York (1983)

Seising, R.: Pioneers of vagueness, haziness, and fuzziness in the 20th century. In: Nikravesh, M., Kacprzyk, J., Zadeh, L.A. (eds.) Forging New Frontiers: Fuzzy Pioneers I, pp. 55–81. Springer, Heidelberg (2007a)

Seising, R.: The Fuzzification of Systems. The Genesis of Fuzzy Set Theory and its Initial Applications – Developments up to the 1970s. Springer, Berlin (2007b)

Siler, W., Buckley, J.J.: Fuzzy Expert Systems and Fuzzy Reasoning. Wiley, Hoboken (2005)

Słowiński, R. (ed.): Intelligent Decision Support – Handbook of Applications and Advances of the Rough Sets Theory. Kluwer, Dordrecht (1992)

Słowiński, R., Greco, S., Matarazzo, B.: Rough set approach to decision support. In: Hryniewicz, O., Kacprzyk, J., Kuchta, D. (eds.) Issues in Soft Computing. Decisions and Operations Research, pp. 87–135. EXIT Publ., Warsaw (2005)

Stachowiak, A.: Trust propagation – Cardinality-based approach. In: Proc. Inter. Multiconference on Computer Science and Information Technology, Mrągowo, Poland, pp. 125–129 (2009)

Stachowiak, A.: Trust propagation based on group opinion. In: Hüllermeier, E., Kruse, R., Hoffmann, F. (eds.) Information Processing and Management of Uncertainty in Knowledge-Based Systems. Theory and Methods. Communications in Computer and Information Science, vol. 80, pp. 601–610. Springer, Heidelberg (2010)

Szmidt, E., Kacprzyk, J.: Group decision making under intuitionistic fuzzy preference relations. In: Proc. 7th Inter. Conf. on Information Processing and Management of Uncertainty in Knowledge-Based Systems (IPMU1998), Paris, pp. 172–178. Editions EDK, Paris (1998)

Szmidt, E., Kacprzyk, J.: Decision making in an intuitionistic fuzzy environment. In: Proc. Joint 4th Meeting of the EURO Working Group on Fuzzy Sets and 2nd International Conference on Soft and Intelligent Computing (EUROFUSE-SIC'99), Budapest, pp. 292–297 (1999)

Szmidt, E., Kacprzyk, J.: Entropy for intuitionistic fuzzy sets. Fuzzy Sets and Systems 118, 467–477 (2001)

Szmidt, E., Kacprzyk, J.: A new similarity measure for intuitionistic fuzzy sets: Straightforward approaches may not work. In: Proc. 2007 IEEE Conf. on Fuzzy Systems, London, pp. 481–486 (2007)

Szmidt, E., Kukier, M.: Classification of imbalanced and overlapping classes using intuitionistic fuzzy sets. In: Proc. 3rd Inter. IEEE Conf. on Intelligent Systems, London, pp. 722–727. IEEE Computer Society Press, Los Alamitos (2006)

Szmidt, E., Kukier, M.: A new approach to classification of imbalanced classes via Atanassov's intuitionistic fuzzy sets. In: Wang, H.-F. (ed.) Intelligent Data Analysis: Developing New Methodologies Through Pattern Discovery and Recovery, pp. 85–101. IGI Global (2008)

Świtalski, Z.: Transitivity of fuzzy preference relations – An empirical study. Fuzzy Sets and Systems 118, 503–508 (2001)

Šostak, A.P.: Fuzzy cardinals and cardinalities of fuzzy sets (in Russian). In: Algebra and Discrete Mathematics, pp. 137–144. Latvian State University, Riga (1989)

Takeuti, G., Titani, S.: Intuitionistic fuzzy logic and intuitionistic fuzzy set theory. J. Symb. Logic 49, 851–866 (1984)

Tamburini, G., Termini, S.: Some fundamental problems in the formalization of vagueness. In: Gupta, M.M., Sanchez, E. (eds.) Fuzzy Information and Decision Processes, pp. 161–166. North-Holland, Amsterdam (1982)

Tanaka, K.: An Introduction to Fuzzy Logic for Practical Applications. Springer, New York (1997)

Termini, S.: Aspects of vagueness and some epistemological problems related to their formalization. In: Skala, H.J., Termini, S., Trillas, E. (eds.) Aspects of Vagueness, pp. 205–230. Reidel, Dordrecht (1984)

Trillas, E.: Sobre funciones de negación en la teoría de conjuntos difusos. Stochastica 3, 47–84 (1979)

Victor, P., Cornelis, C., De Cock, M., Pinheiro da Silva, P.: Gradual trust and distrust in recommender systems. Fuzzy sets and Systems 160, 1367–1382 (2009)

Vlachos, I.K., Sergiadis, G.D.: Subsethood, entropy, and cardinality for interval-valued fuzzy sets – An algebraic derivation. Fuzzy Sets and Systems 158, 1384–1396 (2007a)

Vlachos, I.K., Sergiadis, G.D.: Intuitionistic fuzzy image processing. In: Nachtegael, M., Van der Weken, D., Kerre, E.E., Philips, W. (eds.) Soft Computing in Image Processing, pp. 383–414. Springer, Heidelberg (2007b)

Wagenknecht, M., Kalinina, E.: Some remarks on crisp and fuzzy numbers averaging and aggregating. In: Proc. 13th Zittau Fuzzy Colloquium, Zittau, Germany, pp. 190–201 (2006)

Weber, S.: A general concept of fuzzy connectives, negations and implications based on t-norms and t-conorms. Fuzzy Sets and Systems 11, 115–134 (1983)

Wu, J.K., Ang, Y.H., Lam, P., Loh, H.H., Desai Narasimhalu, A.: Inference and retrieval of facial images. Multimedia Systems 2, 1–14 (1994)

Wu, J.-Z., Zhang, Q.: Multicriteria decision making method based on intuitionistic fuzzy weighted entropy. Expert Systems with Applications 38, 916–922 (2011)

Wygralak, M.: A new approach to the fuzzy cardinality of finite fuzzy sets. Bull. Stud. Exch. on Fuzziness and its Appl (BUSEFAL) 15, 72–75 (1983a)

Wygralak, M.: Fuzzy inclusion and fuzzy equality of two fuzzy subsets, fuzzy operations for fuzzy subsets. Fuzzy Sets and Systems 10, 157–168 (1983b)

Wygralak, M.: A supplement to Gottwald's note on fuzzy cardinals. Kybernetika 20, 240–243 (1984)

Wygralak, M.: Fuzzy cardinals based on the generalized equality of fuzzy sets. Fuzzy Sets and Systems 18, 143–158 (1986)

Wygralak, M.: Rough sets and fuzzy sets – Some remarks on interrelations. Fuzzy Sets and Systems 29, 241–243 (1989)

Wygralak, M.: Generalized cardinal numbers and their ordering. In: Bouchon-Meunier, B., Zadeh, L.A., Yager, R.R. (eds.) IPMU 1990. LNCS, vol. 521, pp. 183–192. Springer, Heidelberg (1991a)

Wygralak, M.: A look at metrics and norms through the Łukasiewicz logic. Jour. Applied Non-Classical Logics 1, 77–81 (1991b)

Wygralak, M.: Powers and generalized cardinal numbers for HCH-objects. Basic notions. Math. Pann. 3, 91–115 (1992)

Wygralak, M.: A cardinality theory for vaguely defined objects – Problems of inequalities and applications. Acta Appl. Math. 30, 1–33 (1993a)

Wygralak, M.: Generalized cardinal numbers and operations on them. Fuzzy Sets and Systems 53, 49–85 (and (1994): Erratum. Ibid., 62, p. 375) (1993b)

Wygralak, M.: A general cardinality theory for vaguely defined objects. In: Proc. First European Congress on Fuzzy and Intelligent Technologies (EUFIT'93), Aachen, Germany, pp. 1265–1271 (1993c)

Wygralak, M.: Cardinal aspects of vaguely defined objects. In: Bouchon-Meunier, B., Yager, R.R., Zadeh, L.A. (eds.) IPMU 1994. LNCS, vol. 945, pp. 187–192. Springer, Heidelberg (1995)

Wygralak, M.: Vagueness and cardinality – A unifying approach. In: Bouchon-Meunier, B., Yager, R.R., Zadeh, L.A. (eds.) Fuzzy Logic and Soft Computing, pp. 310–319. World Scientific, Singapore (1995)

Wygralak, M.: Vaguely Defined Objects. Representations, Fuzzy Sets and Nonclassical Cardinality Theory. Kluwer, Dordrecht (1996a)

Wygralak, M.: Cardinality of objects with respect to many-valued logic – A synthetic approach. In: Proc. Inter. Conf. Fuzzy Logic in Engineering and Natural Sciences, Zittau, Germany, pp. 475–481 (1996b)

Wygralak, M.: Cardinalities of fuzzy sets evaluated by single cardinals. In: Proc. 7th International Fuzzy Systems Association World Congress (IFSA'97), Prague, pp. 73–77. Academia, Praha (1997a)

Wygralak, M.: On the best scalar approximation of cardinality of a fuzzy set. Inter. Jour. of Uncertainty, Fuzziness and Knowledge-Based Systems 5, 681–687 (1997b)

Wygralak, M.: Vagueness and its representations: A unifying look. Mathware and Soft Computing 5, 121–131 (1998a)

Wygralak, M.: From sigma counts to alternative nonfuzzy cardinalities of fuzzy sets. In: Proc. 7th Inter. Conf. on Information Processing and Management of Uncertainty in Knowledge-Based Systems (IPMU'98), Paris, pp. 1339–1344. Editions EDK, Paris (1998b)

Wygralak, M.: Questions of cardinality of finite fuzzy sets. Fuzzy Sets and Systems 102, 185–210 (1999a)

Wygralak, M.: Triangular operations, negations, and scalar cardinality of a fuzzy set. In: Zadeh, L.A., Kacprzyk, J. (eds.) Computing with Words in Information/Intelligent Systems, Vol. 1 – Foundations, pp. 326–341. Physica-Verlag, Heidelberg New York (1999b)

Wygralak, M.: Scalar cardinalities of fuzzy sets with triangular norms and conorms. In: Proc. Joint 4th Meeting of the EURO Working Group on Fuzzy Sets and 2nd Inter. Conf. on Soft and Intelligent Computing (EUROFUSE-SIC'99), Budapest, pp. 322–327 (1999c)

Wygralak, M.: A generalizing look at sigma counts of fuzzy sets. In: Bouchon-Meunier, B., Yager, R.R., Zadeh, L.A. (eds.) Uncertainty in Intelligent and Information Systems, pp. 34–45. World Scientific, Singapore (2000a)

Wygralak, M.: An axiomatic approach to scalar cardinalities of fuzzy sets. Fuzzy Sets and Systems 110, 175–179 (2000b)

Wygralak, M.: Fuzzy sets with triangular norms and their cardinality theory. Fuzzy Sets and Systems 124, 1–24 (2001)

Wygralak, M.: Variants of defining the cardinalities of fuzzy sets. In: Grzegorzewski, P., Hryniewicz, O., Gil, M.Á. (eds.) Soft Methods in Probability, Statistics and Data Analysis, pp. 178–185. Physica-Verlag, Heidelberg (2002)

Wygralak, M.: Cardinalities of Fuzzy Sets. Springer, Heidelberg (2003a)

Wygralak, M.: Cardinalities of fuzzy sets with triangular norms. In: Proc. 3rd Conf. European Society for Fuzzy Logic and Technology (EUSFLAT2003), Zittau, Germany, pp. 691–696 (2003b)

Wygralak, M.: I-fuzzy sets with triangular norms, their hesitation areas and cardinalities. In: Atanassov, K.T., Kacprzyk, J., Krawczak, M., Szmidt, E. (eds.) Issues in the Representation and Processing of Uncertain and Imprecise Information. Fuzzy Sets, Intuitionistic Fuzzy Sets, Generalized Nets, and Related Topics, pp. 397–408. EXIT Publ., Warsaw (2005a)

Wygralak, M.: Fuzziness and cardinality. In: Hryniewicz, O., Kacprzyk, J., Kuchta, D. (eds.) Issues in Soft Computing. Decisions and Operations Research, pp. 137–145. EXIT Publ., Warsaw (2005b)

Wygralak, M.: Representing incomplete knowledge about fuzzy sets. In: Proc. 5th EUSFLAT Conf., Vol. II, pp. 287–292. Ostrava, Czech Rep (2007)

Wygralak, M.: Hesitation and cardinality in the I-fuzzy set theory – A general approach. In: Atanassov, K.T., Bustince, H., Hryniewicz, O., Kacprzyk, J. (eds.) Developments in Fuzzy Sets, Intuitionistic Fuzzy Sets, Generalized Nets and Related Topics. Foundations, pp. 377–385. EXIT Publ., Warsaw (2008a)

Wygralak, M.: Model examples of applications of sigma f-counts to counting in fuzzy and I-fuzzy sets. In: Atanassov, K.T., Hryniewicz, O., Kacprzyk, J., Krawczak, M. (eds.) Advances in Fuzzy Sets, Intuitionistic Fuzzy Sets, Generalized Nets and Related Topics, Vol. II: Applications, pp. 187–196. EXIT Publ., Warsaw (2008b)

Wygralak, M.: On nonstrict Archimedean triangular norms, Hamming distances, and cardinalities of fuzzy sets. Int. Jour. Intelligent Systems 24, 697–705 (2009)

Wygralak, M.: Hesitation degrees as the size of ignorance combined with fuzziness. In: Hüllermeier, E., Kruse, R., Hoffmann, F. (eds.) Information Processing and Management of Uncertainty in Knowledge-Based Systems. Theory and Methods. Communications in Computer and Information Science, vol. 80, pp. 629–636. Springer, Heidelberg (2010)

Wygralak, M.: Scalar and fuzzy cardinalities – tools for intelligent counting under information imprecision. In: Atanassov, K.T., Baczyński, M., Drewniak, J., Kacprzyk, J. (eds.) Recent Advances in Fuzzy Sets, Intuitionistic Fuzzy Sets, Generalized Nets and Related Topics. Vol. I: Foundations, pp. 237–243. IBS PAN – SRI PAS Publ., Warsaw (2011)

Wygralak, M., Pilarski, D.: FGCounts of fuzzy sets with triangular norms. In: Hampel, R., Wagenknecht, M., Chaker, N. (eds.) Fuzzy Control – Theory and Practice, pp. 121–131. Physica-Verlag, Heidelberg (2000)

Yager, R.R.: A new approach to the summarization of data. Information Sciences 28, 69–86 (1982)

Yager, R.R.: Quantifiers in the formulation of multiple objective decision functions. Information Sciences 31, 107–139 (1983)

Yager, R.R.: On ordered weighted averaging aggregation operators in multicriteria decision making. IEEE Trans. On Systems, Man and Cybernetics 18, 183–190 (1988)

Yager, R.R.: Interpreting linguistically quantified propositions. Inter. Jour. of Intelligent Systems 9, 541–569 (1994)

Yager, R.R.: On the fuzzy cardinality of a fuzzy set. Int. J. of General Systems 35, 191–206 (2006)

Yager, R.R., Filev, D.P.: Essentials of Fuzzy Modeling and Control. Wiley, New York (1994)

Yager, R.R., Ford, K.M., Cañas, A.J.: An approach to the linguistic summarization of data. In: Bouchon-Meunier, B., Zadeh, L.A., Yager, R.R. (eds.) IPMU 1990. LNCS, vol. 521, pp. 456–468. Springer, Heidelberg (1991)

Ye, J.: Fuzzy decision-making method based on the weighted correlation coefficient under intuitionistic fuzzy environment. Europ. J. Oper. Research 205, 202–204 (2010)

Zadeh, L.A.: Fuzzy sets. Inform. and Control 8, 338–353 (1965)

Zadeh, L.A.: Similarity relations and fuzzy orderings. Information Sciences 3, 177–200 (1971)

Zadeh, L.A.: A rationale for fuzzy control. Measurement and Control 34, 3–4 (1972)

Zadeh, L.A.: The concept of a linguistic variable and its applications to approximate reasoning – I. Information Sciences 8, 199–249 (1975a)

Zadeh, L.A.: The concept of a linguistic variable and its applications to approximate reasoning – II. Information Sciences 8, 301–357 (1975b)

Zadeh, L.A.: The concept of a linguistic variable and its applications to approximate reasoning – III. Information Sciences 9, 43–80 (1975c)

Zadeh, L.A.: Fuzzy logic and approximate reasoning. Synthese 30, 407–428 (1975d)

Zadeh, L.A.: Fuzzy sets as a basis for a theory of possibility. Fuzzy Sets and Systems 1, 3–28 (1978a)

Zadeh, L.A.: PRUF – A meaning representation language for natural languages. Inter. Jour. Man-Machine Stud. 10, 395–460 (1978b)

Zadeh, L.A.: A theory of approximate reasoning. In: Hayes, J.E., Michie, D., Mikulich, L.I. (eds.) Machine Intelligence, vol. Vol. 9, pp. 149–194. John Wiley & Sons, New York (1979)

Zadeh, L.A.: Test-score semantics for natural languages and meaning representation via PRUF. In: Rieger, B.B. (ed.) Empirical Semantics, pp. 281–349. Brockmeyer, Bochum (1981a)

Zadeh, L.A.: Fuzzy probabilities and their role in decision analysis. In: Proc. 4th MIT/ONR Workshop on Command, Control and Communications, pp. 159–179. MIT Press, Cambridge (1981b)

Zadeh, L.A.: A computational approach to fuzzy quantifiers in natural languages. Comput. and Math. with Appl. 9, 149–184 (1983a)

Zadeh, L.A.: A theory of commonsense knowledge. Mem. No. UCB/ERL M83/26 (1983b)

Zadeh, L.A.: The role of fuzzy logic in the management of uncertainty in expert systems. Fuzzy Sets and Systems 11, 199–228 (1983c)

Zadeh, L.A.: Fuzzy logic, neural networks, and soft computing. Comm. of the ACM 37, 77–84 (1994)

Zadeh, L.A.: Toward a theory of fuzzy information granulation and its centrality in human reasoning and fuzzy logic. Fuzzy Sets and Systems 90, 111–127 (1997)

Zadeh, L.A.: From computing with numbers to computing with words – From manipulation of measurements to manipulation of perceptions. IEEE Trans. on Circuits and Systems 45, 105–119 (1999a)

Zadeh, L.A.: Fuzzy logic = Computing with words. In: Zadeh, L.A., Kacprzyk, J. (eds.) Computing with Words in Information/Intelligent Systems, Vol. 1 – Foundations, pp. 3–23. Physica-Verlag, Heidelberg (1999b)

Zadeh, L.A.: Outline of a computational theory of perceptions based on computing with words. In: Sinha, N.K., Gupta, M.M. (eds.) Soft Computing and Intelligent Systems: Theory and Applications, pp. 3–22. Academic Press, London (2000)

Zadeh, L.A.: A new direction in AI. Toward a computational theory of perceptions. AI Magazine 22, 73–84 (2001)

Zadeh, L.A.: Toward a perception-based theory of probabilistic reasoning with imprecise probabilities. Jour. of Statistical Planning and Inference 105, 233–264 (2002)

Zadeh, L.A., Kacprzyk, J. (eds.): Computing with Words in Information/Intelligent Systems, Vol. 1 – Foundations. Physica-Verlag, Heidelberg (1999a)

Zadeh, L.A., Kacprzyk, J. (eds.): Computing with Words in Information/Intelligent Systems, Vol. 2 – Applications. Physica-Verlag, Heidelberg (1999b)

Zadrożny, S., Kacprzyk, J.: FQUERY for Access: towards human consistent querying user interface. In: Proc. ACM Symp. on Applied Computing (SAC '96), Philadelphia, USA, pp. 532–536 (1996)

Zadrożny, S., Kacprzyk, J.: Linguistically quantified thresholding strategies for text categorization. In: Proc. 3rd EUSFLAT Conf., Zittau, Germany, pp. 38–42 (2003)

Zadrożny, S., Kacprzyk, J.: Computing with words for text processing: An approach to the text categorization. Information Sciences 176, 415–437 (2006)

Index